EAI/Springer Innovations in Communication and Computing

Series Editor

Imrich Chlamtac, European Alliance for Innovation
Ghent, Belgium

The impact of information technologies is creating a new world yet not fully understood. The extent and speed of economic, life style and social changes already perceived in everyday life is hard to estimate without understanding the technological driving forces behind it. This series presents contributed volumes featuring the latest research and development in the various information engineering technologies that play a key role in this process. The range of topics, focusing primarily on communications and computing engineering include, but are not limited to, wireless networks; mobile communication; design and learning; gaming; interaction; e-health and pervasive healthcare; energy management; smart grids; internet of things; cognitive radio networks; computation; cloud computing; ubiquitous connectivity, and in mode general smart living, smart cities, Internet of Things and more. The series publishes a combination of expanded papers selected from hosted and sponsored European Alliance for Innovation (EAI) conferences that present cutting edge, global research as well as provide new perspectives on traditional related engineering fields. This content, complemented with open calls for contribution of book titles and individual chapters, together maintain Springer's and EAI's high standards of academic excellence. The audience for the books consists of researchers, industry professionals, advanced level students as well as practitioners in related fields of activity include information and communication specialists, security experts, economists, urban planners, doctors, and in general representatives in all those walks of life affected ad contributing to the information revolution.

Indexing: This series is indexed in Scopus, Ei Compendex, and zbMATH.

About EAI - EAI is a grassroots member organization initiated through cooperation between businesses, public, private and government organizations to address the global challenges of Europe's future competitiveness and link the European Research community with its counterparts around the globe. EAI reaches out to hundreds of thousands of individual subscribers on all continents and collaborates with an institutional member base including Fortune 500 companies, government organizations, and educational institutions, provide a free research and innovation platform. Through its open free membership model EAI promotes a new research and innovation culture based on collaboration, connectivity and recognition of excellence by community.

Hamed Taherdoost

E-Business Essentials

Building a Successful Online Enterprise

Springer

EAI
RESEARCH MEETS INNOVATION

Hamed Taherdoost
Department of Arts, Communications and Social Sciences
University Canada West
Vancouver, BC, Canada

Hamta Business Corporation
Vancouver, BC, Canada

ISSN 2522-8595 ISSN 2522-8609 (electronic)
EAI/Springer Innovations in Communication and Computing
ISBN 978-3-031-39625-0 ISBN 978-3-031-39626-7 (eBook)
https://doi.org/10.1007/978-3-031-39626-7

This Springer imprint is published by the registered company Springer Nature Switzerland AG
The registered company address is: Gewerbestrasse 11, 6330 Cham, Switzerland

To my father, Reza Taherdoust, a stalwart in the realm of industry for 50 years

Foreword

In today's fast-paced and interconnected world, the corporate landscape has undergone a remarkable transformation fueled by the advent of the Internet. The rise of e-commerce has revolutionized how businesses communicate, collaborate, and conduct transactions with customers and suppliers across the globe. As technology continues to shape and reshape the digital economy, it becomes increasingly crucial for individuals and organizations to comprehend the multifaceted nature of e-business.

This book serves as a comprehensive guide, offering a deep dive into the diverse elements that constitute the realm of e-business. Its overarching objective is to equip readers with a thorough understanding of this dynamic domain, from its fundamental concepts and significance within contemporary business environments to the intricate models and strategies that drive success in the digital era.

Recognizing the inherent opportunities and challenges inherent in the e-business environment, this book explores a wide spectrum of topics. Starting with a primer on e-business and its indispensable role in today's corporate landscape, it proceeds to unravel the myriad models and tactics that businesses can employ to thrive in the digital economy. Addressing the ever-evolving nature of the e-environment, this book delves into crucial aspects such as the growth and utilization of the Internet, as well as the paramount significance of security and control to safeguard online assets effectively.

As businesses embark on the transformational journey of implementing an e-business model, they encounter various challenges. This book navigates through these hurdles, emphasizing the vital role of change management in the successful adoption of e-business practices. Furthermore, it delves into the analysis and assessment of e-business activities, underscoring the significance of key performance indicators (KPIs) as essential tools for monitoring and optimizing company performance in the digital realm.

Rounding out its comprehensive coverage, this book concludes with an in-depth exploration of the legal, regulatory, and ethical dimensions that emerge within the e-business landscape. Privacy, data protection, intellectual property rights, and other critical concerns must be skillfully addressed by businesses operating in the digital economy. Through its meticulous examination of these topics, this book equips readers with the necessary knowledge to navigate the legal and ethical challenges of the e-business realm.

Written with e-business enthusiasts of all levels in mind, including students, professors, and practitioners, this book empowers readers to navigate the intricacies of the digital economy with confidence. Its aim is to foster a profound understanding of the various facets of e-business, enabling readers to flourish in the ever-evolving landscape. By offering valuable perspectives and insights into the realm of e-business, this book serves as a trusted companion for those seeking success in the digital economy.

Embrace the boundless possibilities of e-business, and let this book be your guide on the transformative journey ahead.

Hamta Business Corporation
Fairleigh Dickinson University
Vancouver, Canada

Mitra Madanchian

Preface

In an age where the digital landscape is reshaping the very fabric of business, e-commerce has emerged as a powerful force propelling companies into new frontiers. If you're intrigued by the remarkable transformations unleashed by the Internet and the boundless potential of e-business, then this book is your essential companion on an enlightening expedition.

Drawing from extensive research and practical insights, this comprehensive guide unveils the intricate tapestry of e-business, empowering readers with the knowledge and strategies needed to navigate this ever-evolving realm. With captivating storytelling and real-world examples, the book takes you on a captivating journey from the foundations of e-business to its cutting-edge models and tactics, revealing how forward-thinking organizations harness the digital economy to achieve unprecedented success.

The goal of this book is to offer a thorough understanding of the many aspects of e-business. It covers a broad range of subjects, beginning with an introduction to e-business and its importance in today's corporate environment. The rest of the book delves into the numerous models and tactics that firms might use to thrive in the digital economy. Because the e-environment brings both possibilities and problems, this book delves into these aspects, as well as the growth and uses of the internet in e-business. Security and control are critical components in the e-business environment, and this book discusses the different methods that firms may take to protect their online assets in detail.

Billing and payment systems, supply chain management, and procurement are critical aspects of e-business, and the book covers them all, including the use of technology to speed up the supply chain process. The book also covers digital marketing, customer relationship management, and business intelligence, revealing how organizations may use these features to improve their e-business operations. The rest of the book delves into the challenges that firms encounter while implementing an e-business model, as well as the importance of change management in the process. Furthermore, the book delves into the analysis and performance assessment of e-business activities, as well as the usage of key performance indicators (KPIs) to monitor company performance. The book finishes with a detailed study of the legal, regulatory, and ethical issues that arise in the e-business context. Businesses must handle privacy, data protection, intellectual property rights, and other legal and ethical challenges while operating in the digital economy.

This book is written for e-business aficionados of all levels, including students, professors, and practitioners. It gives a thorough understanding of the numerous aspects of e-business, allowing readers to successfully navigate the e-business environment. This book aims to help readers flourish in the digital economy by delivering e-business perspectives and information.

Vancouver, BC, Canada Hamed Taherdoost

Acknowledgement

I would like to express my deepest gratitude to my incredible wife, Mitra, whose unwavering love and support have been my anchor throughout this journey. To my precious daughter, Hamta, and my son, Kiasha, both of whom have brought immense joy to my life, thank you for being my constant inspiration.

I am also immensely grateful to my colleagues at Hamta Business Corporation and the esteemed faculty at University Canada West for their guidance, collaboration, and invaluable contributions that have shaped the path of my success.

Contents

Chapter 1
Introduction to E-Business

Abstract The rise of the digital age has transformed businesses, giving rise to e-business as an alternative to traditional operations. E-business encompasses online transactions, supply chain management, customer communication, and financial transactions. Digital technology has expanded businesses' reach, enabling global market access and fostering growth and innovation. This chapter provides a comprehensive overview of e-business, covering its historical evolution, critical technologies, trust establishment, and marketing strategies. It begins with a historical overview, tracing e-business from the Internet's introduction to mobile technology advancements. This context enhances understanding of the current state of e-business. Key technologies utilized in e-business implementation, such as the Internet, e-commerce platforms, digital payments, and mobile devices, are explored. These technologies facilitate secure transactions, customer communication, and supply chain management. Building trust is vital in e-business, and mechanisms for establishing trust with customers are examined. Security measures, product/service reliability, and excellent customer service contribute to trust and brand reputation. The chapter concludes by analyzing marketing strategies and platforms in e-business. Customized marketing approaches tailored to online customers' needs are emphasized, including social media platforms, search engine optimization, and digital marketing techniques.

1 Explanation of the Concept of E-Business

E-business has become an increasingly important aspect of the global economy, with more and more businesses turning to digital channels to reach customers and streamline their operations. By using electronic means such as websites, mobile applications, and social media, businesses can offer customers the convenience of shopping anywhere and anytime. They can also collect valuable data on customer behavior and preferences. E-business also allows businesses to reduce costs by automating processes and eliminating the need for physical storefronts or offices.

H. Taherdoost, *E-Business Essentials*, EAI/Springer Innovations in Communication and Computing, https://doi.org/10.1007/978-3-031-39626-7_1

Moreover, e-business facilitates faster and more efficient communication between businesses and their partners, enabling them to collaborate more effectively and make quicker decisions. As a result, e-business has transformed how many industries operate and is expected to grow in importance.

E-business has brought about a revolution in the world of commerce, enabling businesses to reach a wider audience and conduct transactions seamlessly across the globe. With the advent of the Internet, businesses no longer need to rely on traditional brick-and-mortar stores to sell their products and services. Instead, they can leverage various digital channels to reach potential customers, build brand awareness, and drive sales. E-commerce, which involves selling goods and services online, is one of the most popular activities within e-business. E-marketing and e-advertising are other crucial activities that enable businesses to promote their offerings online and attract customers to their websites or social media pages. Meanwhile, e-payment, e-procurement, e-logistics, and e-service all provide different functionalities to help businesses manage their online transactions, source goods and services, and support customers. By combining these different e-business activities, businesses can create a comprehensive online presence and provide a seamless user experience to their customers.

E-commerce has revolutionized how businesses operate in the digital age, allowing them to expand their reach to a global audience and operate around the clock. With the rise of online marketplaces like Amazon, eBay, and Alibaba, e-commerce has become more accessible to small- and medium-sized businesses, enabling them to compete with larger companies. Additionally, using electronic payment systems such as PayPal, Stripe, and Square has made it easier and more secure for consumers to purchase online, leading to increased online sales. Furthermore, advancements in logistics technology and shipping infrastructure have enabled e-commerce businesses to deliver goods to customers quickly and efficiently, further contributing to the industry's growth. Overall, e-commerce has become essential to modern business, transforming how companies sell and consumers buy goods and services.

In today's digitally connected world, e-marketing and e-advertising have become essential components of e-business. These techniques help businesses reach a global audience, allowing them to promote and sell their products and services online. One of the most popular techniques is search engine optimization (SEO), which involves optimizing a website's content to rank higher in search engine results pages. Social media marketing is another effective strategy businesses can use to engage with customers and build their brand. Businesses can establish a loyal following and increase their online presence by creating compelling social media content and engaging with customers. Email marketing is another valuable tool for businesses to promote their products and services. Businesses can build relationships with their audience and generate leads by sending targeted emails to potential customers. E-payment and e-procurement have revolutionized the way businesses conduct financial transactions. E-payment has simplified online transactions, eliminating the need for cash payments, checks, and money orders. This has made it easier for businesses to reach a global customer base, as customers from different countries can easily make payments. E-wallets have also become popular, with users able to store their debit and credit card information in one convenient location. This saves time, as users no longer need to enter their payment information repeatedly for each transaction [1].

Similarly, e-procurement has enabled businesses to streamline their purchasing process by utilizing digital platforms to source and purchase goods and services from global vendors. This has resulted in cost savings, improved efficiency, and reduced lead times, as businesses can source products from anywhere in the world with just a few clicks. E-procurement also offers greater transparency in the procurement process, enabling businesses to track orders and manage their inventory in real time. As a result, e-payment and e-procurement have become integral components of e-business and are crucial in driving growth and success for enterprises.

E-service has revolutionized how businesses handle customer support, providing a faster, more accessible way to assist clients around the clock. With e-service, businesses can offer immediate assistance to customers, regardless of location, through various digital channels. This helps companies save money by reducing the need for extensive customer support teams while still providing customers with a high level of service. Additionally, e-logistics has transformed how businesses manage their supply chain, making it more efficient and effective. Through digital platforms, companies can track inventory, monitor shipments, and manage logistics in real time, reducing delivery times and ensuring that products reach customers in a timely and cost-effective manner. E-business has opened up numerous opportunities for organizations of all sizes, enabling them to reach a global audience, streamline operations, and provide better customer service [2].

Another significant advantage of e-business is its ability to operate 24/7 without being limited by traditional business hours or geographical restrictions. E-businesses can provide services and sell products around the clock, enabling them to generate revenue even when their physical stores are closed. Moreover, e-businesses can target specific customer segments more effectively by using data-driven insights to tailor their marketing efforts and offerings to match their customers' preferences and needs. This capability allows businesses to improve customer satisfaction, build brand loyalty, and increase customer retention. Overall, e-business provides numerous benefits for companies seeking to expand their customer base, streamline their operations, and increase their revenue streams.

The dependence on digital technology in e-business is limited to using digital tools and platforms and extends to managing and storing data. E-businesses generate and store vast amounts of data on their customers, transactions, and other business activities. To manage this data effectively, e-businesses use digital systems such as customer relationship management (CRM) software and enterprise resource planning (ERP) systems. These systems enable e-businesses to store, manage, and analyze their data, providing valuable insights that can be used to make informed business decisions.

Moreover, e-businesses also need to comply with various legal and regulatory requirements to operate effectively. For instance, they should comply with data protection and privacy regulations, such as the EU's General Data Protection Regulation (GDPR) and the USA's California Consumer Privacy Act (CCPA). They also need to comply with taxation laws in different jurisdictions, which can be complex, primarily when operating across borders. Failure to comply with these regulations can result in hefty fines, legal penalties, and damage to the business's reputation. Therefore, e-businesses should stay updated on the latest regulations and comply with all applicable laws.

Furthermore, e-businesses often benefit from lower variable costs as well. They can leverage digital tools and automation to streamline their supply chain management, order processing, and inventory control, resulting in greater efficiency and reduced labor expenses. E-commerce platforms also offer a variety of cost-effective marketing and advertising options, allowing e-businesses to target their audiences precisely and monitor the effectiveness of their campaigns in real time. Consequently, e-businesses can allocate resources more effectively and efficiently to improve their bottom line.

In the highly competitive world of e-commerce, customer experience is a critical factor in building brand loyalty and increasing sales. E-businesses understand the importance of offering customers an easy and enjoyable shopping experience. They focus on creating user-friendly interfaces that make it easy for customers to navigate their websites and find what they want. E-businesses also offer flexible payment methods to accommodate their customers' preferences, such as accepting credit cards, e-wallets, and other digital payment systems. Fast and reliable shipping and delivery services are another crucial aspect of the customer experience. E-businesses strive to provide timely and hassle-free delivery of products to customers, as this can be a decisive factor in the customer's decision to make future purchases from the company. Additionally, online reviews and ratings are commonly used to measure customer satisfaction and provide insights into areas that need improvement, allowing e-businesses to improve their offerings and stay ahead of the competition continuously.

The use of data and analytics is crucial to e-businesses in many ways. First, it allows them to understand better their customer's behavior, preferences, and needs. E-businesses can analyze customer data, such as purchase history and website interactions, to personalize the customer experience, recommend products and services, and target marketing efforts. Second, data and analytics can help e-businesses optimize their operations, from inventory management to supply chain efficiency. By analyzing data on sales, shipping, and delivery times, e-businesses can identify areas for improvement and streamline their processes, ultimately leading to increased profitability. Lastly, data and analytics enable e-businesses to monitor and evaluate their performance over time, making adjustments and improvements where necessary. By tracking metrics such as website traffic, conversion rates, and customer retention, e-businesses can gauge the success of their strategies and make data-driven decisions [2].

E-business and e-commerce are crucial concepts in the modern business environment. With the advent of the Internet and the rise of digital technologies, more and more businesses are exploring the opportunities presented by these concepts. E-business and e-commerce provide companies with new and innovative ways to reach customers, expand their market share, reduce costs, and improve their overall performance. Therefore, it is essential to understand the differences between e-business and e-commerce and how they operate in the digital business environment. This study explores the fundamental differences between e-business and e-commerce and their respective scopes. It focuses on how businesses can leverage these concepts to improve their operations and increase their competitiveness.

E-commerce has become essential to the global economy, with millions of people conducting daily transactions. E-commerce systems provide a convenient and efficient way for customers to purchase products and services from the comfort of their homes, regardless of location. E-commerce platforms are designed to be user-friendly, secure, and reliable, with features like online shopping carts, payment gateways, and real-time inventory tracking. Additionally, e-commerce allows businesses to reach a wider audience, expand their customer base, and increase revenue streams. With the rise of mobile commerce and the increasing use of social media platforms as shopping channels, the scope and potential of e-commerce are continuously expanding, creating new opportunities for businesses and consumers alike.

E-business includes all online activities using digital technology, which goes beyond just buying and selling goods or services. E-business is an umbrella term that encompasses a wide range of business activities, including digital marketing, advertising, payment processing, procurement, logistics, customer service, and other operations that are conducted through digital channels. This allows businesses to connect with customers and partners worldwide using various online tools and platforms. The use of digital technology in e-business provides numerous benefits, including faster transaction times, lower costs, improved efficiency, and enhanced customer experience. Businesses can leverage various digital channels such as websites, social media, mobile applications, email, and messaging platforms to conduct e-business operations effectively.

One of the primary distinctions between e-business and e-commerce is their focus. E-commerce primarily involves buying and selling products and services online, whereas e-business encompasses a broader range of operations beyond commercial transactions. E-business aims to use digital technology to facilitate all aspects of business operations, from marketing to logistics to customer support. By incorporating digital tools and platforms, companies can reduce costs, streamline processes, and improve customer experience.

Another distinction between e-business and e-commerce is their scope. E-commerce is a subset of e-business focused on facilitating transactions between companies and customers or between companies. On the other hand, e-business encompasses a much broader range of activities, including marketing, sales, supply chain management, and customer support. Through e-business, companies can manage and optimize all digital operations, allowing for greater control and efficiency.

One of the critical distinctions between e-business and e-commerce is the degree of integration between them. E-commerce platforms are standalone systems enabling online transactions between buyers and sellers. Such platforms are designed to be easy to use and can often be set up quickly, even by small businesses. However, these platforms are typically separate from a company's digital tools and platforms, such as customer relationship management (CRM), supply chain management, and marketing automation software. In contrast, e-business involves the integration of various digital technologies and platforms to create a seamless and integrated business environment. For example, an e-business may use a CRM system to manage customer data and interactions, a supply chain management system to manage inventory and logistics, and marketing automation software to execute targeted campaigns across various digital channels. This integration enables businesses to streamline operations, increase efficiency, and improve customer experience [2].

The complexity of e-business and e-commerce systems is essential when deciding which to adopt. E-commerce systems are generally more straightforward since they are focused on facilitating transactions between buyers and sellers. They typically involve a relatively small number of digital tools and platforms, such as an online store and a payment gateway. On the other hand, e-business systems encompass a more comprehensive range of activities, including marketing, advertising, customer service, and supply chain management. Each activity may require different digital tools and platforms, creating a more complex and challenging environment. However, the benefits of e-business systems often outweigh the complexity, as they can lead to greater efficiency, reduced costs, and improved customer experiences. Figure 1.1 provides a helpful summary of the differences between e-business and e-commerce systems, which can help organizations make informed decisions about which approach to adopt.

Furthermore, e-commerce is relatively more uncomplicated than e-business, mainly involving buying and selling products and services facilitated by electronic payment systems. On the other hand, e-business includes a more extensive range of digital activities such as e-marketing, e-advertising, e-payment, e-procurement, e-logistics, and e-service in addition to e-commerce. E-commerce transactions are usually conducted between firms and customers or between enterprises, whereas e-business covers all kinds of business activities carried out online using digital

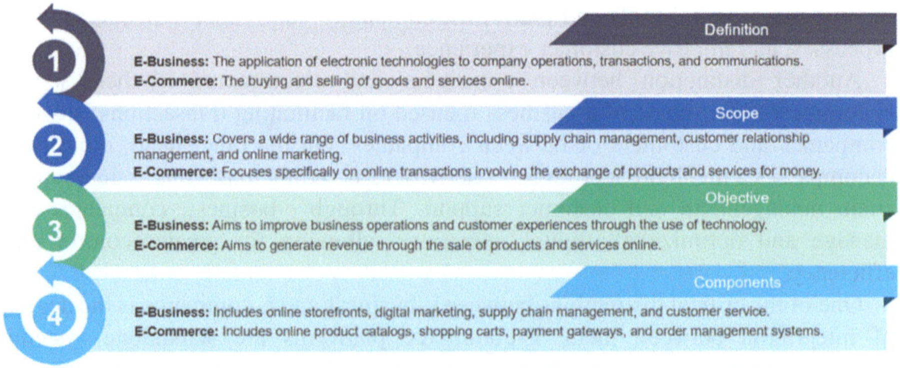

Fig. 1.1 Critical differences between e-business and e-commerce

technology. In summary, while e-commerce is an essential part of e-business, e-business encompasses many more digital tools and platforms, making it a more complex and integrated approach to conducting business in the digital age.

2 Historical Overview of E-Business

The electronic business has a brief history, dating back to the early days of the Internet in the 1990s. While the concept of electronic commerce had existed for decades before the emergence of the Internet, the widespread adoption of the web made e-business a viable option for companies of all sizes. Initially, e-business centered on the online buying and selling products and services, with pioneers like Amazon and eBay leading the way. As e-commerce gained popularity, businesses began to explore other areas of e-business, such as e-marketing and e-advertising. Google and Yahoo! emerged as early leaders in e-marketing, using search engines and targeted advertising to help companies reach new audiences and increase online sales.

With the advent of digital technologies, businesses expanded beyond e-commerce and e-marketing. They began to explore new types of e-business, such as e-procurement, e-logistics, and e-service, to streamline their operations and enhance the customer experience. This led to new e-business models, such as the software-as-a-service (SaaS) model, which provided organizations with access to powerful e-business tools and platforms without requiring significant in-house IT resources.

The evolution of e-business continued with the spread of data analytics and artificial intelligence technologies in the late 2000s and early 2010s. These technologies enabled companies to collect vast amounts of data on their customers, sales, and operations and use them to guide business decisions and improve performance. E-business also incorporated machine learning algorithms and predictive analytics to automate and improve business operations, resulting in cost savings and increased productivity. Recent years have seen the emergence of new technologies, such as blockchain, virtual reality, and the Internet of Things (IoT), which have further expanded the scope and potential of e-business. These technologies have enabled companies to create innovative e-business models, such as blockchain-based supply chain management systems, virtual reality retail experiences, and IoT-enabled innovative logistics platforms (Fig. 1.2) [1].

Overall, the history of e-business has been marked by a steady expansion of its capabilities and functions, driven by the growing complexity of digital technology and the increasing demand for online services and experiences. As e-business continues to mature and evolve, it will undoubtedly play an even more significant role in the global economy, providing companies of all types and sizes with new opportunities to connect with customers, streamline operations, and promote growth and innovation.

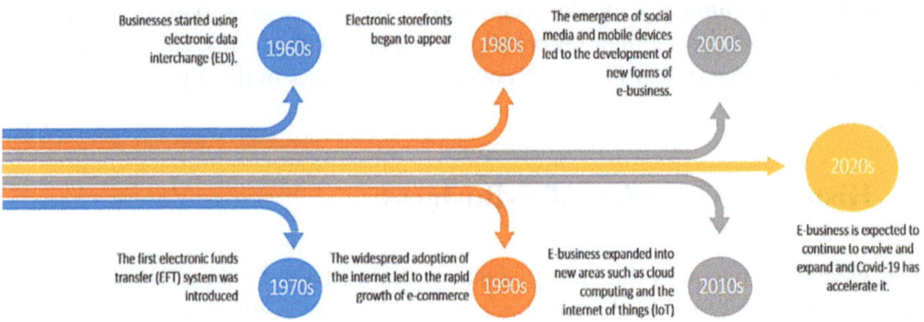

Fig. 1.2 Historical evolution of e-business

Internet's emergence and expansion have dramatically affected the creation and expansion of e-business, or electronic business. The Internet has established a worldwide platform that enables companies to connect with consumers, partners, and suppliers around the globe, hence allowing new kinds of trade, communication, and cooperation. In the 1960s, the US Department of Defense started creating a decentralized communication network, ARPANET. This network was created to facilitate information sharing and project collaboration between military and academic researchers regardless of location. Throughout the 1980s and 1990s, the Internet's initial function as a military and academic communication network started to change. The advent of the World Wide Web, or WWW, in the early 1990s, allowed companies to construct websites that anybody with an Internet connection could visit, thus creating new potential for e-commerce [1].

The expansion of the Internet has had many effects on e-business. One of the most significant effects has been the emergence of a global marketplace that enables firms to connect with clients and partners worldwide. This has allowed new types of e-commerce, such as online marketplaces and digital storefronts, which enable firms to offer items and services to clients without a physical shop or sales staff. The Internet has also allowed new means of communication and cooperation, including email, instant messaging, and video conferencing, revolutionizing how organizations communicate with their consumers, partners, and suppliers. This has resulted in the rise of new types of e-business, such as e-service and e-procurement, which use digital technology to expedite corporate operations and enhance the customer experience. New business models, such as the software-as-a-service (SaaS) model, have emerged due to the Internet's influence on e-business. This strategy enables enterprises of all sizes to engage in the digital economy by providing access to robust e-business tools and platforms without requiring substantial internal IT resources.

Overall, the development and expansion of the Internet have profoundly affected e-business, allowing organizations to access new markets, simplify operations, and develop new kinds of trade, communication, and cooperation. As the Internet develops and matures, it will undoubtedly play a more significant role in the global economy, spurring economic development and innovation in the digital era.

E-business has grown substantially in recent years for several reasons that have increased its significance in the global economy. These reasons include technological advancements, changes in customer behavior, and new business models, all of which have generated new opportunities for firms to compete and thrive in the digital age. The accessibility and affordability of digital technology, such as the Internet, mobile devices, and cloud computing, are some of the primary reasons for the rise and significance of e-business. These technologies have made it easier for organizations to create an online presence, communicate with consumers and partners, and streamline their processes, enabling new growth and innovation prospects.

Consumers increasingly use digital platforms to research products and services, compare pricing, and make purchases, which is a significant factor in the expansion of e-commerce. This shift in consumer behavior has created new opportunities for firms to connect and interact with consumers through digital marketing channels, such as social media, email marketing, and search engine optimization. In addition, the rise of new business models, such as the sharing economy and subscription-based services, has given firms new ways to monetize their assets and compete in new markets. These models have allowed companies to leverage digital technology to provide consumers new value, such as access to pooled resources or personalized goods and services [2].

The increasing globalization of the economy, which has opened new opportunities for companies to access customers and partners worldwide, also contributes to the growth and significance of e-business. E-business has helped enterprises to develop a global presence, expand their customer base, and reduce the costs and risks of international trading. Ultimately, the growth and significance of e-business can be attributed to several factors that have provided new competitive and profitable opportunities for firms in the digital age. As digital technologies expand and mature, e-commerce is poised to become an increasingly significant global economic growth and innovation engine.

E-business has emerged as a successful business model, offering a range of benefits that enable organizations to compete and thrive in the digital era. These benefits include increased efficiency, reduced expenses, wider reach, and enhanced customer engagement. One of the primary benefits of e-business is increased efficiency. By using digital technology, organizations can simplify their operations, automate repetitive processes, and eliminate manual processes, resulting in quicker and more efficient processes. This can help businesses increase productivity, reduce costs, and enhance their bottom line. Another benefit of e-business is reduced expenses. By eliminating the need for storefronts, businesses can save money on rent, electricity, and other costs associated with maintaining a physical presence. Additionally, e-business enables companies to reach a larger audience without costly marketing activities, reducing the cost of customer acquisition. Figure 1.3 demonstrates the advantages and disadvantages of e-business.

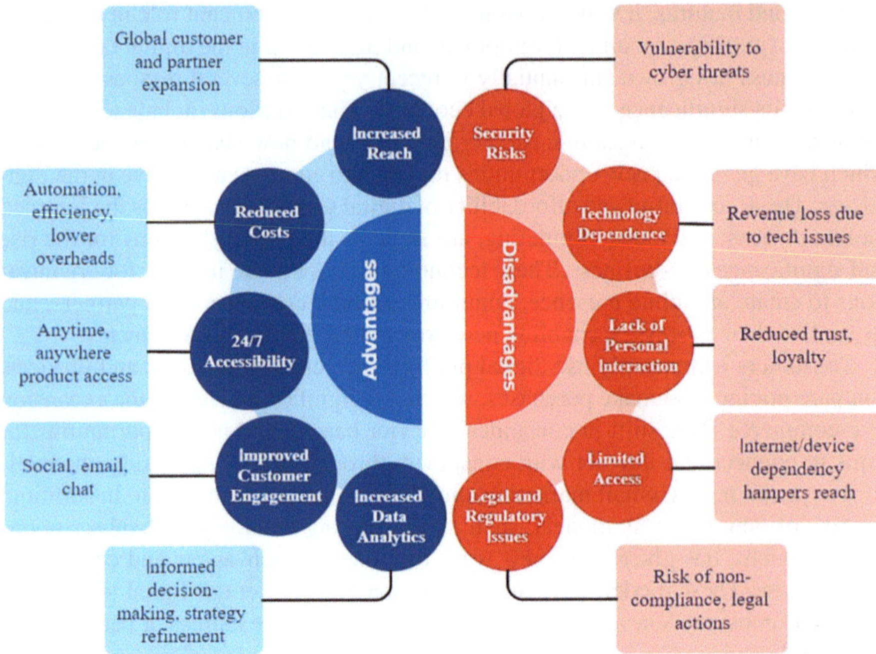

Fig. 1.3 Advantages and disadvantages of e-business

3 Overview of Key Technologies Used in E-Business

E-business refers to the application of digital technologies in commercial transactions and processes. This has led to the developing of various technologies utilized in e-business models and tactics. This section will overview the most critical e-business technologies (Fig. 1.4).

- E-commerce Platforms: E-commerce platforms provide businesses with an online storefront to display and sell their products and services. They typically offer shopping carts, payment gateways, and order management systems.
- Customer Relationship Management (CRM) Systems: CRM systems allow firms to manage customer interactions and connections. This includes monitoring customer contacts, evaluating consumer data, and overseeing customer service.
- Enterprise Resource Planning (ERP) Systems: ERP systems provide firms with an integrated platform for managing multiple facets of their operations, including inventory management, supply chain management, and financial management.
- Cloud Computing: Cloud computing enables organizations to store and access data and applications online instead of on local PCs and servers. This affords businesses greater adaptability, scalability, and cost-effectiveness.

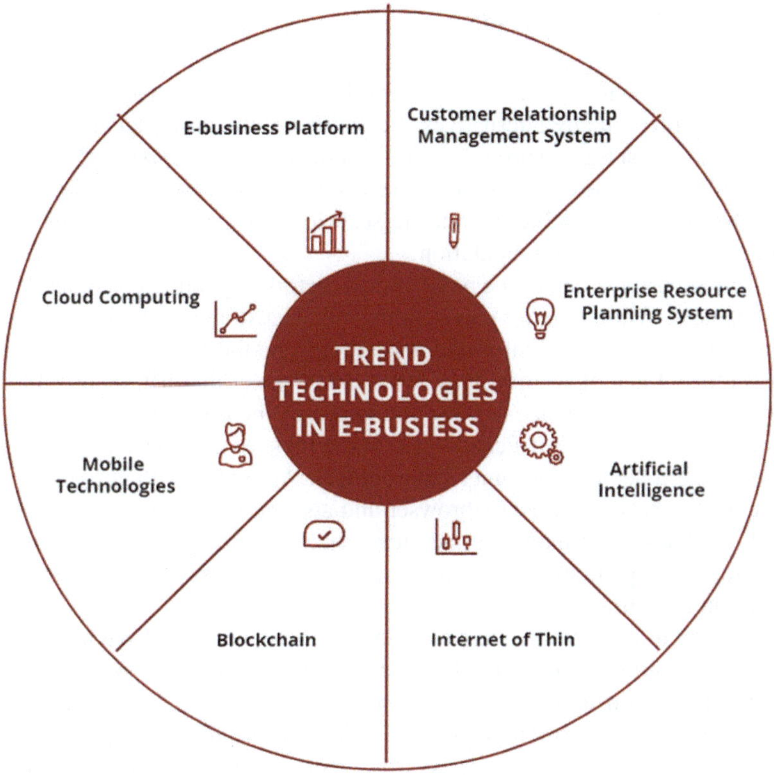

Fig. 1.4 Trend technologies in e-business

- Mobile Technologies: Mobile technologies such as smartphones and tablets have become integral to e-commerce, allowing firms to reach customers on the go and offer a seamless mobile experience.
- Artificial Intelligence (AI): AI technologies like machine learning, natural language processing, and chatbots are widely employed in e-business to automate tasks, deliver personalized experiences, and enhance consumer interaction.
- Blockchain: Blockchain is a decentralized digital ledger that enhances corporate transactions and processes' security, transparency, and traceability.
- Internet of Things (IoT): IoT technologies, including sensors, wearables, and smart devices, are utilized in e-business to provide enterprises with real-time data and insights, improve supply chain management, and enhance consumer experiences.

E-business has been made possible by multiple essential technologies that have revolutionized how firms function and connect with their customers. These technologies have enhanced firms' efficiency, productivity, and competitiveness in a world that is becoming increasingly digital and interconnected.

4 Trust in an E-Business

Establishing trust with clients is critical for the success of any e-business in today's digital era. With so many alternatives accessible online, customers are looking for companies they can trust to provide a safe and secure buying experience. As a result, e-commerce companies should emphasize establishing trust with their consumers via various measures, including website security, transparency, customer reviews, support, and a strong brand reputation.

Website security is one of the most critical factors in gaining customers' trust for any e-business. While making transactions online, customers should have faith that their personal and financial information is secure, and the e-business should provide this assurance. Lack of security may result in consumer distrust, financial loss, and data breaches, all of which can harm the image of the e-business. Getting an SSL (Secure Sockets Layer) certificate is one of the most critical steps e-businesses can take to create a secure consumer environment. An SSL certificate encrypts data communicated between a user's browser and an e-business server, making it more difficult for hackers to intercept and steal data. Customers can recognize an SSL certificate by the padlock symbol in their browser address bar and trust that their information is secure.

In addition to SSL certificates, e-businesses may use secure payment gateways and two-factor authentication as additional security precautions. Two-factor authentication requires consumers to provide a verification number, which may be texted to their phone, to access their account or make a transaction. Doing this adds more protection and unauthorized access to client accounts is more complicated. Customers can also make purchases securely using secure payment gateways like PayPal or Stripe without directly providing their financial information to the online store. E-businesses should routinely check their website for security flaws and apply any necessary security fixes and upgrades. It is essential to keep up with security updates to secure client information since hackers constantly seek methods to exploit website security flaws. E-businesses may also use technologies like firewalls and intrusion detection systems to recognize and stop assaults [2].

To enhance consumer confidence, e-businesses should adhere to legislation and standards like the General Data Protection Regulation (GDPR) and the Payment Card Industry Data Security Standard (PCI DSS). Following these guidelines demonstrates a company's commitment to data security and privacy.

The data breach at the credit reporting agency Equifax in 2017 is a real-world example of website security and creating a secure environment for clients. The breach exposed millions of consumers' names, social security numbers, dates of birth, and credit card information. The Equifax data leak served as a wake-up call for organizations and consumers, highlighting the critical importance of website security and data privacy. In the aftermath of the hack, Equifax received widespread criticism for its lack of proper security measures, including the use of outdated software and failure to fix known vulnerabilities.

As a result of the breach, Equifax took several security measures to improve the security of its website, including implementing two-factor authentication, performing monthly security updates and patches, and using a more secure payment gateway. To better secure client information, the organization also strengthened its compliance with applicable standards and laws, such as the PCI DSS and the GDPR. Although the Equifax data breach was a severe setback for the company and its consumers, it was also a valuable lesson in the significance of website security and the need for e-businesses to take proactive steps to secure client data. By establishing comprehensive security measures and staying informed about evolving threats and vulnerabilities, e-businesses can provide their customers with a safe and secure environment and build brand trust.

Transparency is a fundamental feature of e-business, which requires a company's policies, procedures, and data-gathering systems to be transparent and honest. Transparency facilitates the development of credibility and confidence among consumers, suppliers, and other stakeholders. Transparency in the context of e-business may relate to numerous topics, including:

- Privacy policies: E-businesses should have clear and thorough privacy policies that define how they gather, utilize, and secure client information. This includes giving consumers control over their personal information and advising them of any data sharing with third parties.
- Product descriptions: E-businesses should give accurate and comprehensive descriptions of their goods and services, including information about their features, benefits, and restrictions. This aids in preventing consumer misunderstanding or dissatisfaction and decreases the possibility of returns or conflicts.
- Pricing: Online firms should be upfront about their prices, including all applicable fees, taxes, and shipping expenses. This enables consumers to make educated purchase choices and eliminates unpleasant surprises at checkout.
- Reviews and feedback: E-businesses should let consumers give evaluations and comments about their goods or services, as this may aid in establishing trust and credibility. Also, they should reply to any unfavorable evaluations openly and helpfully, addressing any concerns or difficulties.
- Ethical and social responsibility practices: E-businesses should disclose their ethical and social responsibility practices, such as their supplier chain, labor practices, and environmental impact. This helps establish credibility with clients who appreciate socially responsible companies [1].

Transparency is essential for establishing trust, credibility, and loyalty with consumers, workers, and other stakeholders in e-business. By being transparent and honest about their policies and procedures, e-businesses may show their dedication to ethical and responsible business operations, enhancing their reputation and expanding their clientele.

Outdoor apparel and equipment shop Patagonia is a real-world example of a corporation demonstrating openness in e-business. Patagonia is recognized for its dedication to sustainability and ethical business methods, and its website and marketing materials reflect this. Patagonia's website provides comprehensive information on its supply chain and manufacturing processes, including the factories and mills it collaborates with and the materials used to produce its goods. Additionally, the firm provides details on its environmental and ethical activities, including its Fair Trade Certified program and its commitment to utilizing recycled products. Patagonia's dedication to ethical business methods and commitment to openness has helped create consumer confidence and established the firm as a pioneer in a sustainable fashion. Patagonia has established a devoted consumer base that appreciates the quality of its goods and the company's commitment to social and environmental responsibility by being upfront and honest about its policies and practices [2].

Customer reviews and ratings are a critical component of e-commerce. When making online purchases, users often rely on reviews and ratings from other customers to make informed decisions. These evaluations and ratings can provide insight into product or service quality, customer service, and overall consumer satisfaction. Customer reviews and ratings can also be vital in developing client trust and a favorable reputation for online businesses. Positive reviews and high ratings can help attract new customers, while negative reviews may have the opposite effect.

Several online companies have built systems that allow customers to provide feedback and score their experiences to promote reviews and ratings. Sites like Amazon and TripAdvisor have built-in review systems allowing users to comment and rate products or services on a one-to-five-star scale. Some e-businesses may collect feedback through email surveys, social media, or third-party review systems. E-businesses should regularly monitor customer reviews and ratings to ensure they provide high-quality goods and services and address any issues or complaints. Negative reviews can provide valuable feedback and indicate areas for improvement, but they should be responded to promptly and professionally. E-businesses can also monitor customer reviews and ratings as a marketing strategy. Positive reviews and high ratings can be displayed on the company's website or social media channels to demonstrate customer satisfaction and inspire confidence among potential customers. Some e-businesses may also offer discounts or incentives to customers who submit reviews, which can encourage feedback and enhance the company's image.

Overall, customer evaluations and ratings play a vital role in e-commerce and can influence the success and reputation of a company. By providing excellent goods and services and addressing customer complaints, e-businesses can create a loyal customer base and an excellent reputation.

Amazon, the world's largest online retailer, is an example of a company that effectively uses customer feedback and ratings. Amazon displays user reviews and ratings on its product pages, allowing customers to review the thoughts and experiences of others who have purchased and used the same product. Customers can rate items on a five-star scale and write reviews that include information about the product's features, quality, and performance. Reviews can also include images and videos, providing buyers with additional information. Amazon's use of user reviews and ratings has effectively established consumer trust and credibility. According to a 2018 poll by Feedvisor, 97% of Amazon customers check product reviews before making a purchase, and products with higher ratings and more reviews are more likely to be purchased. In addition to providing buyers with vital information, customer reviews and ratings also provide Amazon and its sellers with valuable feedback. In addition to highlighting areas where products need improvement, reviews can reveal consumer preferences and demand trends. This feedback can guide product development and marketing initiatives and enhance customer care and support.

Customer service is essential to e-business, referring to the support clients receive before, during, and after a transaction or encounter with an online firm. Good customer service is crucial for establishing and sustaining client relationships, enhancing customer happiness, and boosting sales and revenue. Practical customer assistance in e-business depends on various factors, including timeliness, accessibility, and personalized communication. E-businesses should guarantee that they provide a variety of customer care channels, including email, phone, chat, and social media, so that consumers may choose how to reach them. Additionally, they should respond promptly to client inquiries and complaints, ideally within 24 h, to demonstrate that their problems are handled seriously [2].

Personalization is another essential component of efficient customer service. E-businesses should aim to offer individualized contacts, such as addressing consumers by name and customizing solutions to their particular requirements and problems. This may increase consumer trust and loyalty and provide a great experience. Moreover, e-businesses may use technology such as chatbots, virtual assistants, and automated replies to provide clients with prompt and efficient solutions to their inquiries. These technologies are particularly effective for delivering customer service around the clock, minimizing wait times, and addressing common questions. Effective customer service is crucial for the success of e-businesses in today's competitive online economy. By offering timely, accessible, and customized help, e-businesses may strengthen client connections, boost customer happiness, and ultimately promote their growth and profitability.

Zappos, an online shoe and clothing shop, is a real-world example of a corporation providing rapid and dependable customer service in e-business. Zappos has earned a reputation for providing superior customer service, and its website and practices reflect this. Zappos' customer service channels, including phone, email, and live chat, are available around the clock. The company has a staff of customer service professionals trained to provide polite and valuable support to clients, whether they have questions about items, want to help place orders, or have issues with returns and exchanges. Moreover, Zappos has a liberal return policy that enables buyers to return items for free within a year of purchase. This approach and the company's prompt customer service have helped establish consumer trust and loyalty. The dedication of Zappos to providing dependable and prompt customer service has been crucial to the company's success. Zappos has established itself as a pioneer in e-commerce customer support by providing several avenues for assistance and ensuring that consumers receive friendly and helpful service.

In e-business, brand reputation refers to customers' view of a specific brand based on their interactions, experiences, and impressions. It is crucial to the success of any e-business since it influences customer behavior, loyalty, and confidence. To develop a strong brand image in e-commerce, businesses should prioritize some crucial elements. These include delivering high-quality goods and services, being open and honest about their company operations, giving outstanding customer service, and having a solid web presence.

Online reviews and ratings are among the most critical aspects influencing brand reputation in e-business. When making purchase choices, consumers depend significantly on the opinions of other customers, and unfavorable evaluations may dramatically harm a brand's reputation. Therefore, businesses should actively monitor and respond to consumer comments to maintain a favorable reputation.

Cybersecurity is an additional aspect that might affect brand reputation in e-business. As the number of cyberattacks aimed at e-businesses rises, people are more concerned about their personal and financial data safe. Thus, businesses should proactively safeguard client information and preserve their confidence. A strong brand image in e-business can be a potent tool that helps organizations acquire new clients, maintain current ones, and set themselves apart from rivals. By concentrating on essential criteria such as product quality, customer service, online reviews, and cybersecurity, e-businesses may establish a trustworthy brand reputation that encourages long-term success [1].

Uber is a real-world illustration of the significance of brand reputation in e-business. A series of disputes and scandals involving the ride-hailing company tarnished its brand image. In 2017, the firm received criticism and boycotts when a former employee detailed her experiences with sexual harassment and discrimination at the workplace in a blog post. This resulted in widespread outrage and a campaign to uninstall the Uber app. In addition to the sexual harassment issue, Uber was criticized for its aggressive business methods, which included a disregard for local legislation and the unfair treatment of its drivers. These incidents resulted in losing client confidence and loyalty, inadequate media attention, and a decrease in the company's market value. Uber undertook several improvements to strengthen its brand image and recover consumer confidence in response to these concerns. The corporation recruited a new chief executive officer, Dara Khosrowshahi, who apologized for the company's previous errors and committed to promoting safety and ethical business practices. Uber also incorporated additional safety measures, such as in-app emergency help and background checks for drivers, and introduced new rewards and incentives to enhance its connection with drivers. The company's attempts to enhance its image and recover consumer confidence highlight the significance of brand reputation in e-business. By proactively addressing problems and enhancing customer satisfaction, e-businesses can safeguard and strengthen their brand reputation, which is crucial for long-term success.

By effectively leveraging social media and digital marketing channels, Airbnb has established itself as a trusted and reliable platform for booking travel accommodations. Social media and digital marketing platforms are essential for establishing trust in e-commerce. Social media platforms and digital marketing channels allow companies to engage with their consumers, increase brand recognition, and advertise their goods and services. By utilizing these channels successfully, firms may generate consumer trust and a favorable brand image. Creating meaningful and exciting content is one method for using social media and digital marketing platforms to develop trust. These include blog entries, videos, and social media updates that give beneficial information to clients and demonstrate the business's expertise. By delivering high-quality content, companies may establish themselves as trustworthy source of information and earn consumers' confidence.

Social media and digital marketing platforms may be used further by actively connecting with clients. This may include swiftly responding to customer comments and messages, resolving customer problems and complaints, and giving helpful information or resources. Organizations may show dedication to customer service and establish audience trust by responding to and interacting with consumers. Social media and digital marketing platforms allow companies to present their goods or services genuinely and aesthetically attractive. With high-quality photographs and videos, firms may display their products in a visually appealing and motivating manner. This may generate enthusiasm and trust among prospective consumers, thus encouraging them to purchase [2].

Airbnb is an example of a company that uses social media and digital marketing channels to develop trust in its online business. The firm engages with its consumers and promotes its brand using social media platforms such as Facebook, Twitter, and Instagram. Airbnb has a robust social media presence, routinely publishing user-generated content and showcasing unusual and intriguing homes that are available for rent. The organization also utilizes social media to engage with clients, assisting and addressing issues quickly and flexibly. In addition, Airbnb employs digital marketing channels such as email marketing and search engine advertising to contact new consumers and establish credibility. The company delivers customized emails to consumers based on their search and booking history, providing personalized suggestions and deals. Additionally, they use search engine advertising to target individuals searching for vacation accommodations, ensuring their brand appears at the top of search results. Airbnb has established itself as a trustworthy platform for booking holiday lodgings using social media and digital marketing channels efficiently.

Search engine optimization (SEO) is a crucial trust-building approach for e-businesses. It involves leveraging relevant keywords and content to boost a website's exposure and rating on search engine results pages (SERPs). The higher a website ranks on SERPs, the more likely prospective consumers will visit it. Thus, SEO is vital for creating client trust and reputation in the online marketplace. By optimizing their websites for search engines, e-businesses can demonstrate their sector knowledge and provide their target audience with information that meets their demands. Consumers are likelier to trust and buy from firms at the top of search engine results pages. In addition, SEO can assist e-businesses in creating trust by improving the user experience of their websites. By focusing on website speed, mobile compatibility, and other technical characteristics that contribute to a great user experience, e-businesses can enhance client trust and trustworthiness. Furthermore, SEO can help e-businesses establish themselves as thought leaders by providing relevant information and insights through their website content. This can attract and retain clients who see the company as a reliable source of information.

SEO is critical to e-business trust-building, enabling e-businesses to demonstrate their knowledge, enhance the user experience, and position themselves as industry thought leaders. By investing in SEO tactics and keeping abreast of the latest trends and best practices, e-businesses can create consumer trust and credibility and thrive in the competitive online marketplace.

Airbnb is an example of a firm that uses search engine optimization (SEO) to develop client trust. As an online marketplace for short-term rentals, Airbnb has invested substantially in SEO to ensure its website ranks well in search engine results for relevant keywords, establishing user confidence. By optimizing its website content and structure for search engines, Airbnb has attracted a larger audience of prospective consumers and established itself as a reputable and dependable platform for short-term rentals. In addition to search engine optimization, Airbnb has leveraged various digital marketing channels, such as social media and email marketing, to establish consumer trust. For instance, the firm maintains a significant presence on social media sites like Instagram and Twitter. It publishes user-generated content and highlights good experiences from its community of hosts and guests. This helps to develop social proof and trust among first-time users who may need to be more open to using the site.

Paid advertising in e-business refers to Internet advertising where companies pay to display their ads on various online platforms to their target audience, including search engines, social media platforms, and other ad-supported websites. One of the primary advantages of paid advertising is that it can help establish credibility and trust with potential customers by enhancing brand exposure. Customers are more likely to view a company as credible and trustworthy if it appears at the top of search results or on a reputable website. However, there are potential trust issues associated with sponsored advertising. For instance, buyers may be suspicious of organizations that rely heavily on paid advertising, seeing it as a way to buy their way to the top rather than achieving success through the quality of their products and services. Moreover, some organizations may engage in false or misleading advertising, damaging their reputation and undermining consumer confidence.

To establish credibility through paid advertising, companies should ensure that their ads are transparent, truthful, and relevant to their target audience. Moreover, their advertising practices, such as disclosing sponsored content or paid endorsements, should be transparent. By using sponsored advertising wisely and intelligently, businesses can leverage its power to build credibility and trust with their target audience.

Airbnb is an example of a company that uses paid advertising to build trust. This online marketplace connects tourists with hosts for short-term lodging and uses paid search advertising on Google to rank at the top of search results for phrases such as "vacation rentals" and "places to stay." By ranking highly in search engine results, Airbnb can establish credibility with potential customers and position itself as a trustworthy supplier of vacation lodging. Additionally, Airbnb uses Facebook ads to reach potential customers based on their interests, behavior, and demographics, thus enhancing its online presence and reputation.

5 E-Business Marketing Strategies and Platforms

In today's digital era, e-business marketing is a crucial component of every organization that aims to succeed. The introduction of the Internet has provided companies with a new channel for efficiently reaching their target audience. Businesses should establish a thorough e-business marketing plan to accomplish this. In this book, we will explore the most critical parts of designing an e-business marketing strategy, including the importance of digital channels, assessing the efficacy of e-business marketing activities, and best practices for planning and managing e-business marketing campaigns.

Creating a thorough e-business marketing strategy requires the identification of target audiences, the establishment of goals, and the creation of an action plan. This plan should account for the unique features of e-business marketing, such as the need for targeted marketing and the use of digital platforms to reach clients. To identify target markets, market research should be conducted to comprehend the wants and preferences of prospective clients. This data is used to develop a profile of the target client, including demographic, psychographic, and behavioral characteristics. Establishing objectives entails identifying targets for the e-business marketing plan, such as increasing website traffic, generating leads, and boosting conversion rates. The action plan explains the activities that should be followed to attain these goals, including using digital channels such as websites, social media, email, and mobile devices [1].

In e-business marketing, digital channels such as websites, social media, email, and mobile applications should be balanced. These channels provide organizations with a cost-effective approach to contacting their target audience and cultivating consumer connections. Websites offer a venue for companies to promote their goods and services, while social media platforms enable businesses to interact with consumers and increase brand recognition. Email marketing is an excellent method for reaching clients directly, while mobile applications enable firms to give targeted information and offers.

Assessing the efficacy of e-business marketing initiatives is crucial to achieving corporate goals. E-business marketing initiatives are evaluated using web analytics and key performance indicators (KPIs). Web analytics inform organizations about website traffic, bounce, and conversion rates. Organizations may use key performance indicators (KPIs) such as client acquisition costs, customer lifetime value, and return on investment (ROI) to evaluate the effectiveness of their e-business marketing efforts.

Among the best practices for developing and maintaining e-business marketing campaigns are creating landing pages, optimizing conversion rates, and nurturing leads. Landing page design entails building a page optimized for conversions and encourages users to perform the intended action, such as completing a purchase or submitting a contact form. Optimizing conversion rates requires evaluating several parts of the e-business marketing campaign, including the message, design, and call to action (CTA), to increase the conversion rate. Lead nurturing involves connecting prospective consumers via targeted marketing and supplying them with helpful information and offers [2].

In conclusion, building a thorough e-business marketing plan is essential for every contemporary company seeking success in the digital era. This plan should consider the particular aspects of e-business marketing, such as the role of digital channels, best practices for designing and managing e-business marketing campaigns, and assessing the efficacy of e-business marketing activities. By adhering to these best practices, companies may establish a solid online presence, reach their target audience, and accomplish their e-business marketing goals.

Summary

Explanation of the Concept of E-Business

- Definition of E-business: E-business involves conducting business activities and transactions electronically over the Internet or other digital platforms.
- Scope of E-business: It encompasses various activities such as online retail, electronic banking, online advertising, and electronic supply chain management.
- Benefits of E-business: E-business offers advantages such as increased global reach, reduced operational costs, improved customer convenience, and enhanced scalability.

Historical Overview of E-Business

- Early Developments: E-business can be traced back to the 1960s when electronic data interchange (EDI) systems were introduced for electronic document exchange between businesses.
- Internet Revolution: The widespread adoption of the Internet in the 1990s paved the way for the growth of e-business, enabling secure online transactions and the emergence of e-commerce platforms.
- Evolution of Technologies: The evolution of technologies like mobile devices, cloud computing, and artificial intelligence has further expanded the possibilities and reach of e-business.

Trust in an E-Business

- Security Measures: E-business relies on robust security measures, including encryption, firewalls, and secure authentication protocols, to protect sensitive data and build customer trust.
- Privacy Protection: Businesses must adhere to privacy regulations and ensure the secure handling of customer data, fostering trust and maintaining customer loyalty.
- Online Reputation Management: Maintaining a positive online reputation through customer reviews, ratings, and transparent business practices is crucial in establishing trust in e-business.

E-Business Marketing Strategies and Platforms

- Digital Marketing Channels: E-business leverages various digital marketing channels such as search engine optimization (SEO), social media marketing, email marketing, and content marketing to reach and engage customers.

- E-Commerce Platforms: Businesses can utilize popular e-commerce platforms like Shopify, Magento, or WooCommerce to set up online stores and showcase their products or services.
- Social Media Platforms: Social media platforms like Facebook, Instagram, and LinkedIn provide opportunities for businesses to engage with their target audience, build brand awareness, and drive traffic to their e-business platforms.

Case Study

Hamta Business Corporation, comprised of seasoned international business and industry professionals, has established a significant presence in key international markets. They offer innovative business solutions to assist entrepreneurs and business proprietors in establishing, optimizing, or improving their operations. Hamta Business Solution Sdn Bhd REG. NO. 1196225-M, founded in 2011, expanded its operations to Malaysia under Hamta Business Solution Sdn Bhd. In September 2020, the corporate headquarters of Hamta Business Corporation were relocated to Vancouver, Canada, with the registration number BC1267956. The company's primary objective is to enable its customers to achieve their business goals and gain a competitive advantage in the market.

1. How has Hamta Business Corporation utilized its international and industry expertise to provide innovative business solutions to entrepreneurs and business owners?
2. What prompted the relocation of Hamta Business Corporation's headquarters from Malaysia to Vancouver? How has this migration affected the company's operations and market position?
3. How does Hamta Business Corporation aid its clients in attaining their business goals? Can you provide examples of specific firm strategies or services?
4. How does Hamta Business Corporation differentiate itself from other business solution providers? What unique value proposition does the business offer its customers?
5. Please describe the industries or fields in which Hamta Business Corporation has demonstrated expertise. How has the organization effectively navigated the constraints and complexity of these industries to develop effective business solutions?

Discussions

1. What is e-business?
2. How does e-business differ from traditional business models?
3. What are some examples of e-business in action?
4. What are the benefits of e-business for organizations?
5. How has e-business evolved?

6. What role do technology and infrastructure play in e-business?
7. What are some of the critical advantages of e-business?
8. How can e-businesses increase efficiency for organizations?
9. What are some of the challenges that organizations may face when implementing e-business strategies?
10. How can e-business impact customer experiences?
11. What are some of the risks associated with e-business?
12. What steps can organizations take to mitigate these risks?
13. Why is trust important in e-business?
14. What factors contribute to building trust in an e-business environment?
15. What are some of the consequences of a lack of trust in e-business?
16. How can organizations build trust with customers in the e-business sphere?
17. What role does security play in building trust in e-business?
18. How can organizations address customer concerns around data privacy and security in e-business transactions?
19. What kind of technology is needed to support e-business operations?
20. How does cloud computing play a role in e-business infrastructure?
21. What are the critical components of e-business infrastructure, such as payment gateways and supply chain management systems?
22. What are some of the challenges that organizations may face when implementing e-business infrastructure and technology?
23. How can organizations ensure that their e-business infrastructure is secure and reliable?
24. What role do IT professionals play in supporting e-business infrastructure and technology?

Multiple-Choice Questions

1. Which of the following best defines e-business?
   (a) Traditional brick-and-mortar businesses
   (b) Online business transactions only
   (c) The use of electronic methods and technologies for business processes
   (d) Selling products and services using social media platforms
2. E-business can be traced back to which era?
   (a) Industrial Revolution
   (b) Renaissance period
   (c) Information Age
   (d) Ancient civilization
3. Which of the following is *not* a key technology used in e-business?
   (a) Cloud computing
   (b) Artificial intelligence

(c) Blockchain
(d) Morse code

4. What is the role of trust in e-business?

(a) It is irrelevant in the online business environment.
(b) Trust is essential for building customer relationships and encouraging online transactions.
(c) Trust is only important for physical stores.
(d) Trust is the responsibility of the customers, not the businesses.

5. Which of the following is an example of an e-business marketing strategy?

(a) Cold calling
(b) Distributing flyers
(c) Social media advertising
(d) Door-to-door sales

6. Which platform is commonly used for business-to-business (B2B) e-business transactions?

(a) eBay
(b) Facebook
(c) Alibaba
(d) Twitter

7. What does SEO stand for in the context of e-business marketing?

(a) Search engine optimization
(b) Social engagement optimization
(c) Sales enhancement operations
(d) Secure electronic ordering

8. Which of the following is an example of a customer-to-customer (C2C) e-business platform?

(a) Amazon
(b) eBay
(c) Shopify
(d) Walmart

9. Which of the following technologies enables secure online payments in e-business?

(a) Cryptocurrency
(b) Virtual reality
(c) Augmented reality
(d) Nanotechnology

10. What is the primary goal of e-business marketing?

(a) Increasing customer loyalty
(b) Reducing operating costs

(c) Maximizing profit margins
(d) Eliminating competition

11. Which of the following is an example of a business-to-consumer (B2C) e-business model?

    (a) Dropbox
    (b) LinkedIn
    (c) Airbnb
    (d) Microsoft Azure

12. What is the purpose of using big data analytics in e-business?

    (a) To track customer behavior and preferences
    (b) To create fake online reviews
    (c) To manipulate search engine rankings
    (d) To eliminate privacy concerns

13. Which marketing strategy focuses on sending personalized emails to potential customers?

    (a) Content marketing
    (b) Affiliate marketing
    (c) Email marketing
    (d) Influencer marketing

14. Which of the following platforms is primarily used for professional networking in e-business?

    (a) Facebook
    (b) Instagram
    (c) LinkedIn
    (d) Snapchat

15. What is the advantage of using social media platforms for e-business marketing?

    (a) Limited reach to potential customers
    (b) Higher advertising costs
    (c) Ability to engage with a large audience
    (d) Lack of real-time communication options

References

1. Turban, E., et al., *Electronic commerce 2018: a managerial and social networks perspective*. 2018: Springer
2. Traver, C.G. and K.C. Laudon, *E-commerce 2017: Business, Technology, Society*. 2017: Pearson

Chapter 2
E-Business Models and Strategies

Abstract E-business has revolutionized how businesses operate and engage with customers, suppliers, and partners. To stay competitive, organizations must grasp various e-business models, revenue models, and development strategies. The first section of this chapter provides an overview of e-business models. These models outline how organizations create, deliver, and capture value through electronic technologies. Examples include business-to-business (B2B), business-to-consumer (B2C), consumer-to-consumer (C2C), consumer-to-business (C2B), business-to-government (B2G), government-to-business (G2B), and mobile commerce models. The second section focuses on revenue models for e-businesses. These models describe how organizations generate revenue from their products or services. Common examples include subscription-based, pay-per-use, advertising-based, and transaction-based models. Selecting an appropriate revenue model is vital for an e-business's success, as each model presents unique benefits and challenges. The third section discusses e-business strategy development. This involves creating a comprehensive plan to achieve an organization's e-business objectives. An effective e-business strategy considers the organization's e-business model, revenue model, competitive landscape, target market, and technology infrastructure. This chapter aims to provide readers with a comprehensive understanding of e-business models, revenue models, and development strategies. By grasping these essential components, organizations can create effective strategies that ensure competitiveness in the evolving digital landscape.

1 Overview of E-Business Models

The Internet and other digital technology development have changed how businesses operate recently. The development of e-commerce and e-business models has given enterprises new ways to connect with consumers and grow their businesses. E-business models provide companies with a tactical framework for managing their online operations and generating income. Organizations utilize e-business models

H. Taherdoost, *E-Business Essentials*, EAI/Springer Innovations in Communication and Computing, https://doi.org/10.1007/978-3-031-39626-7_2

as strategic frameworks for managing their online business operations. An e-business model describes a company's tactics, procedures, tools, and frameworks to produce, distribute, and profit from its online activities. These strategies provide businesses with the ability to maximize their online presence and online income sources. E-business models fall into the following categories:

The business-to-consumer (B2C) model is an e-business paradigm in which businesses sell directly to individual customers through Internet channels. Using that use, a B2C model often creates online storefronts or marketplaces where customers may explore and buy items or services. B2C e-commerce has risen significantly during the last decade, spurred by technological breakthroughs and shifting customer behavior. Thanks to the advent of mobile devices and social media, consumers now have more methods to find and connect with brands online, making it more straightforward for businesses to reach and engage with their target audiences. The B2C model enables businesses to use the Internet to reach many prospective clients without requiring physical retail locations or conventional advertising tactics. This may result in considerable cost savings, enhanced enterprise efficiency, and increased customer convenience and access. The B2C model, on the other hand, introduces new problems and dangers, such as greater competition and the necessity for strong online marketing and customer service methods. Moreover, worries about data privacy and security may impact customer trust and desire to connect with firms online. The B2C model is a popular and efficient approach for businesses to engage with customers and expand their enterprises via online channels [1].

The business-to-business (B2B) model is an e-commerce paradigm in which businesses offer goods or services to other businesses using Internet channels. Companies in a B2B model generally interact with other businesses and assist transactions by using digital marketplaces, online directories, or other e-commerce platforms. B2B e-commerce has grown in popularity recently as more organizations seek to optimize their supply chains and save expenses via online sourcing and purchasing. B2B organizations may save time and money using online channels instead of conventional procurement techniques such as phone calls, faxes, and face-to-face meetings. The sales process in a B2B model often requires longer lead times and higher buy quantities than in a B2C model. This necessitates a new marketing and sales strategy, focusing on long-term relationship development and providing value to corporate clients. B2B e-commerce has obstacles and dangers, such as sophisticated pricing and payment structures and the need to interact with other company systems and procedures. Nonetheless, the potential advantages of B2B e-commerce, such as enhanced efficiency, cost savings, and access to new markets, make it a popular and critical e-business model for many businesses [1].

The consumer-to-consumer (C2C) model is an e-business paradigm in which individuals or businesses sell directly to other individuals or businesses via online platforms. C2C e-commerce is enabled by online marketplaces, auction sites, and other e-commerce platforms that link consumers and merchants. With the emergence of online marketplaces like eBay, Craigslist, and Etsy, C2C e-commerce has become increasingly popular in recent years. Without needing a physical storefront or other traditional retail infrastructure, these platforms enable individuals and small businesses to sell goods and services to a global audience. In a C2C model, customers and vendors typically negotiate prices and terms of sale directly. Typically, the online portal provides a marketplace and facilitates the transaction in exchange for a commission on the sale price. C2C e-commerce presents unique challenges and risks, including managing reputation, trust, and the possibility of fraudulent activity. However, the advantages of C2C e-commerce, such as increased access to global markets, greater flexibility and autonomy for merchants, and reduced prices for purchasers, make it an attractive and significant e-business model for numerous individuals and small businesses [2].

Consumer-to-business (C2B) is an e-commerce strategy in which individual consumers or small companies sell items or services to larger businesses through Internet channels. Individuals or small companies provide their goods or services to prospective customers, who assess and purchase them if they suit their requirements under a C2B model. In recent years, C2B e-commerce has grown in popularity, notably in fields such as freelancing, consulting, and content production. Individual sellers may provide unique or specialized items to companies via online platforms such as Amazon or eBay. At the same time, freelance writers or designers may offer their talents to businesses through online markets or directories. Individuals or small companies work as suppliers or service providers to more significant firms in a C2B model, with the conventional roles of buyers and sellers reversed. Individuals and small enterprises may benefit from improved freedom and autonomy and enhanced access to new markets and business prospects [3].

The business-to-government (B2G) model is an e-commerce paradigm in which companies sell goods or services to government agencies through Internet channels. Businesses in a B2G model often offer goods or services to government agencies using digital procurement processes, online bidding platforms, or other e-commerce channels. B2G e-commerce has grown in importance in recent years as governments across the globe have sought to expedite procurement procedures and reduce costs via online channels. Government agencies may save time and money by embracing online platforms instead of conventional procurement techniques, including paper-based bids and manual procurement procedures. The sales process in a B2G model sometimes entails complicated bidding and tendering procedures, as well as tight laws and compliance requirements. This necessitates a new marketing and sales strategy, focusing on developing relationships with government customers and proving the capacity to satisfy their specific demands. B2G e-commerce has its hurdles and hazards, such as navigating complicated government procurement legislation and compliance standards and establishing credibility and confidence with government buyers. However, the potential advantages of business-to-business e-commerce, such as enhanced efficiency, cost savings, and access to new markets, make it an essential and rising e-business model for many organizations [2].

The government-to-business (G2B) model is a business paradigm in which businesses interact with government entities for various purposes, such as procurement, licensing, regulatory compliance, and taxation. The G2B model provides businesses with a single point of contact for their dealings with the government, streamlining processes and improving transparency. This model has become increasingly important in recent years as governments worldwide seek to reduce costs and expedite procurement procedures through digital channels. While the G2B model has several advantages, it has some potential drawbacks. For example, navigating the regulations and compliance requirements involved in G2B transactions can be complicated. Businesses should focus on building relationships with government customers and demonstrating their ability to meet their needs. Despite these challenges, the benefits of the G2B model are significant. By embracing digital platforms for procurement and other transactions, government agencies can save time and money while improving transparency and efficiency. Businesses navigating the complexities of the G2B model can access new markets and achieve significant cost savings, making it an essential e-business model for many organizations [1].

Mobile commerce, often known as m-commerce, is an e-business model that involves purchasing and selling products and services using mobile devices such as smartphones and tablets. M-commerce, which allows customers to make purchases from anywhere, at any time, using their mobile devices, has grown in importance in the e-commerce market. M-commerce allows firms to reach out to clients in novel ways, such as via mobile applications, mobile-optimized websites, and mobile payment systems. Mobile devices also give companies many data and insights, allowing them to customize their marketing and sales efforts depending on client behavior and preferences [4].

Advances in mobile technology, such as higher connection rates, improved mobile devices, and more secure payment mechanisms, have fueled the expansion of e-commerce. M-commerce has also been spurred by the rising popularity of mobile shopping and customers' global use of mobile devices. The m-commerce model introduces new obstacles and dangers, such as the need to optimize mobile sites for various device kinds and screen sizes and the necessity to secure the security of mobile payment systems. Nonetheless, the potential advantages of m-commerce, such as more significant consumer interaction, convenience, and accessibility, make it a significant and rapidly increasing e-business model for many companies. Table 2.1 summarizes the pros and cons of each e-business model.

There are several examples of businesses that employ these various e-business concepts. Amazon.com, for example, is a well-known corporation that employs the B2C model to sell items and services directly to customers. Nike.com is another example of a B2C corporation, where consumers may buy sports shoes and gear online. The business-to-business (B2B) model is also popular, with firms such as Alibaba.com, IBM.com, and Salesforce.com employing it to offer goods and services to other businesses. These organizations provide various tools and services to assist firms in streamlining their processes and improving their performance. C2C firms such as eBay.com, Etsy.com, and Craigslist.org enable people to purchase and sell items and services directly to one another.

In contrast, C2B companies such as Fiverr.com, Upwork.com, and Freelancer.com enable freelancers to offer their services to businesses. Government consulting services are provided by B2G organizations like Accenture.com, Deloitte.com, and

Table 2.1 Pros and cons of different e-business models

E-business model	Advantages	Disadvantages
Business-to-consumer (B2C) model	Large addressable market Direct customer feedback Opportunities for personalization	Competition can be intense High customer acquisition costs High levels of customer churn
Business-to-business (B2B) model	High transaction values Repeat business opportunities Potential for long-term partnerships	Long sales cycles Complex decision-making processes Limited addressable market
Consumer-to-consumer (C2C) model	Low overhead costs Large addressable market Reduced barriers to entry	Limited ability to control product quality Limited ability to scale Lack of trust among buyers and sellers
Consumer-to-business (C2B) model	More efficient use of resources Increased access to specialized skills Reduced administrative costs	Limited addressable market Need for effective reputation management Need for effective branding
Business-to-government (B2G) model	Large, stable customer base Potential for long-term contracts Potential for repeat business	Complex procurement processes Intense competition for contracts Need to comply with government regulations
Government-to-business (G2B) model	Access to government contracts Increased opportunities for revenue Reduced competition	Need to comply with government regulations Long procurement cycles Complex bidding processes
Mobile commerce (m-commerce) model	Increased convenience for customers More opportunities for personalization Reduced overhead costs	Limited screen size can make browsing difficult Limited ability to showcase products Technical challenges related to mobile optimization

Boozallen.com, whereas G2B companies such as SAM.gov, FedBid.com, and GSA.gov offer a business platform to bid on government contracts. Mobile commerce (m-commerce) is growing in popularity, with firms like Amazon and Uber providing mobile applications enabling people to buy items or services with only a few clicks. Starbucks also has an app that enables users to pre-order and pay for their beverages, making the buying procedure more accessible.

Finally, there are several e-business models accessible, and businesses should choose the one that best meets their company requirements. Amazon, Nike, and eBay are successful businesses that use these e-business concepts.

2 Revenue Models for E-Businesses

E-business revenue models are the numerous ways an Internet firm produces revenue or makes money. A company's revenue model is determined by the sort of e-business it conducts, the industry in which it competes, the target audience, and the competitive environment. The following are some of the most frequent revenue models used by e-businesses:

Advertising-based revenue models are extensively used by e-businesses, particularly those that offer free content or services to consumers. In this business model, companies generate income by selling advertising space on their website, mobile application, or other digital platforms. The advertising-based revenue model allows companies to display advertisements to users using the company's website or mobile application. These can be display ads, native ads, or video ads and can be tailored to specific users based on their interests, behavior, or demographics. This model generates revenue based on the number of times the advertisement is displayed (impressions) or the number of times it is selected (clicks). The cost of the advertisement is determined by the degree of competition for advertising space and the selection of targeting options. Google, Facebook, Instagram, and Twitter rely on advertising for revenue. These businesses generate substantial revenue by providing their platforms to advertisers who wish to reach a large, engaged audience. One of the benefits of the advertising-based revenue model is that it enables e-businesses to provide users with free content or services, which can help to attract and retain a large audience. The potential for ad saturation, which can negatively impact the user experience, and the difficulty of preserving a balance between user privacy and accumulating data for ad targeting purposes are disadvantages of this model [2].

A subscription-based revenue model is one in which customers are charged a recurring fee for access to a product or service. E-businesses that offer digital content, software as a service (SaaS), online courses, or other subscription-based services frequently use this model. Customers often pay a monthly or yearly charge for continuing access to the product or service under this arrangement. In return, they may get unique material or features, tailored suggestions, or other incentives. One of the benefits of the subscription-based revenue model is that it can provide e-businesses with a consistent and predictable source of revenue.

Furthermore, it can aid in developing a loyal customer base, as customers who have invested in the product or service are more likely to continue paying for access. Netflix, Spotify, Microsoft Office 365, and Amazon Prime are examples of companies that use the subscription-based revenue model. These businesses have built a large and loyal customer base by providing high-quality products and services that customers are willing to pay for regularly. However, there are some potential drawbacks to the subscription-based revenue model. Customers, for example, may be hesitant to commit to a recurring payment if they are unsure of the product's or service's value. Furthermore, e-businesses should constantly innovate and improve their offerings to justify the ongoing cost to customers [1].

A transaction-based revenue model is one in which an e-business makes income by facilitating transactions between buyers and sellers. The e-business may charge a commission or transaction fee for each successful transaction on its platform under this model. This revenue model is widely employed by e-commerce companies that offer products or services to consumers directly. For example, online marketplaces such as Amazon, eBay, and Etsy earn money by charging sellers transaction fees for each item sold on their sites. The transaction-based revenue model has the advantage of providing a scalable source of revenue for e-businesses. The e-business may produce more money as more transactions occur on the platform.

Furthermore, the transaction-based revenue model may provide a low barrier to entry for new customers, as they may only be required to pay a fee when they make a purchase. However, there are some potential drawbacks to the transaction-based revenue model. For example, buyers and sellers may be discouraged from using the platform if the transaction fee is too high. Furthermore, e-businesses using this revenue model may face increased competition from platforms with lower transaction fees. Payment processing companies like PayPal and Square, which generate revenue by charging transaction fees for each payment processed on their platform, are other examples of e-businesses that use the transaction-based revenue model. Similarly, ride-sharing companies such as Uber and Lyft generate revenue by charging drivers transaction fees for each ride facilitated through their platforms.

Affiliate marketing revenue models are those in which an e-business gets a commission for promoting and selling the goods or services of another firm. In this approach, the e-business operates as the other company's affiliate or partner, promoting its goods or services to its clients or audience. The e-business receives a commission from the sale when a consumer clicks on the affiliate's link or ad and makes a purchase. The commission rate varies according to the product or service being advertised and the agreement between the e-business and the affiliate firm. E-commerce enterprises, bloggers, and influencers with a significant audience or following often adopt this income model. For example, an e-commerce company selling outdoor gear may collaborate with an affiliate marketing network that promotes outdoor adventure activities like rock climbing or kayaking. The e-commerce company may then advertise the affiliate program's items to its audience and earn a commission on every transaction through its unique affiliate link. One of the benefits of the affiliate marketing income model is that it may give e-businesses a relatively passive cash stream. After establishing the affiliate network and the items or services are marketed, the e-business may earn sales commissions without actively maintaining inventory or fulfilling orders [3].

Moreover, this strategy may help e-businesses diversify their income sources by offering complementary items or services. However, there are several possible drawbacks to the affiliate marketing income strategy. For example, to produce substantial income, e-businesses may need to devote considerable time and money to advertising the affiliate program to their target audience. Moreover, specific affiliate networks may provide low compensation rates, making it difficult for e-businesses to make considerable income from this model alone. Amazon Associates, which enables e-businesses to earn a commission on Amazon product purchases, and

Commission Junction, which gives access to a broad selection of affiliate programs across multiple sectors, are two examples of e-businesses that employ the affiliate marketing income model.

The freemium revenue model combines the "free" and "premium" income types. The primary product or service is free under this model, whereas the premium version is provided for a cost. This strategy aims to provide clients with a free sample of the product or service before enticing them to upgrade to the premium version by providing extra features and perks. A free product or service often has limited functionality, while the premium version has more sophisticated features and capabilities. Users are expected to like the free version enough to see the value in subscribing to the premium version. Software, mobile applications, and gaming firms often adopt freemium models. Popular productivity tools such as Trello and Evernote, for example, provide free basic versions while charging a charge for more sophisticated features and capabilities [1].

Similarly, famous gaming applications like Candy Crush give out a limited number of lives and features for free but charge a price for infinite lives and advanced capabilities. The freemium model has the benefit of attracting many customers since the free version is readily available and needs no commitment. This may lead to increased brand awareness and client loyalty. Moreover, the strategy enables businesses to earn from free and paying clients. However, the freemium model has its drawbacks. One possible disadvantage is that converting free users to paid clients might require more work. Moreover, providing a free version may diminish the premium version's value in clients' eyes, which may be less eager to pay for it if they believe they can receive much of what they need for free. Lastly, providing a free version might be expensive since resources are required to maintain and support the free and premium versions of the product or service.

The revenue sharing model, also known as the commission-based revenue model, is a business model in which a firm makes income by getting a portion of sales made via its platform or service. This strategy is widespread in e-commerce, where third-party vendors offer their items via an online marketplace, and the platform operator receives a fee on each transaction. In this model, the platform operator works as a middleman between buyers and sellers, charging a portion of the transaction value as a commission fee. The commission fee varies according to the product or service sold and the degree of service offered by the platform operator. Amazon, for example, operates on a commission-based income model. Third-party vendors may sell their items on Amazon, and Amazon receives a percentage on each sale. The commission rate varies according to the product category, with some categories having more excellent commission rates than others.

The commission-based revenue model allows the platform operator to earn money without incurring the risks and expenses of manufacturing or selling items or services. Furthermore, the model incentivizes the platform operator to provide high-quality service because sellers only earn money if selling their products through the platform. However, the commission-based revenue model has its drawbacks. To remain competitive, the platform operator may face competition from other platforms offering similar products or services and may need to innovate and improve their service continuously. Furthermore, the model may be vulnerable to fraud and abuse because some sellers may engage in unethical or illegal practices to generate sales and commissions. The licensing revenue model offers permission to utilize intellectual property such as patents, copyrights, and trademarks in return for a fee or royalty. E-businesses may use this strategy to monetize their intellectual property and create cash by licensing their goods, software, or technology to other companies. Microsoft is one example of an e-business that employs the license revenue model. Microsoft offers many software products, including the Windows operating system and the Office suite, which it licenses to other firms for a price. Another example is Spotify, a music streaming business that obtains licenses from record companies and pays them a part of its earnings.

The licensing revenue model offers various benefits, including establishing consistent cash streams for the e-business and lowering the risk of product development and marketing expenditures. Moreover, licensing may be practical for e-businesses looking to extend their reach and enter new markets. However, there are several disadvantages to the licensing income model. E-businesses, for example, may need to spend significant resources establishing and safeguarding their intellectual property, which may be expensive and time-consuming. Moreover, licensing agreements may be complicated and require legal skills to negotiate and handle successfully. Lastly, e-commerce companies that depend primarily on licensing income may encounter difficulties if their intellectual property becomes obsolete or is infringed upon by rivals.

Crowdfunding is a revenue model that relies on raising funds from many individuals or organizations, typically through an online platform. E-businesses can use crowdfunding to raise capital for new products, projects, or initiatives. Backers typically receive rewards or equity in the e-business in exchange for contributions. Kickstarter is one example of an e-business that uses the crowdfunding revenue model. Kickstarter is a popular crowdfunding platform allowing e-businesses to raise backers' funds to launch new products and services. Backers can make monetary contributions in exchange for benefits such as early access to products, exclusive merchandise, or other perks. Patreon is another example of an e-business that uses the crowdfunding revenue model. Patreon is a crowdfunding platform that allows content creators like artists, musicians, and writers to receive funding from their fans and followers. Backers receive exclusive content, early access to new works, and other perks. The crowdfunding revenue model has several advantages, including providing e-businesses with funding otherwise unavailable from traditional sources such as banks or venture capitalists [4].

Furthermore, crowdfunding can assist e-businesses in developing a community of loyal supporters and creating buzz around new products or initiatives. However, there are some disadvantages to the crowdfunding revenue model. To attract backers and achieve their funding goals, e-businesses may need to invest significant resources in marketing and promotion. Furthermore, crowdfunding can be risky because there is no guarantee that a project will be successful or that backers will receive the promised rewards. Finally, e-businesses may be required to give up some ownership or control in exchange for crowdfunding, which can impact their long-term prospects. Table 2.2 summarizes the pros and cons of each revenue model for e-business.

Table 2.2 Pros and cons of revenue models for e-business

Revenue model	Advantages	Disadvantages
Advertising-based	Can generate significant revenue No direct cost to users	Can be intrusive and annoying for users Revenue may be unpredictable
Subscription-based	Provides predictable revenue streams Users have an ongoing commitment to the product or service	May limit the user base to those willing to pay Users may be hesitant to commit to long-term subscriptions
Transaction-based	Revenue directly tied to sales Minimal overhead costs	Revenue may be variable and unpredictable May require significant marketing and advertising costs to drive traffic
Affiliate marketing	Low overhead costs Revenue generated through third-party advertising No need to produce or maintain products	Revenue may be unpredictable Dependence on third-party advertising sources
Freemium	Can attract a large user base Provides opportunities for revenue growth through premium offerings	Difficult to find the right balance between free and paid features May limit revenue potential
Commission-based	Generates revenue through partnerships and collaborations Minimal overhead costs	Revenue may be unpredictable Dependence on third-party businesses
Licensing	Provides predictable revenue streams Revenue not tied to sales volume	Limited growth potential Dependence on a limited number of licensing agreements
Crowdfunding	Provides initial capital for startups Validates demand for products or services	Limited revenue potential beyond the initial funding Success is dependent on market acceptance and enthusiasm

3 E-Business Strategy Development

Developing an e-business strategy is identifying and implementing a plan for how an organization will use electronic resources and technology to achieve its goals. Developing an e-business strategy requires multiple steps, including defining objectives, analyzing the industry and competition, identifying target customers, selecting appropriate technology platforms and tools, creating a marketing plan, and measuring and adjusting the strategy over time. Developing an e-business strategy is essential for businesses seeking success in the digital age. A well-developed e-business strategy aligns the company's technology with its business goals and objectives. This ensures that technology investments are made strategically, as opposed to haphazardly.

By leveraging technology to create new products or services, enhance the customer experience, or reduce costs, e-business strategies can help businesses gain a

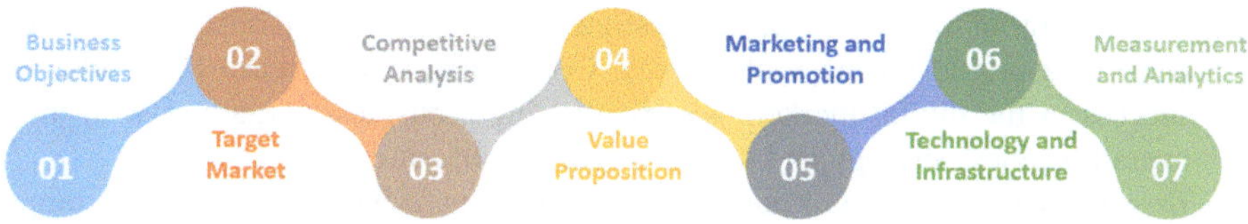

Fig. 2.1 Key elements in an e-business strategy

competitive edge. A good e-business strategy will prioritize the customer experience, ensuring that the organization meets the needs and expectations of its target market. This can increase customer loyalty and engagement. By automating processes, reducing the need for manual intervention, and minimizing errors, e-business strategies can also help businesses reduce costs. In addition, a well-developed e-business strategy can help businesses improve their internal and external communication and collaboration. Developing an e-business strategy is crucial for businesses to remain competitive and thrive digitally. By aligning technology with business objectives and concentrating on the customer experience, businesses can gain a competitive advantage, reduce expenses, and enhance collaboration and communication. Developing a successful e-business strategy is essential for success in the online marketplace. A company's e-business strategy uses digital technologies and platforms to accomplish its business objectives. Developing an e-business strategy necessitates taking into account several crucial factors, including (Fig. 2.1):

- *Business Objectives*: Setting precise business objectives is crucial to any e-business strategy. A company's e-business initiatives are designed to attain specific business objectives defined by its business objectives. These objectives should be aligned with the overall business strategy and incorporate the company's mission, values, and long-term objectives. When establishing e-business objectives, businesses should consider customer requirements, market trends, and the competitive landscape. The objectives of a business may include increasing sales, enhancing customer satisfaction, broadening market penetration, or enhancing operational efficiency. These goals should be quantifiable, attainable, pertinent, and time-bound.
- Additionally, the e-business strategy should identify the key performance indicators (KPIs) that will be used to track progress toward these goals. KPIs include website traffic, conversion, customer retention, and revenue growth. Setting attainable KPIs can help businesses evaluate the success of their e-business initiatives and make necessary adjustments.
- In addition, e-business strategies should consider the impact of emerging technologies on the organization's business objectives. For instance, the rise of mobile devices and social media has provided businesses new opportunities to reach and interact with customers. Therefore, businesses should consider leveraging these technologies to achieve their e-business objectives.

- Developing an effective e-business strategy requires the establishment of specific, measurable business objectives. Companies can increase their odds of success in the digital marketplace by aligning e-business initiatives with their overall business strategy and defining specific outcomes.
- *Target Market:* The target market is crucial to any e-business strategy. A target market is a specific group of customers a business intends to reach with its products or services. Identifying the target market is crucial for any e-business because it facilitates the development of a targeted marketing and sales strategy that speaks directly to the needs and preferences of the target audience. Here are some essential steps for identifying and defining your target market as part of your e-business strategy (Fig. 2.2):

  1. Define your business niche: Begin by defining the products or services you provide and the specific benefits they provide to potential customers. This will assist you in determining your niche and intended audience.
  2. Conduct market research: Investigate your prospective customers' demographics, psychographics, and purchasing behavior. This will assist you in determining their preferences and needs.

Fig. 2.2 Steps for identifying the target market

3. Segment your target market: Divide it into segments based on their shared characteristics, such as age, gender, income, location, interests, and behavior. This will help you target the appropriate audience with your marketing messages.
4. Analyze your competition: Analyze your competitors and identify their target audience. This will help you understand the gaps in the market and how to position your business.
5. Test and refine: Test your marketing strategies and refine your target market as needed. This will help you continuously improve your e-business strategy.

   Remember, your target market is not set in stone and may change as your business evolves. However, the success of your e-business strategy hinges on a thorough comprehension of your target market.

- *Competitive Analysis:* Analyzing the competition is crucial to any e-business strategy. It entails identifying and analyzing your competitors to understand their strengths and vulnerabilities and how your business can differentiate itself in the marketplace. Here are some essential stages for conducting a competitive analysis as part of your e-business strategy (Fig. 2.3):

  1. Identify your competitors: Identify all your direct and indirect competitors. In contrast to direct competitors, indirect competitors offer alternatives that could compete with your products or services.
  2. Analyze their strengths and weaknesses: Analyze the strengths and vulnerabilities of your competitors. Consider their product offerings, pricing strategies, customer service, and distribution channels. Determine the areas in which your competitors excel and those in which they fall short.
  3. Recognize their target market: Recognize whom your competitors are targeting and how they are attempting to reach them. Analyze their marketing messages, channels, and customer demographics.
  4. Determine their pricing strategy: Analyze the pricing strategy of your competitors, including their pricing model, discounts, and special offers. Determine how your pricing strategy compares to competitors' and whether there is room for price adjustments.
  5. Identify opportunities and threats: Determine opportunities and threats based on your analysis. Consider market voids that your business can occupy as well as potential competitors.
  6. Develop your differentiation strategy: Based on your analysis, create a differentiation strategy that distinguishes your business from its rivals. Determine the unique value proposition that your company can provide that its competitors cannot.

     Remember that competitive analysis is an ongoing procedure and that you should continuously monitor your competitors and modify your e-business strategy as necessary. By analyzing your competitors, you can obtain valuable insights that will assist you in making informed decisions and maintaining a competitive advantage.

There are a variety of methods for conducting a competitive analysis. Here are some of the most effective e-business strategy methods for analyzing your competitors:

Fig. 2.3 Essential stages for conducting a competitive analysis

1. SWOT stands for strengths, weaknesses, opportunities, and threats in a SWOT analysis. A SWOT analysis lets you identify the internal and external factors affecting your business and its competitors. By analyzing your competitors' strengths, weaknesses, opportunities, and threats, you can create strategies that capitalize on your strengths and opportunities while mitigating your weaknesses and threats.
2. Benchmarking involves comparing your business to competitors based on specific metrics, such as pricing, product features, customer service, and marketing strategies. By comparing your business to your competitors, you can identify areas for improvement and opportunities for differentiation.
3. Market share analysis involves analyzing your competitors' market share within your target market. By understanding your competitors' market share, you can determine how much of the market they are capturing and how much room your business has for expansion.
4. Analyzing customer feedback can provide insight into how your competitors are viewed by their customers. By analyzing customer reviews and feedback, you can determine where your competitors excel and where they fall short and devise strategies to capitalize on their weaknesses.
5. Industry reports provide valuable insight into the performance of your competitors and market trends. You can better understand the market and identify opportunities and threats by analyzing industry reports.

   Remember that competitive analysis is an ongoing procedure and that you should continuously monitor your competitors and modify your e-business strategy as necessary. By analyzing your competitors, you can obtain valuable insights that will assist you in making informed decisions and maintaining a competitive advantage.

- *Value Proposition:* A value proposition is a statement that describes the distinctive value that your company provides to its customers. It is a crucial component of any e-business strategy because it helps to differentiate your business from competitors and convey the benefits of your products or services to prospective customers. Here are some essential steps that will assist you in creating a value proposition for your e-business (Fig. 2.4):

  1. Identify your target audience: Identify the specific group of customers you intend to attract. Understanding your target audience will help you determine the most relevant benefits.
  2. Evaluate your rivals: Analyze your competitors' value propositions and identify market gaps. Determine what value your company can provide that its competitors cannot.
  3. Define your unique value proposition: Define your unique value proposition that sets your business apart from your competitors. Your value proposition should emphasize your business's advantages to its customers, such as quality, convenience, affordability, or innovation.
  4. Test and modify: Test and modify your value proposition as necessary. Utilize customer feedback to identify opportunities to enhance your value proposition and better meet the needs of your target audience.

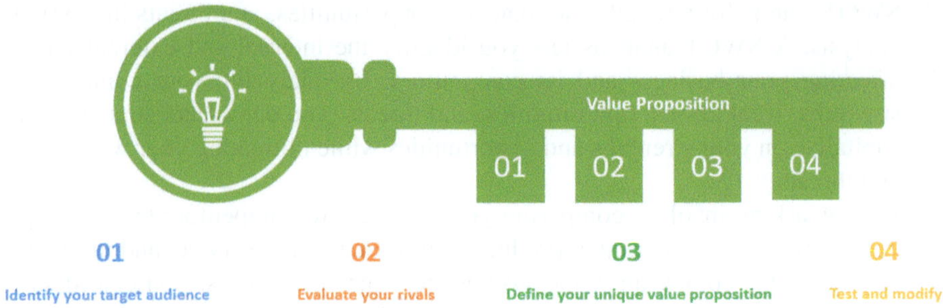

Fig. 2.4 Essential in creating a value proposition for an e-business

Remember that your value proposition should be concise, clear, and straightforward. It should convey the unique value that your business provides to customers and set your business apart from competitors. You can drive the success of your e-business by creating a compelling value proposition that attracts and retains customers.

- *Marketing and Promotion:* Promotion and marketing are essential to any e-business strategy. They aid in increasing brand awareness, attracting new customers, and driving sales. The following steps will assist you in developing a marketing and promotion strategy for your online business (Fig. 2.5):

1. Identify your target audience: Identify the specific customer segment you are pursuing. Understanding your target market will assist you in creating marketing messages that resonate with them.
2. Develop a marketing message: Create a clear, concise, and persuasive marketing message that conveys the unique value your business offers to its customers. Your message should emphasize your products or services' benefits and how they can solve your customers' problems.
3. Determine your marketing channels: Determine the channels that will most effectively reach your target audience. Social media, email marketing, search engine optimization (SEO), paid advertising, content marketing, and influencer marketing may be included.
4. Develop a content strategy: Develop a content strategy that supports your marketing message and assists in attracting and engaging your target audience. This could include blog posts, videos, infographics, and social media posts that provide value to your customers.
5. Establish a budget: Establish a budget for your marketing and advertising efforts. Determine the budget for advertising, content creation, and other marketing costs.
6. Measure your results: Evaluate the efficacy of your marketing and promotion initiatives. Monitor metrics, including website traffic, engagement, conversion, and customer acquisition expenses. Utilize this information to improve your marketing strategy and channels.

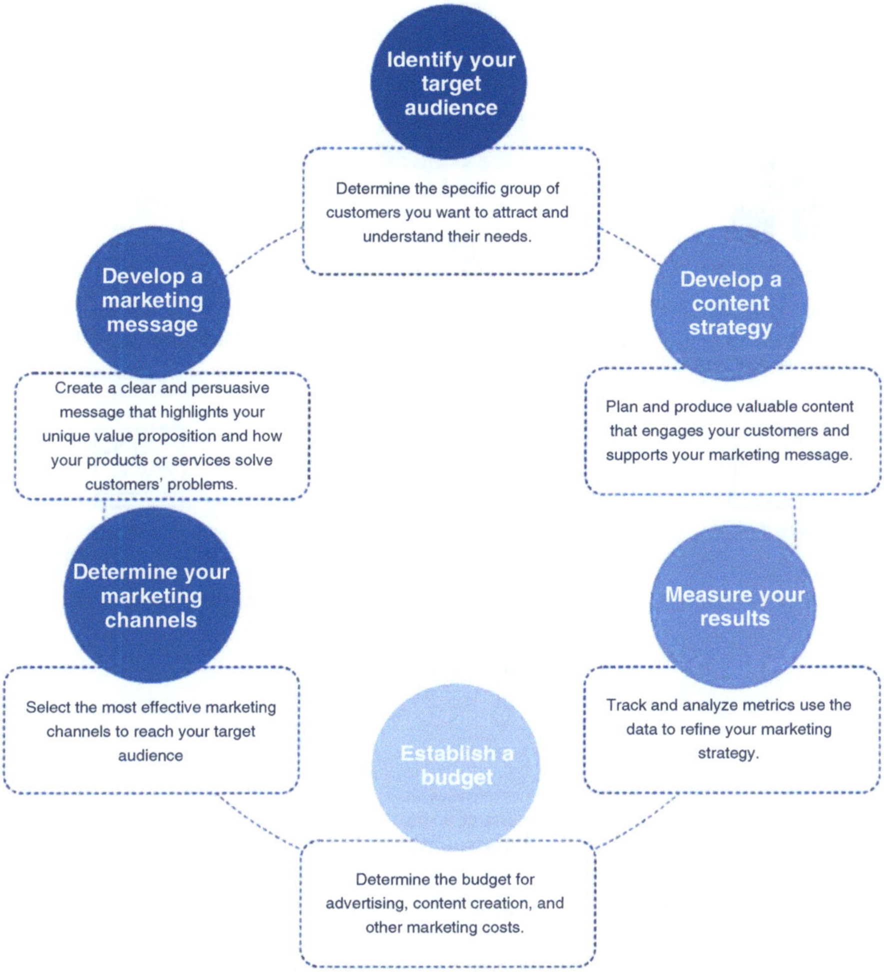

Fig. 2.5 Steps in developing a marketing and promotion strategy for your online business

Remember that effective marketing and promotion require a comprehensive strategy that aligns with the goals and objectives of your e-business. You can attract and retain customers and propel the success of your e-business by developing a clear marketing message, identifying the appropriate channels, and measuring the results.

- *Technology and Infrastructure:* Technology and infrastructure are indispensable elements of any e-business strategy. Developing a technology and infrastructure strategy for your e-business requires careful consideration of the following factors (Fig. 2.6):

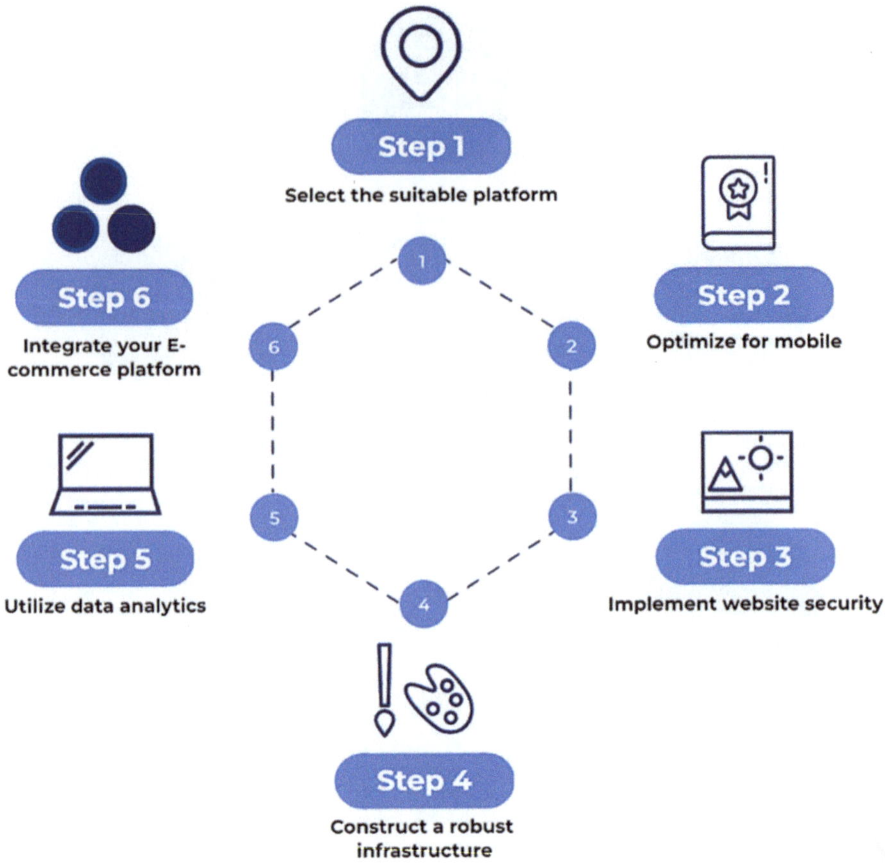

Fig. 2.6 Steps for developing a technology and infrastructure strategy for an e-business

1. Select the suitable platform: Select a scalable, secure, and trustworthy platform. Consider usability, adaptability, personalization options, and integration capabilities. The most popular e-commerce platforms are Shopify, WooCommerce, Magento, and BigCommerce.
2. Optimize for mobile: Ensure your website and e-commerce platform are compatible with mobile devices. Mobile optimization is essential to the success of your e-business, as a growing number of consumers shop online using their smartphones and tablets.
3. Implement website security: It protects your website and customer data from cyber threats. This includes SSL encryption, regular software and plugin updates, and the implementation of strong passwords and two-factor authentication.
4. Construct a robust infrastructure: Construct a robust infrastructure that can support the growth of your e-business. This includes purchasing dependable servers, bandwidth, storage, backup, and disaster recovery solutions.

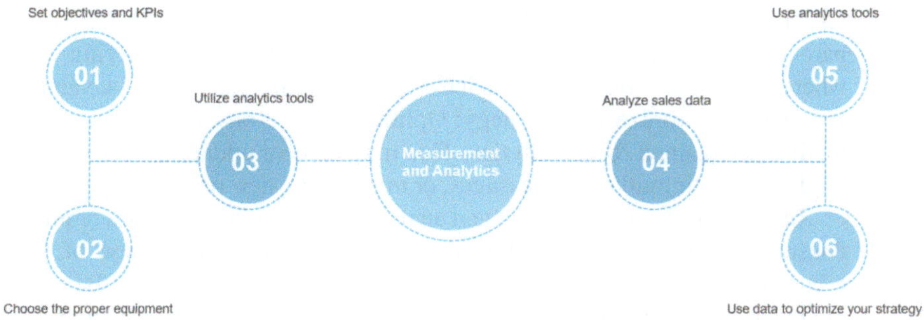

Fig. 2.7 Steps in developing a measurement and analytics strategy for online business

5. Utilize data analytics: Use data analytics tools to obtain insight into customers' behavior and preferences. This will allow you to optimize your website, product offerings, and marketing efforts to satisfy the requirements of your consumers.
6. Integrate your e-commerce platform with other systems, such as your inventory management, accounting, and shipping systems. This will improve your operations and streamline them.

   Keep in mind that technology and infrastructure are essential to the success of your e-business. By selecting the right platform, optimizing for mobile, securing your website, constructing a robust infrastructure, leveraging data analytics, and integrating with other systems, you can provide your customers with a seamless and practical e-commerce experience.

- *Measurement and Analytics:* Measurement and analytics are indispensable elements of any e-business strategy. They aid in monitoring performance, measuring success, and identifying areas for improvement. The following steps will assist you in developing a measurement and analytics strategy for your online business (Fig. 2.7):

1. Set objectives and KPIs: Begin by establishing e-business-aligned objectives and key performance indicators (KPIs). This may include website traffic, conversion rates, average order value, customer acquisition costs, and lifetime value.
2. Choose the proper equipment: Select the appropriate analytics tools that will assist you in monitoring and analyzing your website and customer data. Google Analytics, Adobe Analytics, and Kissmetrics are popular tools.
3. Utilize analytics tools: Utilize analytics tools to monitor customer behavior, including website visits, page views, and click-through rates. This will enable you to comprehend how customers interact with your website and identify areas for enhancement.
4. Analyze sales data: Analyze sales data to identify patterns and trends in customer behavior, such as which products sell the most and which marketing channels generate the most revenue.

5. Use analytics tools: Use analytics tools to track your competition and compare your performance to theirs. This will assist you in identifying areas where you can improve your e-business and maintain a competitive advantage.
6. Use data to optimize your strategy: Optimize your e-business strategy using the insights gained from analytics. This may involve updating your website, modifying your marketing messages, or introducing new products or services.

   Remember that measurement and analytics are ongoing processes requiring constant monitoring and optimization. You can drive the success of your e-business and achieve your objectives by establishing goals and KPIs, selecting the appropriate analytics tools, tracking customer behavior, analyzing sales data, monitoring your competition, and optimizing your strategy with data.

   Developing an effective e-business strategy necessitates a thorough consideration of the factors above. By emphasizing these factors, businesses can develop a comprehensive and successful e-business strategy that drives success in the online marketplace.

4 Case Studies of Successful E-Business Strategies

Amazon is one of the most successful e-commerce companies in the world due to its vast selection of products and services, quick and dependable delivery, and excellent customer service. Their e-business strategy is multifaceted and consists of the following key components: Amazon's e-business strategy is built around a customer-centric focus to provide customers with the best shopping experience possible. They use data analytics to personalize the customer experience and provide targeted product recommendations, and they are renowned for their speedy and dependable shipping. Amazon offers various products, including books, electronics, clothing, and household goods. This enables them to serve a large customer base and maintain a competitive advantage in the e-commerce industry. Amazon has diversified its services by offering Amazon Web Services (AWS) and Amazon Prime Video in addition to its e-commerce platform. This allows them to generate income from various sources and create a more robust business model. Amazon is recognized for its innovation in e-commerce, which emphasizes enhancing the customer experience through technology. They were one of the first e-commerce businesses to offer one-click ordering and have continued to innovate with Amazon Prime and Alexa. Amazon has formed strategic partnerships with other companies to expand its reach and provide consumers with more services. For instance, they have partnered with Whole Foods to provide grocery delivery and with Kohl's to enable in-store returns. Amazon has achieved accelerated growth and become one of the most valuable companies due to these aspects of its e-business strategy. They have disrupted conventional retail models and set the bar for excellence in e-commerce.

Shopify is an e-commerce platform that has become one of the world's most successful e-commerce enterprises. Their e-business strategy focuses on providing a user-friendly platform that enables businesses to create and manage online stores, emphasizing customization and integrations with popular tools. The following are critical components of Shopify's e-business strategy: Shopify's platform is designed to be accessible and user-friendly, even for businesses with limited technical expertise. This has made it an attractive option for small- and medium-sized businesses seeking to expand their online presence. Shopify provides extensive customization options, including various templates and themes, enabling businesses to create a unique and personalized online store. This has helped Shopify distinguish itself from rivals and attract a large customer base. Shopify integrates with various popular tools, such as Facebook, Google, and Instagram, making it simple for businesses to manage their online presence across multiple platforms. Shopify provides various support and resources to help businesses succeed, such as marketing tools, shipping and payment options, and a community forum where businesses can connect and share ideas. Shopify is renowned for its innovation in e-commerce, regularly introducing new features and services to stay ahead of the curve. For instance, they recently introduced a buy-now-pay-later option for customers, enabling businesses to offer flexible payment options.

Warby Parker is an e-commerce startup whose direct-to-consumer business model has revolutionized eyewear. Their e-business strategy centers on providing affordable, high-quality eyewear with an emphasis on customer experience and social responsibility. Critical components of Warby Parker's e-business strategy include the following: Warby Parker sells eyewear directly to consumers via their website and retail locations, eliminating traditional intermediaries and offering lower prices than rivals. Warby Parker's eyewear is offered at affordable prices, beginning at $95 for a pair of glasses. This has increased the number of customers who can purchase their products. Warby Parker offers customers a personalized experience with a virtual try-on tool that enables them to try on glasses from their homes. In addition, they offer a program where customers can try on five frames for free before purchasing. Warby Parker is committed to social responsibility, as evidenced by their Buy a Pair, Give a Pair program, in which they donate one pair of glasses for every pair purchased. Warby Parker is primarily an e-commerce company. However, they have also opened brick-and-mortar stores, which allow customers to try on glasses in person and provide a physical presence in key markets.

Through these components of their e-business strategy, Warby Parker has disrupted the eyewear industry and become a popular option for customers seeking affordable, high-quality eyewear emphasizing customer experience and social responsibility. They have established a loyal customer base and set the standard for the eyewear industry through direct-to-consumer e-commerce. Due to these aspects of their e-business strategy, Shopify has been able to experience accelerated growth and become a popular option for businesses seeking to expand their online presence. They have disrupted the traditional e-commerce market and set the standard for user-friendly and customizable platforms.

Summary

Overview of E-Business Models

- Definition: E-business models refer to the frameworks and structures that define how electronic businesses operate and generate value online.
- Types of e-business models:
  - B2C (business-to-consumer): E-commerce platforms that sell products or services directly to consumers
  - B2B (business-to-business): E-commerce platforms that facilitate business transactions
  - C2C (consumer-to-consumer): E-commerce platforms where consumers can buy and sell products directly to/from other consumers
  - C2B (consumer-to-business): E-commerce platforms where consumers offer products or services to businesses

Revenue Models for E-Businesses

- Definition: Revenue models outline how e-businesses generate income and monetize their products or services online.
- Common revenue models:
  - Sales Revenue: E-businesses generate revenue by selling products or services directly to customers.
  - Advertising Revenue: E-businesses earn revenue by displaying advertisements on their platforms.
  - Subscription Revenue: E-businesses charge users a recurring fee for access to premium content or services.
  - Freemium Model: E-businesses offer essential services for free but charge for additional features or upgrades.
  - Affiliate Marketing: E-businesses earn a commission by promoting and selling other companies' products or services.
  - Data Licensing: E-businesses sell anonymized user data to third parties for market research or advertising purposes.

E-Business Strategy Development

- Definition: E-business strategy development involves planning and implementing a comprehensive approach to achieve business goals in the online marketplace.
- Key components of e-business strategy development:
  - Market Analysis: Research the target market, customer needs, and competitive landscape.
  - Value Proposition: Define a unique value proposition that differentiates the e-business from competitors and attracts customers.
  - Customer Segmentation: Identify and segment the target audience based on demographics, preferences, and behavior.
  - Marketing and Promotion: Develop marketing strategies to reach and engage the target audience through various digital channels.

- Technology Infrastructure: Establish a robust technological infrastructure to support e-business operations and ensure scalability.
- User Experience: Design intuitive, user-friendly interfaces to enhance customer satisfaction and encourage repeat visits.
- Data Analytics: Utilize data analytics to gain insights into customer behavior, optimize business processes, and make data-driven decisions.
- Security and Privacy: Implement robust security measures to protect customer data and build user trust.
- Continuous Improvement: Monitor and evaluate the e-business performance regularly, identify areas for improvement, and adapt strategies accordingly.

Case Study

Company X aspires to disrupt the retail industry by offering a one-of-a-kind online purchasing experience. Using augmented reality technology, the company has developed a smartphone app that allows users to put on clothing and accessories virtually. They have worked with other fashion brands to give a varied selection of things. Company X is still in its early stages of growth and seeks to build an effective e-business strategy.

1. Identify the target market groups that Company X should focus on based on its unique virtual try-on service. How might Company X better target these demographics with its marketing efforts?
2. How does Company X's virtual try-on feature offer value to the purchase experiences of customers when compared to traditional Internet shopping? How can the company communicate and reinforce its value proposition to potential customers?
3. Investigate the competitive climate of the online fashion business. Identify and evaluate the e-business strategies and techniques of Company X's significant competitors. How can Company X set itself apart and get a market competitive advantage?
4. Given the nature of the firm and the technology involved, which revenue model(s) would be suitable for firm X? Justify your choice and explain how it aligns with the company's aims and target market.
5. Research a successful e-business strategy from a similar industry or location. How may Company X adapt and incorporate analogous elements into its e-business plan to increase its chances of success?

Discussion

1. What is the definition of e-business models and how have they evolved over time?
2. What are the different types of e-business models and how do they differ from each other?

3. Can a company use more than one e-business model? If so, how does this impact their operations?
4. How have e-business models changed the way companies do business?
5. What are the key characteristics of B2B and B2C e-business models?
6. How do B2B and B2C e-business models differ from each other in terms of customer behavior, purchasing process, and marketing strategies?
7. What are some advantages and disadvantages of using B2B and B2C e-business models?
8. How do companies successfully implement B2B and B2C e-business models?
9. What are the different types of revenue models for e-businesses and how do they work?
10. How do revenue models impact a company's pricing strategy and customer acquisition?
11. What are the pros and cons of subscription-based, transaction fee-based, and advertising-based revenue models?
12. How do companies choose the best revenue model for their e-business?
13. Why is it important for companies to develop an e-business strategy?
14. What are the key elements of an e-business strategy, and how do they contribute to a company's success?
15. How do companies conduct customer segmentation, develop a value proposition, and perform competitive analysis when developing an e-business strategy?
16. Can you provide examples of companies that have successfully developed and implemented e-business strategies?

Multiple-Choice Questions

1. Which of the following statements best defines e-business models?
   (a) The physical layout of an online store
   (b) The process of selling products and services over the Internet
   (c) The software used to manage online transactions
   (d) The pricing strategies implemented by online businesses
2. Which of the following is not a type of e-business model?
   (a) B2B (business-to-business)
   (b) B2C (business-to-consumer)
   (c) C2C (consumer-to-consumer)
   (d) B2G (business-to-government)
3. Which of the following companies is an example of a B2B e-business model?
   (a) Amazon
   (b) Alibaba
   (c) eBay
   (d) Dell

4. Key characteristics of a B2C e-business model include:
   (a) Selling products/services to other businesses
   (b) Long-term contractual agreements with customers
   (c) Direct interaction with end consumers
   (d) Large-scale bulk transactions
5. What is one of the key differences between B2B and B2C e-business models?
   (a) B2B models focus on individuals as customers, while B2C models focus on businesses.
   (b) B2B models typically involve larger transaction volumes than B2C models.
   (c) B2C models rely on third-party intermediaries, while B2B models do not.
   (d) B2B models are more consumer-driven, while B2C models are more supplier-driven.
6. Which of the following companies uses a B2B e-business model?
   (a) Amazon
   (b) Walmart
   (c) eBay
   (d) Shopify
7. Revenue models for e-businesses refer to:
   (a) The overall profit margin of an e-business
   (b) The marketing strategies used to generate sales
   (c) The different ways an e-business generates revenue
   (d) The pricing models employed by e-businesses
8. Which of the following is an example of a subscription-based revenue model for an e-business?
   (a) Google AdSense
   (b) PayPal
   (c) Netflix
   (d) eBay
9. Which revenue model relies on charging a fee for each transaction facilitated through the e-business platform?
   (a) Subscription-based model
   (b) Advertising-based model
   (c) Transaction fee-based model
   (d) Freemium model
10. Which of the following companies uses an advertising-based revenue model for its e-business?
   (a) Amazon
   (b) Facebook
   (c) eBay
   (d) Netflix

11. Why is developing an e-business strategy important?
    (a) To increase customer satisfaction
    (b) To attract investors
    (c) To differentiate from competitors
    (d) To reduce operational costs
12. Which of the following is a key element of an e-business strategy?
    (a) Marketing budget allocation
    (b) Supplier selection criteria
    (c) Competitive analysis
    (d) Distribution channel optimization
13. Which element of an e-business strategy involves identifying and targeting specific groups of customers?
    (a) Competitive analysis
    (b) Value proposition
    (c) Customer segmentation
    (d) Market research
14. Which of the following is a successful e-business strategy case study?
    (a) Blockbuster
    (b) Kodak
    (c) Amazon
    (d) Toys "R" Us
15. In an e-business strategy, a value proposition refers to:
    (a) The price of products/services offered
    (b) The unique benefits or advantages an e-business provides to its customers
    (c) The target market segment for the e-business
    (d) The promotional activities employed by the e-business

References

1. Timmers, P., *Business models for electronic markets.* Electronic markets, 1998. **8**(2): p. 3–8.
2. Dubosson-Torbay, M., A. Osterwalder, and Y. Pigneur, *E-business model design, classification, and measurements.* Thunderbird International Business Review, 2002. **44**(1): p. 5–23.
3. Al-Debei, M.M. and D. Avison, *Developing a unified framework of the business model concept.* European journal of information systems, 2010. **19**(3): p. 359–376.
4. Turban, E., et al., *Electronic commerce 2018: a managerial and social networks perspective.* 2018: Springer.

Chapter 3
E-Environment

Abstract In the digital era, businesses must adapt to the changing landscape and embrace the e-environment to remain competitive. The e-environment refers to the virtual platform enabling businesses to conduct activities like marketing, selling, and customer service. This chapter provides an overview of the e-environment in e-business and its impact. The e-environment is characterized by speed, scalability, and flexibility. Digital technologies allow businesses to quickly respond to market changes, scale operations, and adapt to trends. It also offers a cost-effective alternative to traditional methods, reaching a wider audience with minimal overhead costs. E-commerce platforms and digital marketing have revolutionized operations and customer interaction. Businesses can now reach global customers 24/7, creating a level playing field for enterprises of all sizes. Consumers benefit from a convenient and personalized shopping experience with access to a broader range of products and services. Managing the e-environment requires a comprehensive e-business strategy. Prioritizing data security and privacy is also crucial to protect customer information. The e-environment has transformed the business landscape, presenting opportunities and challenges. Understanding its characteristics and impact is crucial for businesses to thrive. By adopting a comprehensive e-business strategy, businesses can leverage the e-environment to reach new customers, enhance customer experience, and drive growth.

1 Definition and Explanation of the E-Environment

In today's digital era, the e-environment idea in e-business has grown more critical. It is the digital environment where organizations use digital technologies such as the Internet, mobile devices, and cloud computing. In e-business, the e-environment enables companies to access a worldwide audience, manage operations, and perform transactions online. Several reasons have contributed to this transition toward e-business, including the expanding popularity of e-commerce, the increase of mobile devices, and the rising demand for digital services. E-commerce has become

H. Taherdoost, *E-Business Essentials*, EAI/Springer Innovations in Communication and Computing, https://doi.org/10.1007/978-3-031-39626-7_3

a cornerstone of e-business, allowing firms to extend their reach and enhance profits by selling their goods and services online.

In e-business, the e-environment also includes using digital tools and platforms to manage operations and transactions. E-commerce platforms, customer relationship management software, and digital payment systems are examples of these technologies. Businesses may use these digital technologies to simplify operations, boost productivity, and improve client experience. Another critical feature of the e-environment in e-business is the need for companies to have a solid online presence. You must create a website, social media profiles, and other digital marketing platforms to reach prospective clients. An excellent online presence assists organizations in increasing brand visibility, consumer interaction, and revenue. Despite the various advantages of the e-environment in e-business, there are also hurdles and hazards to functioning in this digital ecosystem. To secure customers' information, businesses must follow local rules, preserve consumer data, and prioritize data security and privacy. Furthermore, firms must remain adaptable and inventive to keep ahead of the competition and adapt to new technology and changing client expectations.

Finally, in today's digital world, the e-environment notion in e-business is a dynamic and ever-evolving concept that has grown more significant. Businesses must prioritize data protection and privacy, be nimble and inventive, and have a robust online presence to compete in this climate. Businesses may leverage digital tools and platforms to simplify processes, boost productivity, and deliver a better customer experience.

The e-environment is a digital ecosystem that is important to e-business success. This ecosystem comprises numerous critical components that allow firms to function online correctly. E-commerce platforms, digital marketing channels, customer relationship management software, and digital payment systems are examples of these components.

E-commerce platforms are an essential part of the e-business landscape. These platforms enable companies to offer their goods and services online, reaching a larger audience and enhancing revenues. E-commerce systems may range from essential online stores to sophisticated platforms that interface with other company software. Businesses may use these systems to manage their online catalog, handle orders, and maintain client information.

Another critical component of the e-environment in e-business is digital marketing channels. Businesses may use these channels to market their goods and services online, attract new consumers, and increase brand recognition. Social media, email marketing, search engine optimization, and content marketing are examples of digital marketing channels. Businesses may reach their target audience and boost traffic to their e-commerce platforms by successfully exploiting these channels.

CRM (customer relationship management) software is also an essential component of the e-environment in e-business. CRM software facilitates the management of customer interactions and relationships by offering a centralized platform for managing customer data, communications, and feedback. This data may enhance customer experiences and guide marketing and sales tactics. Businesses may develop personalized experiences for their consumers and establish long-term connections with them by employing CRM software [1].

Digital payment systems are an essential part of the e-business ecosystem. These systems help companies to safely and effectively accept online payments, giving consumers a smooth checkout experience. Credit card processors, payment gateways, and digital wallets are examples of digital payment systems. Businesses that use these systems may provide a variety of payment choices to their clients, increasing the overall customer experience.

Cloud computing is another critical component of the e-environment in e-business. Businesses may use cloud computing to store and access data and apps online rather than on local servers. Businesses can now access their data and apps from any location, anytime, and on any device. Cloud computing also helps organizations to grow their operations fast and effectively since computer resources can be readily added or removed as required.

Another critical component of the e-environment in e-business is data security. Businesses must secure client data while adhering to local data security and privacy standards. Implementing secure payment methods, encrypting sensitive data, and employing secure communication channels are all part of this. Businesses may develop trust with their consumers and safeguard their brands by prioritizing data security. Using analytics and business intelligence tools is part of the e-environment in e-business. Businesses may use these tools to analyze data and obtain insights into consumer behavior, sales patterns, and other critical performance metrics. Businesses may use analytics and business intelligence technologies to make data-driven choices that enhance operations and promote development.

Finally, the e-environment in e-business includes various critical components such as e-commerce platforms, digital marketing channels, customer relationship management software, digital payment systems, cloud computing, data protection, analytics, and business intelligence tools. By adequately integrating these components, businesses may reach a larger audience, enhance customer experiences, and drive growth. However, to preserve customers' information and retain their reputation, firms must prioritize data security and compliance with local rules [2].

The e-environment in e-business has evolved significantly (Fig. 3.1). Simple Internet storefronts that enabled people to buy things online characterized the early days of e-commerce. The e-environment, however, has gotten considerably more complicated and sophisticated as technology has improved and consumer behavior has altered.

One of the first instances of e-commerce may be found in the 1970s when corporations started adopting electronic data interchange (EDI) to exchange commercial papers electronically. Many supply chain operations, such as ordering and invoicing, might now be automated thanks to this technology. While customers did not directly utilize EDI, it did provide the groundwork for the eventual development of e-commerce platforms. The advent of the Internet in the 1990s ushered in a new age of e-commerce. Online marketplaces like eBay and Amazon arose, enabling customers to buy things from various suppliers. The emergence of secure online payment methods like PayPal made it simpler for customers to make online transactions.

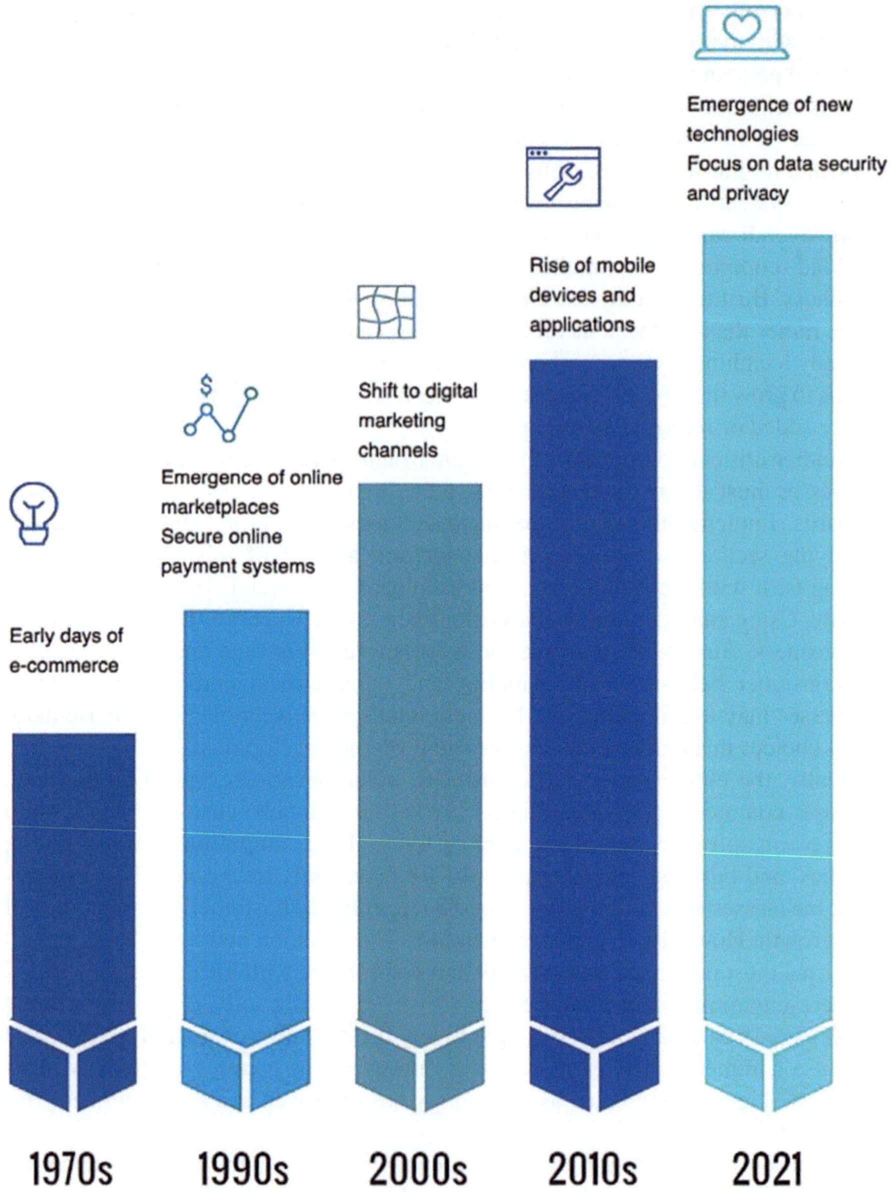

Fig. 3.1 Historical evolution of the e-environment

The e-environment in e-business started to develop beyond standard e-commerce platforms in the early 2000s. Businesses started to use digital marketing platforms to contact clients and advertise their goods and services, such as search engine optimization (SEO), email marketing, and social media. This transition signaled the beginning of a new era in the e-environment, as companies started to use technology to communicate with consumers in novel and creative ways. In the late 2000s and early 2010s, the emergence of mobile devices and apps considerably influenced the e-environment in e-business. As consumers started to purchase online using their smartphones and tablets, companies were forced to optimize their e-commerce platforms and digital marketing channels for mobile devices. This resulted in the development of mobile-first design and the usage of mobile apps to give personalized experiences to consumers.

With the introduction of new technologies such as artificial intelligence (AI) and machine learning in recent years, the e-environment in e-business has continued to grow. These technologies can analyze massive volumes of data and identify patterns and trends that can be utilized to influence corporate plans. AI and machine learning may be used by businesses to personalize product suggestions for particular consumers, optimize pricing tactics, and enhance supply chain management.

The growth of social media and the growing significance of brand reputation have also impacted the e-environment in e-business. Businesses may use social media sites like Facebook, Twitter, and Instagram to engage with consumers and market their goods and services. However, social media can be a double-edged sword since bad reviews and criticism may spread swiftly and harm a company's image.

Finally, the growing relevance of data security and privacy has impacted the e-environment in e-business. Because of the surge in cybercrime and data breaches, companies have been obliged to take precautions to secure their customers' personal and financial information. As a result, new technologies such as blockchain have emerged, which may be used to safeguard online transactions and verify the legitimacy of items.

Finally, the historical growth of the e-environment in e-business has been impacted by many variables, such as technical improvements, changing customer behavior, and the introduction of new technologies such as AI and blockchain. As the e-environment evolves, companies must keep up with the newest technology and trends to remain competitive while protecting their customers' data and privacy [1].

2 Characteristics of the E-Environment

In e-business, information speed and accessibility are critical aspects of the digital environment. Electronic transactions occur at a breakneck pace, and companies must ensure their clients have fast and easy access to the information they need. In this section, we will explore the key factors that affect the speed and accessibility of information in an e-business environment (Fig. 3.2).

The availability and quality of Internet connections heavily influence the speed and accessibility of information in the e-environment. High-speed Internet connections enable fast and efficient data transmission, allowing companies to complete

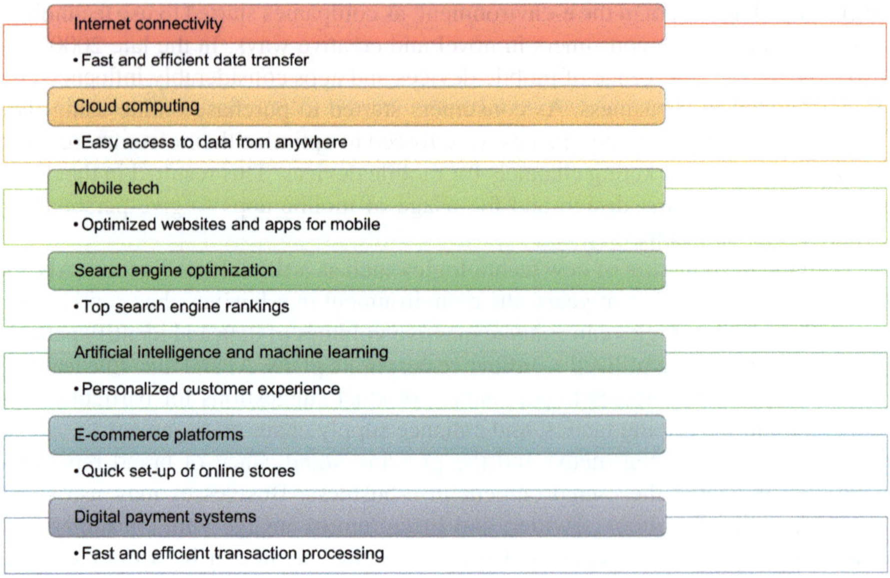

Fig. 3.2 Key components of the e-environment that contribute to the speed and accessibility of information in e-business

transactions swiftly and efficiently. With the advent of 5G technology, Internet access is set to become even quicker and more dependable.

Cloud computing has revolutionized the way organizations store and access data. Businesses can now store data on the cloud instead of physical servers, enabling easy access from anywhere worldwide. This allows employees to work more efficiently and provides customers fast access to product and service information.

The widespread use of mobile technology has also significantly impacted the speed and accessibility of information in the e-environment. As most Internet users now access the Internet via mobile devices, companies must ensure that their websites and apps are mobile-friendly. This allows customers to access information quickly and easily, regardless of location.

Search engine optimization (SEO) is a critical aspect of e-commerce as it ensures that companies' websites and content appear at the top of search engine results pages. By optimizing their websites for search engines, businesses can increase the speed and accessibility of information, making it easier for customers to find what they are looking for.

Machine learning and artificial intelligence (AI) are increasingly being used in the e-environment to improve the speed and accessibility of information. Businesses can tailor the customer experience and provide relevant information in real time by using data analytics and machine learning algorithms. This improves the customer experience and helps organizations increase customer loyalty and retention.

E-commerce platforms are crucial to the speed and accessibility of information in the digital world. Businesses can quickly set up an online shop and start selling products using e-commerce systems such as Shopify, WooCommerce, or Magento. Customers can access information about goods and services on these platforms, making informed purchase decisions easier [3].

Digital payment systems are another critical component of the e-environment, allowing companies to handle transactions quickly and efficiently. Thanks to digital payment systems such as PayPal, Stripe, and Square, businesses can now accept payments from customers worldwide, further improving the speed and accessibility of information in the e-environment.

Ultimately, the speed and accessibility of information in the e-environment are critical to the success of any organization operating in this environment. By using technologies such as cloud computing, mobile technology, SEO, AI, e-commerce platforms, and digital payment systems, businesses can provide customers with fast and efficient access to the information they need, improving the customer experience and increasing corporate success.

The emergence of the e-environment has resulted in a dramatic change in how companies function and connect with their consumers. Because of the environment's worldwide reach and influence, companies can now reach consumers from all over the globe. Customers can purchase from the comfort of their homes, regardless of their location. In this section, we will look at the worldwide reach and influence of the e-business environment and how it has changed how firms function.

The e-environment has allowed firms to reach a worldwide client base, removing geographical constraints that were formerly a key impediment to expansion. Businesses may now quickly sell their goods and services to a worldwide audience by exploiting digital channels such as social media, search engines, and e-commerce platforms, thanks to the growth of the Internet. This has increased competition while providing a significant opportunity for organizations to grow their client base and income streams. The e-environment has also changed how firms function, allowing them to enter new markets and develop abroad. This is aided by the environment's ease of access to worldwide markets, which allows enterprises to set up online storefronts and marketplaces with little overhead expenses. This has resulted in an explosion of e-commerce platforms, which serve as entry points to worldwide marketplaces for firms of all sizes [1].

The e-environment has revolutionized how companies connect with their consumers, allowing them to access worldwide markets. Because of the advent of social media, companies can now interact with their consumers on a more personal level, fostering brand loyalty and trust. This has resulted in a change in marketing techniques, with corporations emphasizing customer connections rather than merely driving sales. The environment's worldwide reach and effect have also increased cross-cultural knowledge and understanding. As organizations develop abroad, they are compelled to adapt to other cultures and norms, resulting in greater awareness and respect for diversity. As a result, new goods and services customized to particular cultural settings have been developed, allowing companies to respond to the distinct demands and preferences of their worldwide consumer base.

The e-environment has also had a tremendous economic influence, propelling development and innovation. Access to global markets has facilitated the growth of new industries and employment possibilities, generating wealth and propelling economic progress. The e-environment has also allowed small- and medium-sized organizations (SMEs) to compete with more prominent corporations, leveling the playing field and fostering innovation. However, the environment's worldwide reach and effect pose substantial hurdles for enterprises. Because of increased competition, counterfeit items have proliferated, and firms must be alert to preserve their intellectual property rights. Furthermore, because of the worldwide nature of the e-environment, firms must comply with various rules and legislation in many countries, posing a considerable compliance problem.

Finally, the worldwide reach and influence of the e-environment associated with e-business have revolutionized how firms function and engage with their consumers. Access to global markets has created massive potential for development and expansion, but it has also introduced new obstacles and hazards. Businesses must adapt to the shifting landscape and adopt new technology and marketing techniques to reach and engage their global client base to compete in the e-environment.

The growth of e-business has transformed how businesses connect with their clients. Businesses have been able to reach a bigger audience and provide individualized experiences that were previously unachievable due to the advent of the e-environment. The capacity for consumers to connect with companies and develop their content is a significant characteristic of the e-environment. This paper will examine the role of interactivity and user-generated material in the e-environment and how it has affected e-business.

Interactivity is a crucial component of the e-environment because it enables consumers to interact with companies in previously impossible ways. Customers may contact companies through numerous digital channels such as social media, email, and chatbots and obtain individualized replies to their questions. This has dramatically enhanced customer service since firms can react swiftly and effectively to client requests. Interactivity has also enabled companies to collect vital client input, which can be utilized to enhance goods and services.

Another key feature of the e-environment is user-generated content (UGC), which enables consumers to produce and share their material linked to a company or brand. Product reviews, social media postings, and videos are examples of this. Customers prefer the views of their peers above conventional advertising. Hence, UGC has grown in popularity in recent years. Businesses have realized the value of UGC and started using it in their marketing plans. Businesses may boost their reputation and attract a larger audience by publishing UGC on their social media channels or websites.

Interactivity and UGC have substantially influenced e-business. Businesses that value interaction and UGC have developed closer connections with their consumers, resulting in greater loyalty and revenue. Businesses may establish a more accurate and approachable brand image by giving individual replies and using UGC in their marketing plans. This may lead to enhanced consumer trust and trustworthiness, increasing profitability.

However, several things could be improved in interaction and UGC in the e-environment. One of the most challenging problems is controlling one's Internet

Table 3.1 Comparison of personalization and customization strategies in e-business

	Personalization	Customization
Definition	The process of tailoring a customer's experience based on their preferences and behaviors	The process of allowing a customer to design and choose their own unique experience
Type of input	Behavioral data and customer history	Active customer input and choice
Level of control	The business has more control over the process	The customer has more control over the process
Benefits	Increased customer loyalty, satisfaction, and engagement	Increased customer satisfaction, sense of ownership, and value perception
Challenges	Requires extensive data collection and analysis—the risk of over-personalization	Requires resources to provide and manage options—risk of inconsistency or complexity

reputation. Businesses must be attentive in maintaining their Internet presence now that consumers may post reviews and express their ideas online. Unwanted reviews or comments may rapidly spread and harm a company's image. Therefore, firms must have methods to monitor and react to unwanted criticism.

Another issue is the potential of consumers sharing incorrect or harmful information. Businesses must have mechanisms to regulate UGC and delete unwanted information. Because this may be time-consuming and expensive, firms must assess the advantages and dangers of using UGC in their marketing plans.

Finally, interactivity and user-generated content are critical elements of the e-environment that have significantly influenced e-business. Businesses may strengthen consumer interactions and boost their reputation and reach by emphasizing interactivity and implementing UGC into their marketing plans. However, difficulties related to interaction and UGC must be appropriately controlled to safeguard a company's image and limit damage. To stay ahead of the competition in the evolving e-environment, organizations must remain agile and imaginative in their approach to interactivity and UGC.

The advent of e-environments in recent years has transformed how companies connect with their clients. The capacity for companies to personalize and tailor the client experience is one of the fundamental advantages of this ecosystem. Personalization and customization are two ideas that have received much attention in the e-commerce market because they allow firms to adjust their offers to specific client demands and preferences (Table 3.1).

Personalization is adapting a product or service to a specific customer's wants and preferences. This may range from suggesting items based on prior purchases to sending personalized emails based on a customer's browsing history. Personalization is often seen as improving the customer experience since it may make consumers feel cherished and understood. In contrast, customization enables consumers to customize a product or service themselves. This might range from picking a product's color to selecting the features of a service. Customization often empowers consumers and gives them ownership over their purchases. Personalization and customization may both be beneficial to organizations in the e-environment. Businesses may boost customer satisfaction and loyalty by adapting their offers to specific consumer demands and preferences. Customers are more likely to remember a company that provides a unique or tailored experience, which may help firms stand out in a competitive market. However, there are several drawbacks to personalization and customization in the e-environment. One of the most challenging difficulties is the necessity for data. Businesses need access to client data to tailor or modify offers, which might cause privacy issues. Businesses must be upfront about their data collecting and usage policies and ensure compliance with applicable rules [2].

Despite these obstacles, personalization and customization are becoming more significant in e-commerce. Businesses not providing personalized and customized experiences may struggle to compete as consumers get increasingly used to them. Personalization and customization assist organizations in providing a more unique and engaging experience for their consumers, resulting in greater loyalty and revenue.

E-commerce is one sector where personalization and customization are very significant. Personalization and customization may help e-commerce enterprises stand out among increasing rivals. E-commerce enterprises may boost the possibility of purchase by providing customized suggestions based on browsing and purchase history. E-commerce enterprises may boost client engagement and pleasure by providing customization choices, such as the opportunity to customize a product [3].

Personalization and customization are very crucial in the field of content marketing. Information marketing entails generating and disseminating quality information to attract and maintain a specific audience. Businesses may use personalization and customization to provide more relevant and exciting content for their target audience. Businesses may improve the efficiency of their content marketing efforts by personalizing information to specific client demands and preferences.

Finally, personalization and customization are two ideas that have received much attention in e-business. Businesses may boost customer satisfaction and loyalty by personalizing services to consumer requirements and preferences. However, there are also drawbacks to personalization and customization, such as the requirement for data and technology. To stay ahead of the competition in the evolving e-environment, organizations must remain adaptive and imaginative in their approach to personalization and customization.

3 Impact of the E-Environment on Businesses and Consumers

The advent of e-commerce transformed the way customers engage with companies. The transition to digital platforms has opened up new avenues for companies to reach a larger audience and provide customized experiences. Consumer behavior and expectations are changing as the e-environment evolves. This paper will examine how changes in customer behavior and expectations in the e-environment affect e-business.

The need for convenience is one of the most fundamental shifts in customer behavior in the e-environment. Consumers expect to be able to access goods and services from anywhere, anytime, thanks to the growth of mobile technology and the Internet. As a result, on-demand services such as ride-sharing and meal delivery have grown in popularity, catering to customers' needs for immediate satisfaction. Businesses that offer a streamlined and efficient shopping experience, from browsing through checkout and delivery, are more likely to attract and keep consumers.

Another significant shift in customer behavior is the growing prominence of social proof. With so much information accessible online, buyers are more inclined to check out other people's thoughts and experiences before purchasing. This has increased online evaluations and user-generated material, which may substantially influence a company's reputation and legitimacy. Businesses promoting openness and welcoming client input are more likely to gain trust and loyalty.

Aside from convenience and social proof, customers are increasingly interested in personalization and customization. Consumers want companies to provide tailored experiences and suggestions based on their specific interests and habits, thanks to the plethora of data accessible online. This has resulted in a rise in targeted advertising and product suggestions, which have the potential to enhance conversion rates and consumer happiness significantly. Businesses that invest in data analytics and customization technology have a better chance of succeeding in the e-commerce market.

Furthermore, customers are becoming more conscious of their environmental effects and are emphasizing sustainability. This has increased eco-friendly goods and sustainable corporate strategies, such as waste reduction and carbon emission reduction. Businesses that emphasize sustainability and communicate their efforts to customers have a better chance of appealing to environmentally sensitive clients and establishing a favorable reputation. The e-environment has changed consumer expectations surrounding customer service. With the ease of access to digital media, customers want companies to respond to their questions and concerns quickly and efficiently. This has resulted in the rise of chatbots and social media customer care, which may give rapid and customized assistance. Businesses that emphasize customer service and invest in customer relationship management systems are likelier to develop strong and long-lasting customer connections.

Finally, changes in customer behavior and expectations in the e-environment have severe consequences for e-business. Businesses creating convenient, customized, and sustainable experiences and focusing on social proof and customer service will do better in the digital era. As the e-environment evolves, firms must stay adaptive and imaginative to satisfy customers' evolving demands and expectations.

The emergence of online grocery shopping is one real-life illustration of changes in customer behavior and expectations in the e-environment connected to e-business. Because of the COVID-19 epidemic, many opted for Internet buying as a safer alternative to in-person shopping. As a result, demand for online grocery delivery and pickup services increased, prompting merchants to alter their e-business strategies to suit evolving customer wants and expectations.

Walmart was one store that rapidly responded. They extended their online grocery services to include curbside pickup and same-day delivery. They also implemented a contactless payment option, allowing shoppers to pay for their goods without touching payment equipment. They also devised a virtual queuing system, which allows consumers to join a digital queue rather than wait in busy shops.

Walmart's e-business strategy adjustments directly reacted to customer behavior and expectation changes. Customers today demand ease, security, and speed in their online purchasing experiences, and Walmart's adaptability enabled them to remain ahead of the competition while retaining their client base.

The growth of the e-environment has presented enterprises with both possibilities and problems. On the one hand, companies can now reach a larger audience and provide individualized experiences that were previously unavailable. However, they must deal with new issues, such as controlling their Internet image and safeguarding consumer data protection. This section will examine some of the essential possibilities and problems for organizations in the e-environment. The capacity for enterprises to access a worldwide audience is one of the essential possibilities provided by the e-environment. Businesses may now offer their goods or services to people all around the globe thanks to the advent of e-commerce platforms. This has the potential to improve sales and profitability substantially. Furthermore, the e-environment enables companies to function around the clock since consumers may purchase or access services at any time of day or night.

Another advantage of the e-environment is the capacity to collect valuable data and insights about client behavior. Businesses may better know their consumers' requirements and preferences using analytics software and other data-collecting techniques. This information may be utilized to customize goods and services to those demands better, enhancing customer satisfaction and loyalty.

However, there are several issues that firms must deal with in the e-environment. One of the most challenging problems is managing one's Internet reputation. Businesses must be attentive in maintaining their Internet presence now that consumers may post reviews and express their ideas online. Unwanted reviews or comments may rapidly spread and harm a company's image. Therefore, firms must have methods to monitor and react to unwanted criticism.

Another issue is the possibility of cyber assaults and data breaches. Businesses grow increasingly exposed to cyber dangers as they acquire and keep more consumer data online. Businesses must invest in effective cybersecurity solutions and remain current on the newest cybercrime risks and trends. Failure to do so may have severe financial and reputational consequences.

Furthermore, the e-environment has made the marketplace more crowded and competitive, with companies from all over the globe competing for consumers' attention and allegiance. Companies must be more original and creative in marketing and consumer interaction strategies. They must also be adaptable to shifting market circumstances and customer preferences. However, the e-environment has also given rise to new tools and technology organizations may use. Social media platforms, for example, provide companies with a reliable tool to communicate with consumers and raise brand recognition. Businesses may boost their exposure and reach a larger audience by building a solid social media presence.

Table 3.2 Opportunities and challenges for businesses in the e-environment

Opportunities	Challenges
Increased reach and access to a broader audience	Managing online reputation
Personalized experiences and customer service	Risk of inappropriate or offensive content
Gathering valuable feedback from customers	Costly and time-consuming UGC moderation
Incorporating user-generated content (UGC)	Keeping up with rapidly evolving technology
Building stronger relationships with customers	Ensuring data privacy and security
Improved marketing and advertising strategies	Adapting to changing consumer behavior and needs
Increased brand credibility and trust	Competing with established e-businesses

To summarize, the e-environment offers both possibilities and problems to enterprises. While technology enables companies to reach a worldwide audience and get valuable insights into consumer behavior, it also requires vigilance to maintain their online image and secure client data. Businesses must remain adaptive and imaginative in their approach as the e-environment evolves while also being sensitive to the dangers and difficulties provided by the digital world. Businesses may position themselves for success in the digital era by doing so. Table 3.2 demonstrates an overview of opportunities and challenges for businesses in the e-environment.

The growth of the e-business environment has transformed company marketing and advertising techniques. Businesses can reach a larger audience and tailor their marketing efforts to particular client categories thanks to the growing variety of accessible digital platforms. This has resulted in a change in how firms handle marketing and advertising, with a greater focus on developing connections with consumers and adding value to their lives. This section will examine the consequences of e-business marketing and advertising tactics, such as the necessity of tailored messages, the role of social media, and the need for data-driven decision-making (Fig. 3.3) [1].

Personalized messaging is an essential component of efficient e-business marketing and advertising tactics. Customers nowadays are bombarded with advertisements and need tailored content that speaks directly to their needs and interests. Businesses may boost engagement and drive conversions by segmenting audiences and delivering tailored marketing using customer data. Email campaigns, social media postings, and targeted advertising are all examples of personalized communications. The idea is to bring value to the consumer by knowing their wants and interests and communicating with them in a way they understand.

Social media has also been essential in advancing e-business marketing and advertising tactics. Businesses may use social media sites like Facebook, Instagram, and Twitter to engage with their consumers and establish connections. Businesses may expand their reach and reputation by providing compelling content and supporting user-generated content. Businesses may also use social media to react to client comments and offer individualized customer care.

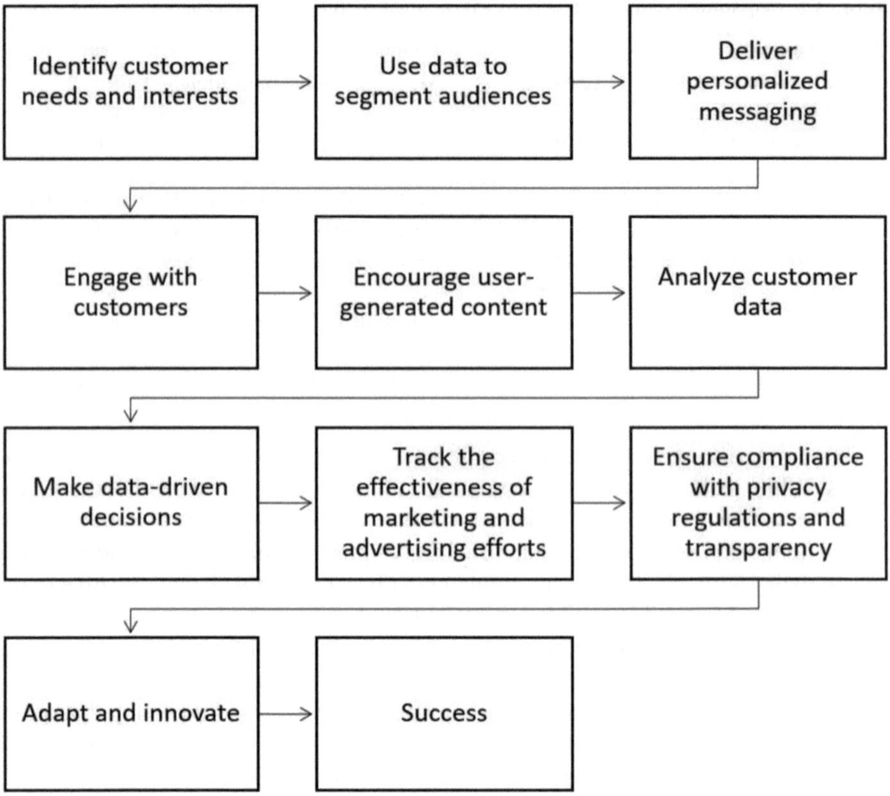

Fig. 3.3 Effective marketing and advertising strategies in the e-business environment

Another critical part of efficient marketing and advertising tactics in the e-business context is data-driven decision-making. Businesses may now make informed judgments regarding their marketing and advertising activities because of the large quantity of data accessible today. Businesses may spot patterns and preferences in consumer data and customize their marketing appropriately. Data-driven decision-making also enables companies to monitor the efficacy of their marketing and advertising activities and make necessary modifications. However, there are several difficulties involved with e-business marketing and advertising. One of the most challenging tasks is managing consumer data and adhering to privacy standards. Businesses must be upfront with their consumers regarding data collection and usage and guarantee that all relevant rules are followed. Another difficulty is adapting and developing in response to changing client preferences and new technology [2].

Finally, successful e-business marketing and advertising tactics emphasize targeted messages, social media interaction, and data-driven decision-making. Businesses may enhance their marketing and advertising activities and drive success by giving value to consumers via tailored messaging, creating connections through social media, and employing data to assist decision-making. On the other hand, businesses must be aware of the issues connected with the e-business environment, such as securing client data and staying adaptive in the face of change.

4 Strategies for Managing the E-Environment

Businesses must utilize techniques that concentrate on online reputation management, consumer interaction, monitoring, analysis, data management, and security to manage the e-environment successfully.

Maintaining an excellent online reputation is critical for the success of any organization in today's e-business climate. Businesses must be careful in monitoring and controlling their online presence as social media and online review platforms become more prevalent. This section will cover best practices for online reputation management in the e-business environment, such as online customer engagement strategies, monitoring and analyzing online behavior, and tools for managing and preserving online data (Fig. 3.4).

Having a good plan for communicating with consumers online is one of the finest practices for online reputation management. This includes developing and maintaining an excellent social media presence and reacting quickly to consumer requests and comments. Businesses may also engage consumers by producing interesting and valuable material, such as blog entries or videos, that provide value to their target audience. Businesses engaging with consumers online may create trust and credibility, eventually improving their online reputation.

Monitoring and analyzing online behavior is another effective method for online reputation management. This includes keeping track of any Internet references of the company and its goods or services and reacting appropriately to nasty comments or complaints. Businesses may spot potential problems before they worsen and proactively resolve them by monitoring Internet behavior. Furthermore, firms may employ analytics tools to analyze the efficacy of their online reputation management initiatives and make data-driven choices.

Businesses must adopt tactics for securing and managing Internet data to manage their online reputation properly. This includes safeguarding consumer data and developing rules and processes for sensitive information. Businesses must also be open and honest with their consumers about how they gather and use data and give clear instructions on managing their online accounts and securing their personal information.

Online reputation management software may help control and secure online data. These technologies assist organizations in tracking and analyzing their online reputation, monitoring brand mentions online, and responding to real-time client feedback. They may also assist firms in identifying possible security issues and mitigating them.

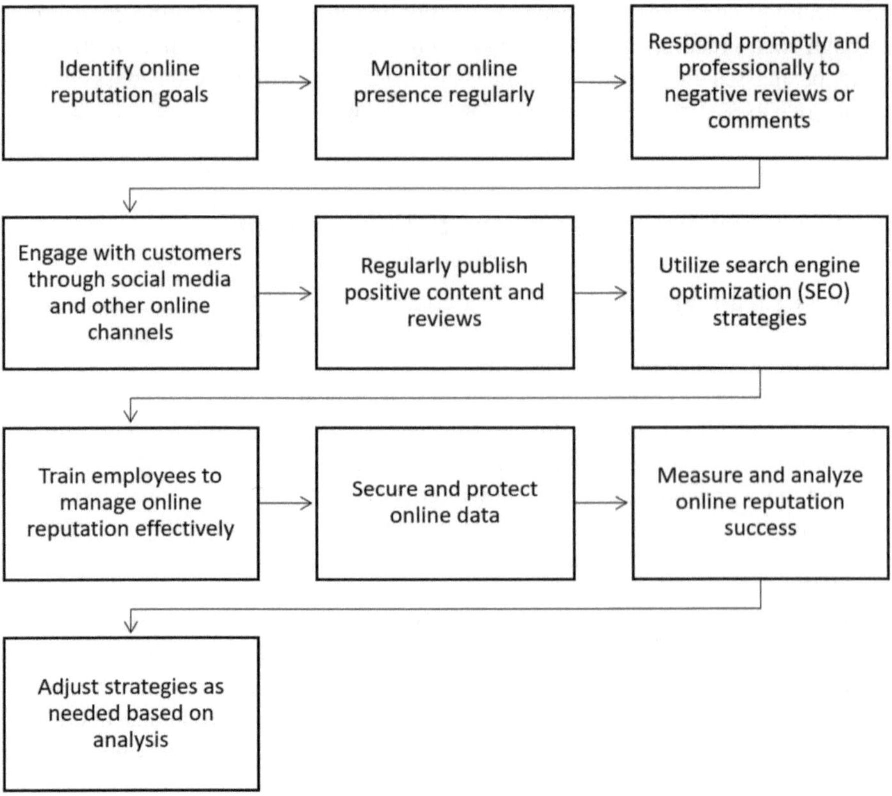

Fig. 3.4 Best practices for online reputation management in e-business

Finally, best practices for online reputation management in the e-business environment include a mix of online consumer engagement tactics, monitoring and analyzing online behavior, and tools for managing and preserving online data. Businesses may successfully manage their online reputation and enhance their chances of success in the competitive e-business environment by actively interacting with consumers, monitoring online behavior, and having robust data protection rules and processes.

Engaging with consumers online is vital to creating connections and achieving success in the e-business world. Businesses can connect with customers in various ways and add value to their lives thanks to the growing number of digital channels available. This paper will look at tactics for engaging consumers online, such as the significance of tailored messages, the usage of social media, and the necessity of delivering outstanding customer service.

Personalized messaging is a critical component of efficient online client interaction. Customers now want messaging that speaks directly to their specific needs and interests. Businesses may boost engagement and drive conversions by segmenting audiences and delivering tailored marketing using customer data. Email campaigns, social media postings, and targeted advertising are all examples of personalized communications. The idea is to bring value to the consumer by knowing their wants and interests and communicating with them in a way they understand.

Social media has also become an essential instrument for online client engagement. Social media sites like Facebook, Instagram, and Twitter allow companies to engage with their consumers and establish connections. Businesses may expand their reach and reputation by providing compelling content and supporting user-generated content. Businesses may also use social media to react to client comments and offer individualized customer care. Responding quickly to client complaints and publicly addressing their problems may help boost customer satisfaction and loyalty.

Another crucial part of online client engagement is delivering outstanding customer service. Customers have high expectations for rapid and efficient service in the e-business era. Excellent customer service may assist organizations in differentiating themselves from their competition and building a loyal consumer base. This may be accomplished in various ways, including giving self-service alternatives on the corporate website, delivering customized customer care through chatbots, and reacting quickly to consumer inquiries and concerns.

In addition to these techniques, companies may use various technologies to communicate with clients online successfully. Customer relationship management (CRM) software is one such product that assists firms in managing client data and interactions across numerous channels. CRM software enables firms to monitor customer contacts, analyze customer data, and customize communications to increase engagement and conversions. Web analytics software, which gives insights into website traffic and user activity, is another tool that firms may utilize. Web analytics may assist firms in identifying patterns and opportunities to increase their online client interaction.

In conclusion, communicating with clients online is crucial for e-business success. Businesses may generate success by deploying tailored messages, leveraging social media, offering exceptional customer service, and utilizing efficient technologies such as CRM software and web analytics. The goal is to understand consumers' requirements and preferences and appropriately adjust interaction techniques. Table 3.3 provides examples of effective online customer engagement strategies.

Businesses must monitor and evaluate their online behavior to remain competitive and maximize their potential in the age of e-business. This entails tracking and measuring consumer interaction, website traffic, social media activity, and other factors. This post will look at some of the most successful strategies for monitoring and evaluating online activity in the e-business environment, including website analytics, social media monitoring, and sentiment analysis.

Website analytics is one of the most effective tools for monitoring and evaluating Internet activity. Businesses may acquire significant consumer behavior and preference insights by measuring website traffic. This contains data such as the number of website visitors, the pages they view, and the time they spend on each page. Website analytics may also inform website users' demographics, such as age, gender, and geography. This data can be used to identify customer behavior trends and patterns, which can then inform marketing and advertising strategies.

Table 3.3 Examples of effective online customer engagement strategies

Online customer engagement strategies	Description
Social media marketing	Use social media platforms to promote and engage with customers, such as creating shareable content, running promotions and contests, and responding to customer comments and messages
Personalized marketing	Delivering personalized messaging to customers based on their preferences, demographics, and past interactions with the brand, such as targeted email campaigns or product recommendations
Live chat support	Providing real-time customer support through live chat on the company website or social media platforms, allowing customers to ask questions and receive immediate assistance
User-generated content	Encouraging customers to create and share their content related to the brand, such as social media posts or product reviews, and leveraging this content to engage with other customers
Loyalty programs	Offering rewards and incentives to repeat customers, such as discounts, free products, or exclusive access to events or content, to build long-term relationships and increase customer retention
Interactive content	Creating interactive content such as quizzes, polls, or games to engage with customers and provide a unique and entertaining experience
Virtual events	Hosting virtual events such as webinars, Q&A sessions, or live streams to connect with customers and provide valuable information or entertainment

Another effective tool for monitoring and evaluating Internet activity is social media monitoring. With billions of users globally, social media platforms such as Facebook, Twitter, and Instagram give companies an unequaled chance to engage with their consumers and increase brand recognition. Businesses may obtain essential insights into client mood, engagement, and preferences by monitoring social media activity. This may be accomplished using various tools and approaches, such as social media listening and sentiment analysis.

Monitoring social media platforms for mentions of a particular business or product is what social media listening entails. This may be done manually or by using specialist software. Businesses may acquire insights into consumer sentiment and preferences by monitoring brand or product mentions. This data may guide marketing and advertising tactics and suggest product and service improvement areas.

Another tool for monitoring and evaluating social media activity is sentiment analysis. This entails analyzing social media messages and categorizing them based on their sentiment using natural language processing and machine learning techniques. Businesses may acquire insights into client attitudes and preferences by studying social media sentiment. This data may guide marketing and advertising tactics and suggest product and service improvement areas.

Aside from website analytics and social media monitoring, organizations may monitor and analyze online activity using various tools and approaches. Examples include search engine optimization (SEO) tools, email marketing analytics, and customer relationship management (CRM) software. Businesses may enhance their marketing and advertising tactics by employing these tools and approaches to understand their customer's behavior and preferences better.

Finally, for organizations functioning in the e-business environment, monitoring and evaluating online activity is critical. Businesses may acquire significant insights into consumer behavior and preferences by adopting website analytics, social media monitoring, and sentiment analysis. This data may guide marketing and advertising tactics and suggest product and service improvement areas. By adopting the correct tools and practices, businesses may remain competitive and realize their potential in the e-business environment. Table 3.4 summarizes the techniques used for monitoring and analyzing online activity in an e-business.

Online data has become one of the most important assets for organizations in today's digital era. As a result, organizations must have the proper tools and strategies to manage and safeguard their online data. This section will explore several tools and approaches organizations may use to manage and safeguard their online data in the e-business environment.

Encryption is a critical tool for managing and securing online data. Encryption is the process of encoding data so that it is unintelligible to anybody who needs the correct decryption key. This is especially true for sensitive information such as credit cards or personal identification numbers. Encryption may be performed on the client side, meaning that data is encrypted before being sent over the Internet or on the server side, which means that data is encrypted while stored on a server. Various encryption technologies, including Pretty Good Privacy (PGP) and Secure Sockets Layer (SSL), are available to safeguard online data.

Firewalls are another critical tool for controlling and securing Internet data. Firewalls are network security systems that monitor and regulate traffic based on predefined security rules. They serve as a line of defense between a company's internal network and the outside world, limiting illegal access to critical data. There are various firewalls, including network, host, and application-based firewalls. Each type has advantages and disadvantages, and the company's specific needs determine the appropriate type of firewall.

In addition to encryption and firewalls, various solutions are available for managing and protecting online data. For example, intrusion detection systems (IDS) and intrusion prevention systems (IPS) may detect and block illegal network access. IDS monitor network traffic for signals of future attacks, while IPS actively prevent assaults by blocking suspicious traffic. Virtual private networks (VPNs) and multi-factor authentication (MFA) are two other techniques that may be used to protect online data.

Many e-businesses now have real-world examples of technologies for managing and preserving online data. For example, companies like Amazon and eBay utilize SSL encryption to secure client data during transactions. Similarly, many banks utilize firewalls, intrusion detection systems, and intrusion prevention systems to prevent unwanted access to consumer data. Furthermore, many businesses use VPNs and MFA to guarantee that only authorized personnel can access critical data. These solutions are crucial for securing online data and guaranteeing the security of

Table 3.4 Techniques for monitoring and analyzing online activity in e-business

Technique	Description	Benefits	Drawbacks	Tools
Social media listening	Monitoring social media platforms to gain insights into customer conversations, industry trends, and brand reputation	Helps identify potential issues early on and respond quickly Provides valuable insights into customer needs and preferences Enables businesses to track competitors and industry trends	It can be time-consuming to manage multiple social media accounts It can be challenging to filter through and analyze large amounts of data May not capture all relevant online conversations	Hootsuite, Sprout Social, Brand24
Web analytics	Tracking website traffic and user behavior to optimize website design and content, measure marketing campaign performance, and improve conversion rates	Provides insights into website performance and user behavior Helps identify areas for improvement and optimization Enables businesses to measure the effectiveness of marketing campaigns	May require technical expertise to set up and interpret data May not capture all user behavior and interactions It can be challenging to determine the cause of certain behaviors	Google Analytics, Adobe Analytics, Piwik
Sentiment analysis	Using natural language processing (NLP) and machine learning algorithms to analyze the sentiment of online conversations and social media posts	Provides a quick and objective way to gauge public opinion and brand reputation Helps businesses understand how customers feel about their products and services Enables businesses to respond to negative sentiment and improve customer satisfaction	Can be limited by the accuracy of NLP and machine learning algorithms May not capture sarcasm, irony, or another nuanced language May be influenced by the context of the conversation	IBM Watson, Lexalytics, Clarabridge
Clickstream analysis	Tracking user clicks and navigation patterns on a website to gain insights into user behavior and preferences	Provides insights into user behavior and preferences Helps businesses identify areas for improvement and optimization Enables businesses to improve conversion rates and user experience	May not capture all user behavior and interactions It can be challenging to determine the cause of certain behaviors May be limited by privacy concerns and regulations	ClickTale, Crazy Egg, Mouseflow
Search engine monitoring	Tracking and analyzing search engine results pages (SERPs) to gain insights into search behavior, competitor rankings, and brand reputation	Provides insights into search behavior and trends Helps businesses track competitors and industry trends Enables businesses to monitor and manage their online reputation	It can be time-consuming to track and analyze SERPs May be influenced by search engine algorithms and personalized search results May not capture all relevant search queries and results	SEMrush, Ahrefs, Moz

Table 3.5 Comparison of tools for managing and protecting online data in e-business

Tool	Description	Benefits	Drawbacks
Encryption software	Software that uses encryption algorithms to convert data into unreadable code	Provides robust security for sensitive information	Can slow down computer performance
Virtual private network (VPN)	A network technology that creates a secure and encrypted connection over the Internet	Protects against cyber threats such as hacking and phishing	It may reduce Internet speed
Anti-virus software	A program that detects and removes viruses and malware from computer systems	Protects against cyber threats	It may not catch all types of viruses or malware
Firewall	A security system that monitors and controls incoming and outgoing network traffic	Protects against cyberattacks and unauthorized access	It can be complex to set up and configure
Multi-factor authentication (MFA)	An authentication method that requires users to provide more than one type of verification to access an account	Provides an extra layer of security against unauthorized access	It can be inconvenient for users to set up and use

e-businesses in the digital age. Table 3.5 provides a summary of tools and their features that are used for managing and protecting online data.

Summary

Definition and Explanation of the E-Environment

- The e-environment refers to the digital landscape encompassing the Internet, online platforms, and virtual spaces where interactions, transactions, and communications occur.
- It includes websites, social media platforms, e-commerce platforms, mobile applications, and other online channels facilitating digital interactions.

Characteristics of the E-Environment

- Global Reach: The e-environment enables businesses and consumers to connect across geographical boundaries, allowing worldwide access and reach.
- 24/7 Availability: Online platforms and services operate around the clock, providing businesses with continuous availability and consumers with convenience.
- Interactivity: The e-environment fosters interactive and dynamic experiences through features like live chats, customer reviews, and personalized recommendations.
- Speed and Efficiency: Digital interactions in the e-environment are characterized by rapid communication, instant transactions, and streamlined processes.

Impact of the E-Environment on Businesses and Consumers

- Expanded Market Opportunities: The e-environment allows businesses to reach a broader audience, tap into new markets, and explore diverse customer segments.

- Enhanced Customer Experience: Consumers benefit from increased convenience, personalized recommendations, and easy access to information and reviews.
- Disruption of Traditional Business Models: The e-environment has disrupted various industries, leading to the emergence of new business models and reshaping market dynamics.
- Increased Competition: The ease of entry in the e-environment has intensified competition, requiring businesses to differentiate themselves and offer unique value propositions.
- Data-Driven Insights: The e-environment generates vast amounts of data, providing businesses with valuable insights for decision-making, marketing strategies, and customer analysis.

Strategies for Managing the E-Environment

- Online Presence and Branding: Businesses need to establish a robust online presence, create a consistent brand image, and optimize their websites and social media profiles.
- Customer Engagement and Relationship Building: Engaging with customers through social media, email marketing, and personalized interactions help build lasting relationships and loyalty.
- E-Commerce and Online Sales: Implementing secure and user-friendly e-commerce platforms and optimizing the online purchasing process to enhance sales and conversions.
- Digital Marketing and Advertising: Utilizing various digital marketing techniques, such as search engine optimization (SEO), social media advertising, and content marketing, to reach and engage target audiences.
- Data Privacy and Security: Safeguarding customer data, complying with data protection regulations, and implementing robust security measures to maintain trust and protect against cyber threats.

Case Study

The multinational e-commerce company XYZ offers a diversity of products online. The company has a robust web presence and relies heavily on the Internet to conduct business. Due to the expanding influence of the e-environment, XYZ Company has witnessed significant shifts in consumer behavior and expectations over the past few years. This case study concentrates on the impact of the e-environment on XYZ Company and the strategies it may employ to effectively manage the challenges and opportunities presented by this digital landscape.

XYZ Company has been in online retail for over a decade and has established a trustworthy brand. In recent months, however, the firm has experienced a decline in client engagement and revenue. It believes that transforming e-environment dynamics and altering consumer behavior contribute to this decrease. In order to resolve these issues, XYZ Company desires to comprehend the impact of the e-environment on its operations and develop strategies for enhancing its online presence, consumer engagement, and overall success.

1. How has the quickness and availability of information in the e-environment impacted XYZ Company's ability to connect and engage customers successfully?
2. What specific shifts in consumer behavior and expectations should XYZ Company be aware of to enhance its online customer experience and increase sales?
3. What opportunities and challenges does the e-environment present for XYZ Company's marketing and advertising initiatives? How can the organization maximize these opportunities while overcoming these obstacles?
4. How can XYZ Company effectively manage its online reputation to maintain consumer trust and credibility in the digital sphere? What are the most effective ways to manage one's online reputation?
5. How does XYZ Company monitor and evaluate Internet activity related to its brand and products? How will the company use this data to make more informed business decisions and improve its online performance?

Discussion

1. What is the e-environment and how does it differ from the physical environment?
2. Why is the e-environment important for businesses and consumers?
3. How has the e-environment changed the way people communicate and access information?
4. What are the main components of the e-environment?
5. How do digital technologies and online communication contribute to the e-environment?
6. How have these components evolved over time and what impact have they had on society?
7. How has the e-environment evolved over time?
8. What were some of the key milestones in the development of the e-environment?
9. How has the historical evolution of the e-environment influenced its current state?
10. What are the main characteristics of the e-environment?
11. How do these characteristics impact businesses and consumers?
12. What are some of the challenges and opportunities associated with these characteristics?
13. How has the e-environment changed consumer behavior and expectations?
14. What are some of the opportunities and challenges for businesses in the e-environment?
15. How have marketing and advertising strategies been impacted by the e-environment?

16. What are some best practices for online reputation management?
17. How can businesses effectively engage with customers online?
18. What are some techniques for monitoring and analyzing online activity?
19. What tools are available for managing and protecting online data?

Multiple-Choice Questions

1. Which of the following best describes the e-environment?
   (a) The physical surroundings of electronic devices
   (b) The digital landscape created by online platforms
   (c) The study of environmental conservation through digital means
   (d) The process of recycling electronic waste
2. What are the key components of the e-environment?
   (a) Social media and search engines
   (b) Print media and television
   (c) Traditional communication methods
   (d) Digital technology and online communication
3. Which statement best describes the historical evolution of the e-environment?
   (a) The e-environment has remained unchanged throughout history.
   (b) The e-environment only emerged in the last decade.
   (c) The e-environment has evolved rapidly over time.
   (d) The e-environment is a recent concept with no historical roots.
4. Which characteristic of the e-environment refers to the speed and accessibility of information?
   (a) Global reach and impact
   (b) Interactivity and user-generated content
   (c) Personalization and customization
   (d) Speed and accessibility
5. What does the term "interactivity and user-generated content" mean in the context of the e-environment?
   (a) The ability to communicate with people from different countries
   (b) The customization of digital products and services
   (c) The sharing and creation of content by online users
   (d) The availability of information from around the world
6. Which characteristic of the e-environment allows businesses to tailor their products or services to individual consumers?
   (a) Speed and accessibility of information
   (b) Global reach and impact

(c) Interactivity and user-generated content
(d) Personalization and customization

7. How has the e-environment impacted consumer behavior and expectations?

(a) Consumers are less reliant on digital technology.
(b) Consumers have higher expectations for personalized experiences.
(c) Consumers prefer traditional communication methods.
(d) Consumers are less connected to global trends and events.

8. What opportunities and challenges does the e-environment present for businesses?

(a) Limited customer reach and engagement
(b) Increased competition and market saturation
(c) Enhanced targeting and customer insights
(d) Decreased need for online marketing and advertising

9. What are the implications of the e-environment for marketing and advertising strategies?

(a) Traditional marketing channels are no longer effective.
(b) Businesses can no longer target specific customer segments.
(c) Personalized and targeted advertising is more feasible.
(d) Online advertising is not cost-effective.

10. What are the best practices for online reputation management?

(a) Ignoring negative feedback and reviews
(b) Removing all user-generated content
(c) Responding promptly and appropriately to feedback
(d) Promoting false-positive reviews

11. What strategies can businesses use to engage with customers online?

(a) Avoiding social media platforms entirely
(b) Implementing one-way communication channels
(c) Encouraging user-generated content and participation
(d) Limiting customer interactions to offline channels

12. What techniques can be used to monitor and analyze online activity?

(a) Ignoring online data and analytics
(b) Using traditional market research methods
(c) Leveraging web analytics tools and social listening platforms
(d) Relying solely on personal intuition and assumptions

13. Which tools can businesses use to manage and protect online data?

(a) Sharing sensitive data on public platforms
(b) Using outdated security measures
(c) Implementing secure data storage and encryption
(d) Ignoring data privacy regulations

14. How has the e-environment impacted businesses' approach to data collection?
    (a) Businesses collect less data due to privacy concerns.
    (b) Data collection methods have remained the same.
    (c) Businesses collect more data to enhance personalization.
    (d) Data collection is no longer necessary in the e-environment.
15. What is the primary purpose of managing the e-environment for businesses?
    (a) To limit customer reach and engagement
    (b) To maintain a negative online reputation
    (c) To improve customer experiences and relationships
    (d) To discourage online interactions with customers

References

1. Morris, N., *Understanding digital marketing: marketing strategies for engaging the digital generation.* Journal of Direct, Data and Digital Marketing Practice, 2009. **10**(4): p. 384–387.
2. Suleiman, M.I., et al., *Benefits and challenges: for marketing strategies on-line.* European Journal of Molecular & Clinical Medicine, 2020. **7**(3): p. 164–179.
3. Constantinides, E. and S.J. Fountain, *Web 2.0: Conceptual foundations and marketing issues.* Journal of direct, data and digital marketing practice, 2008. **9**: p. 231–244.

Chapter 4
Evolution and Applications of the Internet in E-Business

Abstract The Internet has revolutionized business operations, transforming them from local brick-and-mortar establishments to global digital enterprises. This chapter explores the history and evolution of e-business on the Internet, highlighting its uses, benefits, and challenges. It also evaluates successful case studies of businesses that have expanded through Internet marketing. Understanding the current state of e-business requires knowledge of its historical origins and the significant events that have shaped it. The Internet has had a profound impact on businesses, from the days of dial-up modems to the advent of broadband and wireless connectivity. To delve into e-commerce, we must explore online purchasing, digital marketing, social media, and cloud computing. These applications enable businesses to implement innovative strategies, reduce costs, and expand their customer base. However, they also introduce security and privacy concerns that businesses must address to mitigate risks and overcome obstacles. Moreover, we examine notable case studies of companies like Amazon, Alibaba, and Airbnb that have thrived through Internet-based strategies. This chapter concludes with a high-level overview of e-business, analyzing its history, current state, potential applications, and potential hazards. E-business plays a vital role in a company's ability to innovate, expand internationally, and maintain a competitive edge in the twenty-first-century global economy.

1 Overview of the Evolution of the Internet in E-Business

The development of the Internet in the second part of the twentieth century was a turning point in corporate history. Various online marketplaces, search engines, and other Internet-based services were created because business people and visionaries quickly saw the potential of the Internet as an e-commerce tool. Online marketplace eBay is widely mentioned as one of the pioneers of e-commerce. When eBay launched in 1995, it brought people from all over the globe who wanted to buy and sell products online together. Since eBay, other online marketplaces like Amazon and Alibaba have become giants in their respective industries worldwide.

H. Taherdoost, *E-Business Essentials*, EAI/Springer Innovations in Communication and Computing, https://doi.org/10.1007/978-3-031-39626-7_4

The development of search engines was essential for the early stages of e-commerce since it made it easier to find new clients online. Google has become the industry leader since it was released to the public in 1998. Due to Google's algorithm, companies may now improve their websites for search engines, raising their search engine ranks and bringing in more customers. The early years of Internet business have their own unique set of difficulties. Worries about the security and trustworthiness of online transactions have hampered e-commerce's expansion. A common way to send money online is via PayPal, one of the many online payment systems and security procedures created to allay these worries [1].

Online commerce underwent a crucial turning point during the "dot-com bubble" of the late 1990s and early 2000s. Many dot-coms went out of business when the market crashed due to the speculative enthusiasm sparked by the fast growth of Internet-based businesses. The Internet market rebounded and grew as adaptable enterprises adapted to the new setting. Early in the twenty-first century, social media platforms like MySpace and Facebook made it possible for businesses to communicate with their target audiences in new ways, which improved sales and brand recognition. Social media marketing allows companies to reach a wider audience of potential customers.

Technology improvements and shifting customer tastes have played a significant role in the Internet's and e-commerce's remarkable growth over the last several decades. The birth of the World Wide Web, the explosive growth of e-commerce, and the growing role of mobile devices are all covered in this section's discussion of the evolution of the Internet and e-commerce. The launch of the World Wide Web at the beginning of the 1990s signaled a turning point in the history of the Internet and electronic business. The growth of the World Wide Web has opened up new channels for e-commerce and global communication. Two of the most well-known web browsers, Internet Explorer and Netscape Navigator, have simplified the web and its many services. The World Wide Web not only made communication more accessible but also spawned cutting-edge business models like eBay and Amazon.

An important turning point in the history of the Internet and online commerce was the sudden growth of e-commerce in the late 1990s and early 2000s. E-commerce created a new sector of virtual markets and exchanges and allowed customers to purchase online. By offering customers a massive assortment of reasonably priced goods and services, businesses like Amazon and Alibaba have grown to dominate the e-commerce sector. Improvements in online payment options and security protocols have fueled the rapid growth of e-commerce. The increased usage of mobile devices has significantly impacted Internet shopping. A growing number of consumers are adopting mobile computing devices like smartphones and tablets to surf the web and make online purchases due to their widespread availability. New e-commerce opportunities have been opened up by the advent of mobile technology, like mobile payments and smartphone apps. Businesses that invest in mobile solutions have an edge over their competitors because they can reach customers wherever they are and provide a consistent user experience across all their devices [2].

The spread of social media platforms has significantly impacted the Internet and e-commerce. Just a handful of the countless new channels companies can connect with clients and advertise their goods include Facebook, Twitter, and Instagram. The growth of social media has directly led to the emergence of new business models, such as influencer marketing. Consumers are happier and more devoted to businesses because of how simple it is for them to provide feedback and get answers to their questions through social media. The rising importance of data analytics has also significantly impacted Internet use and e-commerce. Data analytics tools make gathering and evaluating a sizable quantity of data on client preferences, habits, and trends feasible. This material could be helpful for marketing professionals, customer service representatives, and operational managers. Data analytics has become a crucial component of the industry as the importance of understanding customer behavior and staying ahead of the competition in e-business increases. Lastly, every organization that does business online is very concerned about the growing significance of cybersecurity. As e-commerce grows and more sensitive data is sent online, businesses are increasingly susceptible to hacks. Encryption, firewalls, and intrusion detection systems are a few key cybersecurity measures that companies must use in the digital era to safeguard their data and the security of their customers [1].

Technology improvements and shifting customer behavior have caused substantial shifts and upheavals in the Internet and e-business. The creation of the World Wide Web, the popularity of e-commerce, the proliferation of mobile devices, the growth of social media platforms, the development of data analytics, and the growth of cybersecurity have all significantly influenced the Internet and e-business. Businesses must keep up with these advances to maintain a competitive edge and succeed in the rapidly evolving world of e-commerce. Figure 4.1 demonstrates key developments in e-business and the Internet over time.

New technical and cultural breakthroughs are significantly impacting the future of Internet commerce. Many factors, such as shifting customer tastes, faster technical development, and global economic swings, will continue to affect the future of e-business. This part will look at some of the most anticipated advancements in online commerce and the Internet in general (Table 4.1).

Artificial intelligence (AI) is expected to play an increasingly important role in Internet businesses shortly. AI has the potential to significantly improve many aspects of e-commerce, including search engine optimization, product suggestions, and customer service/support. As technology advances and becomes more accessible, an increasing number of firms will be able to capitalize on its potential. Amazon, for example, has employed artificial intelligence to enhance its recommendation engine, resulting in more sales and delighted customers.

The rise of mobile commerce is another key trend that will impact the future of e-business. As the number of people who own smartphones and tablets grows, so does the importance of mobile commerce for all companies. In the following years, businesses will emphasize mobile customers by establishing mobile-friendly websites and apps and accepting mobile payments through Apple Pay and Google Wallet. Walmart, for example, has substantially invested in its mobile app to streamline the shopping experience for customers and to encourage customer loyalty. Customization will become more critical in the future of e-commerce. More and more firms are gathering customer information to personalize their services to each

Fig. 4.1 Critical developments in e-business and the Internet over time

individual's interests. Personalized product suggestions, communications, and content are examples of this personalization form. For years, companies like Netflix have used personalized suggestions to enhance the customer experience, and this trend will likely continue.

The increasing incidence of voice searches will influence Internet commerce shortly. Because of the rise of popular virtual assistants like Siri and Alexa, many individuals utilize voice search to investigate items and make online purchases. As a result, companies must optimize their websites and goods for voice inquiries. The Starbucks mobile app, for example, now offers a new feature that lets customers make orders using voice commands. Finally, the continued rise of social media will influence the future of e-business. Facebook and Instagram, among other networks, have matured into incredibly successful commercial marketing and customer service tools. The number of firms that utilize social media to connect with customers

Table 4.1 Future trends and implications for e-business

Future trends and predictions	Implications for e-business
The emergence of artificial intelligence	AI-powered businesses enjoy improved search engine results and recommendation engines, personalized customer service and support, and competitive advantage
Growth of mobile commerce	Increased significance of mobile-optimized websites and applications, increased use of mobile payment platforms, and enhanced consumer convenience and loyalty
Importance of personalization	Personalized product suggestions, personalized communications, and tailored content enhance user experience
Rise of voice search	Businesses that embrace this trend will discover new opportunities arising from the need to optimize websites and products for voice search
Growth of social media	Increased emphasis on developing brand loyalty and engagement; a potent instrument for businesses to communicate with consumers and market their products and services

and establish brand reputation will grow. Nike, for example, has utilized Instagram campaigns like #betterforit to encourage women to talk about their workouts and build a community around the brand.

The e-business and Internet industries will have several prospects in the following years. Artificial intelligence (AI), mobile commerce (m-commerce), customization (personalization), and voice search are examples of new trends and technology that successful firms will foresee and adapt to. Businesses may survive and prosper in today's changing e-commerce industry by focusing on these and other rising trends.

2 Applications of the Internet in E-Business

The numerous benefits of the Internet to businesses have had a significant influence on e-commerce in particular. Businesses may now contact clients all over the globe thanks to e-commerce and online purchases. Thanks to digital marketing and advertising, businesses may contact their target audiences more quickly and efficiently. New customer connection and brand loyalty channels have evolved due to social media and user-generated content. Lastly, cloud computing and software as a service have streamlined corporate operations and lowered costs by allowing businesses to access software and data storage on demand without investing in costly infrastructure and hardware. These e-business Internet apps have already changed how organizations function and will continue to affect the sector in the future.

E-commerce and Internet purchasing have substantially altered the conventional retail scene. The Internet has enabled companies to interact with clients all over the globe, as well as people, to fulfill their shopping needs without ever leaving their homes. E-commerce has become popular over the last decade as many firms move to online sales. We will look at the beginnings and evolution of e-commerce, its benefits and drawbacks, and its impact on the retail business. Online purchase has a long and illustrious history, dating back to the 1970s with early trials in electronic transactions. In the 1990s, with the introduction of the World Wide Web, e-commerce gained significant traction. The first e-commerce website went live in 1994, and by the end of the decade, online purchases had grown from their modest beginnings. E-commerce expanded due to the expansion of high-speed Internet, the development of more foolproof security systems, and other technical improvements. The main benefit of e-commerce and Internet buying is convenience. Businesses are no longer required to have a physical site to sell to customers all over the globe, and consumers may shop from the comfort of their homes. Furthermore, the ease of Internet buying has permitted more customized services. The data acquired from online customer interactions may be utilized to enhance service [3].

There are pros and downsides to e-commerce and Internet purchasing. The issue of safety is paramount. Customers subject themselves to the danger of identity theft and fraud when requested to give financial and personal information online. Online firms may also need help with shipping, refunds, and customer service. Despite these constraints, e-commerce and Internet shopping have significantly influenced retail. Traditional stores have adopted e-commerce systems or developed their online sites to survive. By adopting digital, small firms have been able to compete with industry titans. New technology and business models have arisen due to the growth of the Internet marketplace. Dropshipping, or selling things online without keeping an inventory, has lately gained favor among would-be Internet entrepreneurs. Subscription-based food delivery and online streaming services are becoming more popular.

Overall, e-commerce and online purchasing have changed the retail and corporate landscapes. Portability, versatility, and worldwide accessibility much exceed the dangers and concerns. Due to these changes, retail will continue to innovate and adapt, generating new ideas and growth opportunities. Figure 4.2 provides a view of the e-commerce transaction steps.

The introduction of digital marketing and advertising has transformed how firms reach out to consumers in order to offer them goods and services. A company's digital marketing activities may reach more people as more people utilize the Internet and mobile devices. This section will look at the history and growth of digital marketing, its many forms, and the possibilities and challenges it brings to current enterprises.

The first online banner advertising appeared in 1994, marking the beginning of what would be known as "digital marketing." Over the years, digital marketing has evolved into various subcategories, including search engine optimization (SEO), social media marketing, email marketing, content marketing, and influencer marketing. Businesses may reach their target audience and meet their marketing goals using channel-specific methods and approaches.

One of the digital marketing's most significant benefits is its ability to reach a wide range of people. Thanks to the spread of social media and mobile devices, businesses can now engage with customers anytime and wherever they want. By

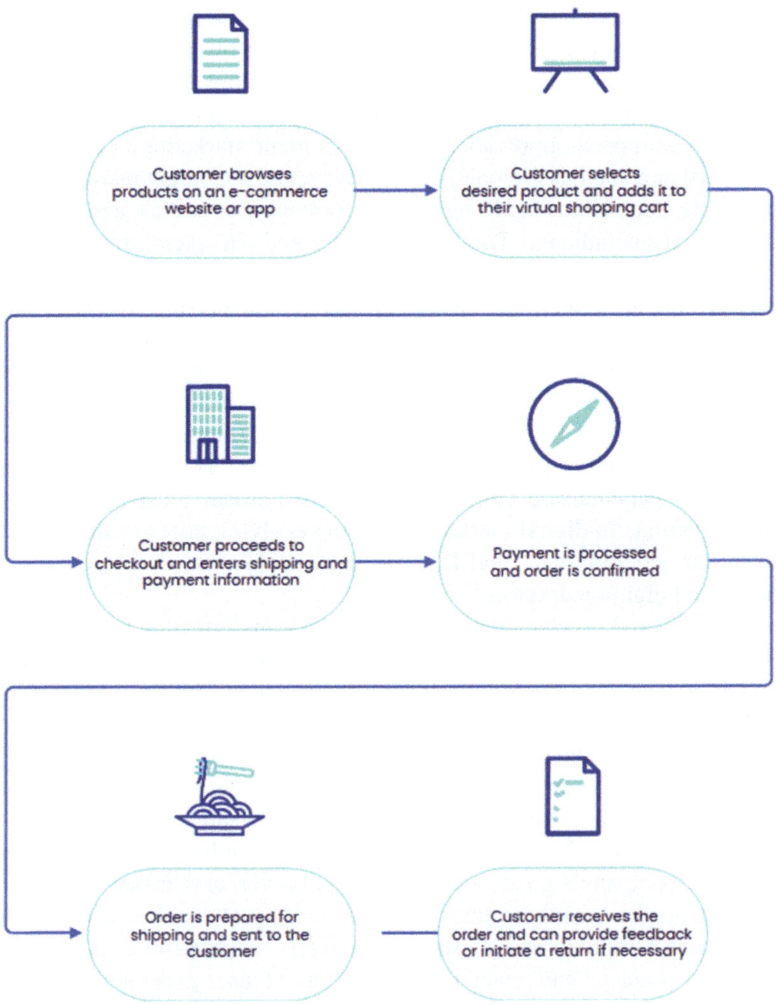

Fig. 4.2 Steps in an e-commerce transaction

allowing firms to engage with specific customer subgroups via tailored content and offers, digital marketing enhances targeting. However, digital marketing has its challenges. The ever-changing nature of digital marketing platforms and algorithms is a challenging task. To succeed, corporate strategies and tactics must adapt to technological advancements and customer preferences. Another barrier is the fight to get noticed in the Internet world. Many organizations compete for the same customers, making it harder to distinguish apart. Businesses may overcome these challenges by building a thorough digital marketing plan that considers their unique goals, consumer base, and available resources. A digital marketing strategy must take into consideration the organization's particular aims and needs in order to be successful.

Social media marketing has quickly emerged as the most popular area of digital marketing. Businesses may reach out to customers, improve their reach, and increase website traffic using social media platforms like Facebook, Twitter, Instagram, and LinkedIn. Sponsored posts, influencer connections, and unpaid content production are all methods to promote on social media. Content marketing is another kind of digital advertising that has grown in popularity recently. Content marketers generate and distribute educational and engaging information to attract and maintain the attention of a given audience. This may be performed via several informative forms, such as blog posts, videos, infographics, eBooks, etc.

Every profitable business has three goals: thought leadership, audience trust, and conversions. Businesses may now promote their goods and services in new ways thanks to digital marketing and advertising. Businesses now have access to a large audience that can be reached via several digital marketing channels because of the widespread usage of the Internet and mobile devices. Despite the challenges, digital marketing offers benefits, including a broader audience reach, more accurate targeting, and more personalized communications. Businesses must keep up with the newest innovations in digital marketing and the evolving tastes of their target consumers to succeed. Figure 4.3 and Table 4.2 provide an overview of a comparison of traditional and digital marketing.

Social media and user-generated content have transformed how businesses conduct business and communicate with customers online. Millions of individuals worldwide use social networking sites like Facebook, Twitter, Instagram, and LinkedIn as their primary sources of information. Consider blog posts, videos, reviews, and social media updates as user-generated content. Thanks to social media and user-generated content, businesses may engage with their customers more personally. Businesses may participate in two-way conversations with their customers through social media. Their comments and insights can be utilized to improve the quality of the company's goods and services. However, user-generated content and social media do not exist in a vacuum. The possibility of negative comments harming a company's reputation is a significant worry. Furthermore, businesses may need help to manage and react to the amount of user-generated material efficiently [2].

Despite these challenges, social media and user-generated content have gained importance in contemporary marketing. Many companies use them to promote goods, communicate with customers, and improve income. Businesses have learned that forming alliances with popular social media users to promote their goods is a successful tactic. Businesses are embracing user-generated material, such as reviews and testimonials from current customers, to develop trust and credibility with new consumers. Businesses increasingly resort to social media contests and campaigns to connect with their target audience and spread the word about their goods and services. It is realistic to expect social media and user-generated content to increase and improve in the following years. AI and virtual reality are two examples of future technology that will allow marketers to engage with their target population more deeply. The relevance of social media as a sales channel is projected to grow as social commerce evolves, allowing users to make purchases directly via social

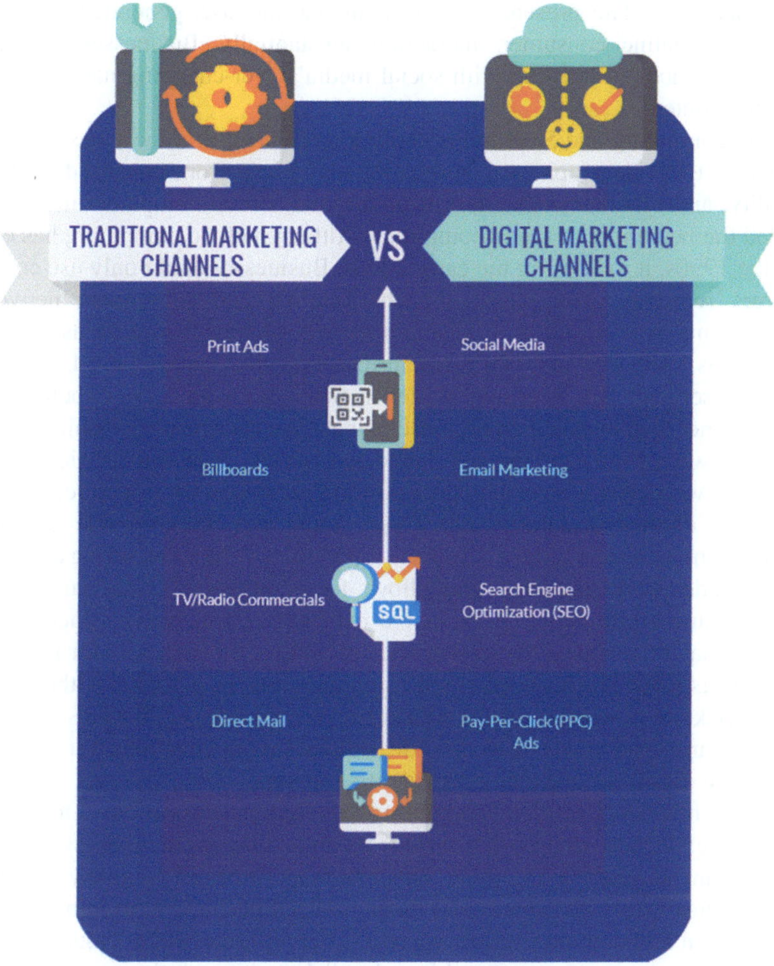

Fig. 4.3 A comparison of traditional and digital marketing channels

Table 4.2 Comparison of traditional marketing vs. digital marketing

Aspect	Traditional marketing	Digital marketing
Reach	Limited to local or regional areas	Global reach
Cost	Expensive, especially for TV and print ads	Lower cost, with options for targeted advertising
Tracking	Difficult to track the effectiveness of campaigns	Easy to track metrics and adjust strategies
Interactivity	Limited interactivity with the audience	Ability to interact with the audience in real time
Speed	Slow process for campaign development and execution	Quick development and execution of campaigns

media platforms. The presence of social media and user-generated content has transformed online consumer interaction substantially. Businesses must adopt inventive methods to keep up with social media's ever-changing nature and user-generated content.

Cloud computing and SaaS are becoming more critical in the e-business industry. Due to these developments, businesses may profit from enhanced scalability, flexibility, and efficiency. The advent of the first virtual computers in the 1960s signaled the beginning of cloud computing. Although cloud computing has existed since the 1990s, it took off in the early 2000s. Businesses could only use computer resources remotely after developing the Internet and subsequent networking improvements. As a result, new business models such as "software as a service," "platform as a service," and "infrastructure as a service" have evolved. SaaS (software as a service) is a licensing model in which software is distributed via the Internet rather than installed on each user's machine. As a result, organizations may access software from any Internet-connected device. Salesforce, Microsoft, and Google are well-known SaaS suppliers offering various business applications.

Because of their scalability, cloud computing and SaaS provide advantages. Businesses may raise or reduce computer capabilities without incurring costly hardware upgrades or downgrades. This allows for rapid reactions to changes in customer demand while avoiding excessive expenses. Furthermore, since many IT functions may be assigned to cloud service providers, the necessity for in-house IT employees is reduced. However, cloud computing and SaaS are not without danger. Security risks arise when sensitive data is kept on unprotected distant servers. Data portability and vendor lock-in may be difficulties for organizations if it is challenging to move data from one cloud provider to another.

Despite these worries, cloud computing and SaaS have contributed to the rise of e-commerce. Because of these advancements, firms may get the processing capability they want without paying a lot. This has made it easier for many enterprises, especially innovators and small- and medium-sized businesses, to join the market. Collaboration and communication are two different areas where cloud computing and SaaS are advantageous. Cloud-based collaboration solutions such as Google Docs and Microsoft Teams have helped to level the playing field for geographically scattered teams.

Furthermore, due to the popularity of cloud computing and SaaS, new business models and possibilities for innovation have evolved. Cloud-based platforms like Uber and Airbnb, for example, have aided the expansion of the contract economy, in which employees are engaged on a temporary or freelance basis. SaaS and cloud computing have transformed the e-business environment by enhancing scalability, flexibility, and efficiency. Despite legitimate worries, the advantages of these technologies cannot be overlooked. As a result of these developments, the e-business sector will continue to expand and adapt, opening up new avenues for innovation and expansion. Figure 4.4 provides an overview of how cloud computing can be implemented in an e-business.

Internet payment options have entirely transformed the sector for doing business online. Manual payment processing procedures such as checks, money orders, and

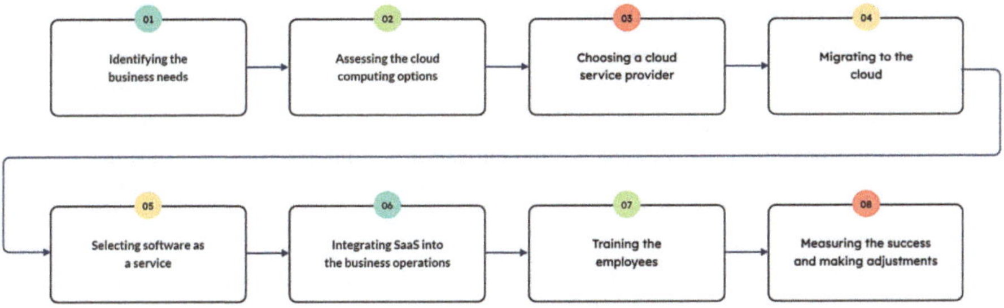

Fig. 4.4 Cloud computing and software as a service in e-business

wire transfers were time-consuming and costly. Nonetheless, the emergence of Internet payment systems has made the electronic transmission of payments between companies and consumers more convenient. Table 4.3 provides an overview and comparison of popular online payment systems.

The first payment gateway was introduced in the early 1990s, signaling the beginning of the age of online payment systems. Online payment methods, on the other hand, became widely used in the late 1990s and early 2000s. Many online payment systems have arisen in response to the growing need for security and efficiency in online business. Credit cards, debit cards, electronic wallets, bank transfers, and mobile phone bill pay are just a few payment methods accessible for online transactions today. Customers' credit and debit card payments, the most common online payment option, are accepted. A customer may use a bank transfer to send payments directly from their bank account to a corporation's bank account. E-wallets like PayPal and Skrill make tracking and moving payments across accounts easy. Apple Pay and Google Wallet are two mobile payment systems that allow users to make purchases using their handsets. The usability of Internet payment systems is a significant benefit. Customers do not need to worry about transferring cash or checks since transactions may be completed online anytime. Businesses may receive payments quickly, saving time and money throughout the payment processing step. Because of the security and reliability they offer while transferring payments, businesses and consumers may confidently use online payment systems [1].

Certain risks accompany the many advantages of Internet payment options. The issue of safety is paramount. Because online payment systems use personal and financial information, they are vulnerable to hijacking and hacks. Both organizations and customers have to safeguard their financial data by using secure payment systems and strong password practices. Furthermore, many online payment systems may charge high transaction costs, reducing a company's profit margin. Despite these difficulties, online payment systems have substantially influenced e-commerce. They have improved corporate operations' effectiveness and efficiency while providing customers with a more accessible and secure way to transact. Because of the widespread usage of online payment systems, new business models such as subscription-based services and mobile payments have emerged. Internet payment systems have transformed e-businesses' function by allowing for safe and easy

Table 4.3 Comparison of popular online payment systems

Payment system	Fees	Payment options	Security	Ease of use	Availability
PayPal	2.9% + $0.30 per transaction	Credit/debit cards, bank transfers, PayPal balance	Two-factor authentication, fraud protection	User-friendly interface	Widely available globally
Stripe	2.9% + $0.30 per transaction	Credit/debit cards, ACH payments, digital wallets	Encryption, PCI compliance	Easy integration, customizable checkout	Available in 39 countries
Square	2.9% + $0.30 per transaction	Credit/debit cards, digital wallets, Square payment card	End-to-end encryption, PCI compliance	User-friendly interface, POS integration	Available in the USA, Canada, Japan, Australia, and the UK
Google Pay	Free	Credit/debit cards, bank transfers	Tokenization, two-factor authentication	Seamless integration with Google services	Available in 30+ countries
Apple Pay	2.9% + $0.30 per transaction	Credit/debit cards, bank transfers, Apple Pay cash	Tokenization, two-factor authentication	User-friendly interface, seamless integration with Apple devices	Available in 50+ countries
Amazon Pay	2.9% + $0.30 per transaction	Credit/debit cards, bank transfers	Fraud protection, payment protection guarantee	Easy integration with Amazon services, user-friendly interface	Available in 170+ countries

financial transactions. The expansion of online payment systems has led to the introduction of many payment methods, each with its own set of benefits and drawbacks. Online payment systems have advantages and disadvantages; companies and customers should know them before committing. Online payment systems will continue to be critical to the success of e-businesses as e-commerce advances and increases. Figure 4.5 demonstrates a simple overview of how online payment systems work.

Supply chain management has become an essential component of modern businesses. The activity of coordinating several activities, including buying, manufacturing, and shipping, to fulfill customer demand in a timely and cost-effective way is known as supply chain management. Over the last several decades, the Internet's expansion has had a tremendous influence on supply chain management, allowing firms to coordinate their activities across the supply chain better. Table 4.4 provides an overview of the comparison between traditional and e-business supply chain management.

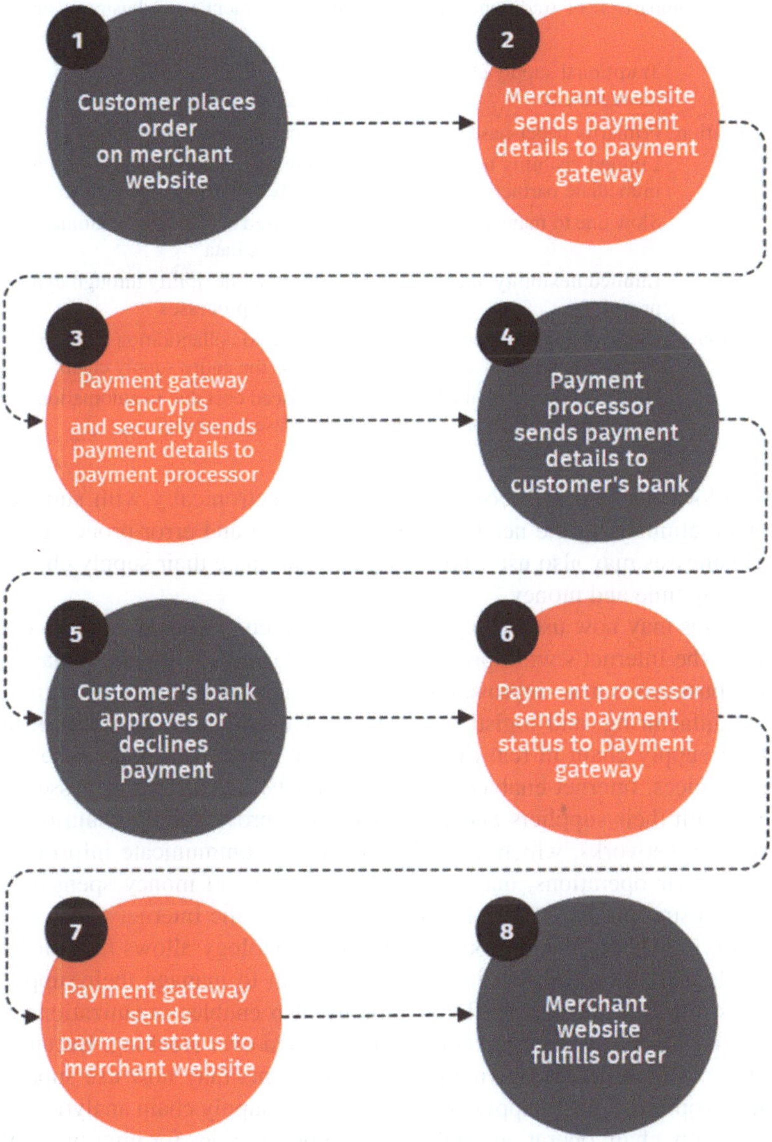

Fig. 4.5 How online payment systems work

E-procurement systems are a classic example of the Internet's potential in supply chain management. Businesses may use these technologies to perform all supplier purchases online, eliminating all paper-based procedures. E-procurement systems' extra features, such as real-time order tracking and inventory level monitoring, contribute to effectively managing supply chain activities. Another use of this technology in supply chain management is using EDI systems through the Internet. EDI

Table 4.4 A comparison of traditional supply chain management vs. e-business supply chain management

Aspect	Traditional supply chain management	E-business supply chain management
Communication	Primarily paper-based	Electronic communication
Visibility	Limited visibility beyond immediate partners	Increased visibility across the entire supply chain
Speed	Slow due to manual processes	Increased speed due to automation and real-time data
Flexibility	Limited flexibility due to rigid processes	Increased flexibility through dynamic and adaptive processes
Collaboration	Limited collaboration beyond immediate partners	Increased collaboration and coordination across the entire supply chain
Cost	High cost due to manual processes and lack of visibility	Reduced cost due to automation and increased visibility

systems allow businesses to communicate data electronically with suppliers and consumers, eliminating the need for time-consuming and error-prone human data input. Businesses may also use EDI systems to automate their supply chain procedures, saving time and money.

Businesses may now use supply chain management software more successfully because of the Internet's widespread availability. Using supply chain management software, businesses may improve the efficiency and effectiveness of their procurement, manufacturing, and logistics processes. This software also allows firms to track their supply chain in real time, providing a competitive edge when making strategic choices. Internet-enabled supply chain networks allow businesses to communicate with their suppliers and customers to improve supply chain operations. Supply chain networks, which allow businesses to communicate information and coordinate their operations, may minimize the time and money spent on supply chain administration. RFID systems that operate over the Internet are an extra tool for simplifying supply chain processes. RFID technology allows firms to monitor their goods and stocks in real time, allowing them to manage their supply chain operations more effectively. RFID technology also enables organizations to keep more precise inventory records, reducing under- and over-stocking. Because of the widespread availability of Internet connectivity, firms may now use supply chain analytics to optimize their supply chain operations. Supply chain analytics analyzes existing supply chain operations and suggests opportunities for improvement using quantitative data. Businesses may improve the efficacy and efficiency of their supply chain operations by identifying inefficiencies and devising remedies using supply chain analytics.

Finally, the Internet has tremendously influenced supply chain management, allowing organizations to handle supply chain activities more efficiently and effectively. Electronic Data Interchange (EDI), supply chain networks, and radiofrequency identification (RFID) are just a few of the Internet's supply chain management applications. These apps have assisted firms in simplifying their supply chain

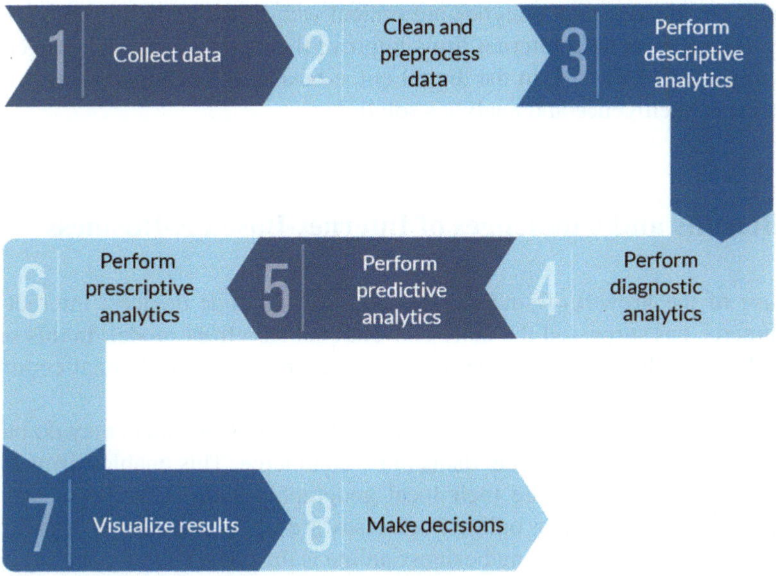

Fig. 4.6 Business intelligence and analytics process flowchart for e-business

operations, resulting in less time and money spent on administration and improving the efficacy of such operations.

The importance of web-based business intelligence (BI) and analytics cannot be overstated. Business intelligence and analytics comprise tools, methods, and procedures for gathering, analyzing, and presenting data, allowing for better corporate decision-making. In today's data-driven business climate, all e-commerce companies must incorporate these technologies to retain a competitive advantage. Figure 4.6 shows a simple process of implementing BI in e-business.

The capacity to derive meaning from massive amounts of data is a valuable asset of Internet business intelligence and analytics. Social media, website traffic, customer interactions, and sales data are just a few data types generated by online companies. Online firms may use business intelligence and analytics to fine-tune their operations, enhance the customer experience, and find neglected market niches. Additionally, online firms may monitor critical metrics such as sales, visits, and conversion rates in real time using BI and analytics dashboards. This allows Internet firms to quickly examine data, adjust plans, and respond to market conditions.

BI and analytics may also assist e-commerce companies in enhancing their marketing tactics. Online firms may learn much about their consumers' interests and behavior using business intelligence and analytics tools, leading to higher engagement, sales, and loyal patronage. Another use of BI and analytics is the optimization of e-business processes. Data analysis may help online firms with supply chain restrictions, appropriate inventory levels, and logistical simplicity. This can decrease costs, shorten changeover times, and increase customer satisfaction. Finally, business intelligence and analytics are critical web-based tools for today's e-businesses. These tools help Internet firms improve their procedures and deliver better customer care. To flourish in the digital era, e-businesses will increasingly depend on business intelligence and analytics solutions.

3 Benefits and Challenges of Internet-Based E-Business

Because of the growth of smartphones and other mobile devices, Internet-based e-commerce has surpassed the more conventional storefront model. In this section, we will look at how e-commerce may be an upgrade over traditional commercial procedures (Fig. 4.7).

The ease of doing business online is a huge benefit. Customers may do business at any time and from any place, thanks to e-commerce. This enables firms to grow their customer bases outside their local geographic areas. Customers may shop whenever convenient since online businesses are open 24 hours a day, 7 days a week. Another benefit of doing business online is the chance to save money. Online enterprises may experience lower expenditures due to needing a physical site. This

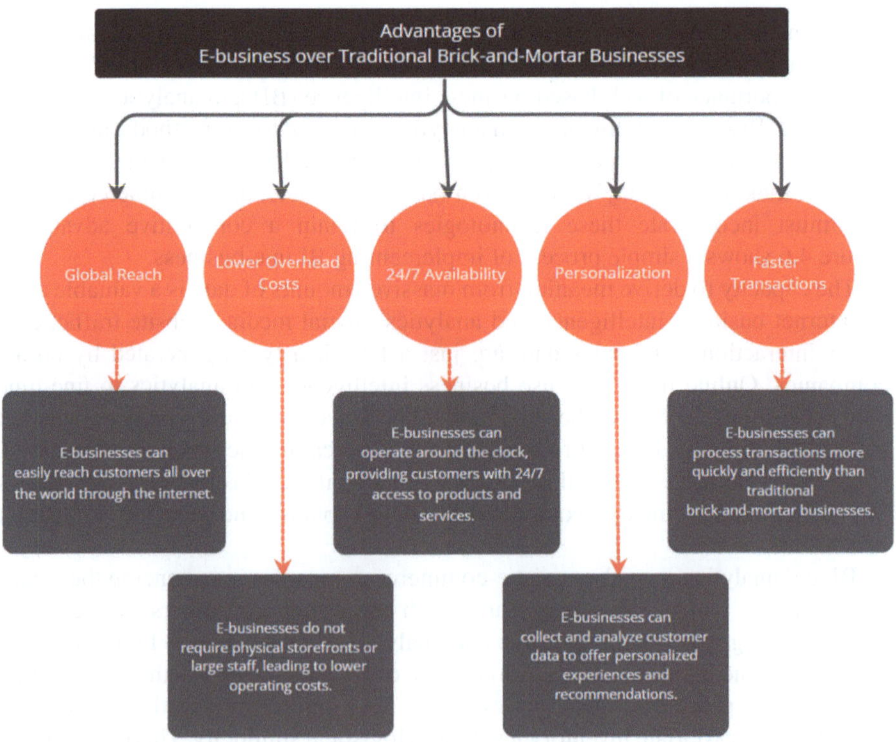

Fig. 4.7 Advantages of e-business over traditional brick-and-mortar businesses

enables firms to provide their goods and services to customers at a lower cost. Furthermore, Internet firms save money since they do not have to spend on expensive promotions such as billboards, print advertisements, and television commercials. Instead, online firms may contact clients using digital marketing tactics such as social media and search engine optimization.

The Internet customer base of a business may grow at an exponential rate. Unlike Internet enterprises, which may reach clients all over the globe, brick-and-mortar stores typically can only sell to people in their local region. Consequently, firms can acquire a broader customer base and boost their bottom line. Furthermore, e-commerce allows enterprises to reach new markets without the time and cost of establishing a physical presence in such areas. Another benefit of doing business online is the capacity to gather and analyze customer information. Online merchants may track the actions and preferences of their customers using several approaches. Businesses may better satisfy the particular demands of their clients with this knowledge. It may also assist firms in making data-driven choices that improve operations and bottom-line outcomes.

Another benefit of running an Internet company is its flexibility. They are versatile and can quickly adjust to new situations. Users may make quick changes to their website, pricing policies, and advertising campaigns based on user input. Businesses that can quickly adjust to shifting circumstances and market conditions have a better chance of success. Scalability is now available to e-commerce ventures. Businesses that operate from a physical location have a more difficult time growing once they reach a specific size. Online firms, on the other hand, may boost output in response to increased demand without investing in costly new facilities. Because a firm does not need to maintain a physical presence, it may grow quickly. E-commerce has various benefits over traditional retail, including better accessibility, lower prices, a broader customer base, data analytics, flexibility, and scalability. Businesses and entrepreneurs seeking to expand their consumer base and bottom line increasingly turn to the Internet.

Today, e-commerce is increasingly essential to the operations of the vast majority of businesses. E-business offers many advantages, but it also has many risks and obstacles. In this section, we will look at the dangers and obstacles online merchants face and the steps they may take to limit their effect. Data security is one of the most severe challenges in Internet commerce. Data security is critical for every online company that wants to secure its clients and itself. Data breaches and invasions are too prevalent in the Internet business world and may negatively impact a company's brand and bottom line. To secure their customers' sensitive information, businesses must employ cybersecurity precautions such as encryption and two-factor authentication. Another challenge of doing business online is obtaining the confidence of customers. Consumer trust is complicated to develop for online firms. Online businesses, unlike traditional brick-and-mortar enterprises, need a more physical presence that customers associate with trust. Businesses may mitigate this risk by developing a solid online brand and reputation, investing in a user-friendly and functional website, providing excellent support services, and supporting secure payment methods.

E-commerce involves logistical and shipping challenges in addition to logistical and shipping challenges. Unlike brick-and-mortar establishments, customers of online shops must depend on a third party's services for delivery. Customer satisfaction necessitates on-time, undamaged product delivery. E-commerce enterprises may alleviate the consequences of this problem by investing in technology such as real-time tracking and monitoring and cooperating with dependable logistics and delivery partners. Another impediment to Internet business is the legal and regulatory environment. The rules and regulations that regulate Internet businesses vary by region and may be difficult. International transactions may be subject to limitations, and data privacy laws must be followed. Businesses may decrease their exposure to this risk by monitoring legislative developments, consulting with lawyers as needed, and following the regulations as they are written.

Finally, online corporations must acknowledge the challenge of maintaining a positive public image. If customers begin to talk negatively about a firm on social media, the company's online reputation may suffer greatly. Businesses may limit their sensitivity to this danger by closely monitoring customer feedback, reacting swiftly and professionally to criticism, and investing in enhancing their online reputation.

To summarize, in order to be successful, Internet firms must overcome several risks and hurdles. The long-term sustainability of any firm requires proactive risk reduction, particularly in cybersecurity, logistics, and regulatory compliance. Despite these challenges, firms may succeed in e-commerce by investing in cutting-edge technology, developing a solid brand and reputation, and emphasizing customer service. Table 4.6 lists common risks and challenges that e-businesses may face.

Internet-based e-commerce has proved critical to many businesses' growth, productivity, and competitiveness. Nonetheless, Internet enterprises must deal with a variety of risks and problems. This category of obstacles includes cybersecurity hazards, regulatory and legal compliance issues, data privacy concerns, and the requirement for successful marketing and customer involvement. To meet these threats and possibilities, e-businesses must create and execute successful strategies. Creating a thorough security program is one method of reducing cyber dangers. Security measures such as firewalls, intrusion detection and prevention systems, and data encryption should be deployed and verified as part of this program.

Furthermore, online organizations must incorporate user authentication and access controls to limit who has access to critical data and infrastructure. Another strategy is to stay in conformity with all relevant rules and regulations. This necessitates being updated on new legislation about online business, such as the EU's General Data Protection Regulation (GDPR) and the California Consumer Privacy Act (CCPA) in the United States. Online shops must have privacy policies that define how consumer information is collected, stored, and used. Data privacy concerns are especially relevant for Internet enterprises. To counteract this problem, online enterprises must implement data protection and privacy policies in compliance with existing laws and regulations. This includes avoiding data breaches and

Table 4.6 Common risks and challenges in e-business

Risk/challenge	Description
Cybersecurity threats	Data security threats to online organizations include hacking, ransomware, and phishing. These may lead to monetary losses, ruined reputations, and legal trouble
Technical issues	Website outages, server failures, and software problems are all examples of technical difficulties that may affect an online company and result in lost revenue and unhappy consumers
Payment fraud	Credit card theft, chargebacks, and identity theft are only some forms of fraud that may occur in online payment systems. Both financial losses and a loss of confidence from customers are possible outcomes
Legal and regulatory compliance	Data privacy, consumer protection, intellectual property, and online advertising are the only areas where e-commerce operations must adhere to specific legal mandates. There may be monetary and legal repercussions for disobedience
Supply chain disruptions	Supply chains for online firms are notoriously complicated due to the need for various suppliers and transportation partners. Customer dissatisfaction and a drop in revenue might result from interruptions like delivery delays, stockouts, or faulty products
Competition	Competition for e-commerce sites comes from both offline stores and other online vendors. To maintain a competitive edge, businesses must constantly explore new approaches to advertising and technology

seeking express authorization from consumers before collecting and utilizing their information.

When there is a strong marketing and customer involvement, e-commerce thrives. Search engine optimization (SEO), email marketing, social media marketing, and customer relationship management (CRM) technology are among the marketing and consumer contact tactics accessible to online firms. E-commerce sites that use these tactics attract more customers, reach more individuals with their message, and enhance brand loyalty. E-businesses may protect themselves from harm and overcome hurdles by taking a proactive approach to risk management. To do so, you must identify possible threats and then create countermeasures. To keep up with the continuously changing e-business market, online enterprises must remain current on technology innovations and trends.

Finally, today's digital world provides several threats and challenges to e-businesses. E-commerce businesses may manage or overcome the risks and difficulties that are specific to them with the help of well-thought-out strategies: implementing successful marketing and customer contact methods, taking a proactive approach to risk management, building thorough security processes, and addressing data privacy issues. Online enterprises should apply these techniques to compete in today's digital market. Table 4.7 and Fig. 4.8 provide an overview of strategies for mitigating risks and addressing challenges of Internet-based e-business.

Many companies have leveraged the Internet's capabilities to grow and thrive in the digital era. Amazon is an example of a successful Internet business. Amazon.com, which started as an online bookshop in 1994, is today an international technology firm providing services such as e-commerce, cloud computing, digital

Table 4.7 Strategies for mitigating risks and addressing challenges of Internet-based e-business

Risk/challenge	Strategy
Cybersecurity threats	Develop a comprehensive security program that includes regular vulnerability assessments, penetration testing, and implementing security controls such as firewalls, intrusion detection and prevention systems, and data encryption Implement user authentication and access controls to prevent unauthorized access to sensitive data and systems
Legal and regulatory compliance	Stay up to date with the latest laws and regulations related to e-business, such as the General Data Protection Regulation (GDPR) in the European Union and the California Consumer Privacy Act (CCPA) in the United States. Implement privacy policies explaining how customer data is collected, stored, and used
Data privacy concerns	Implement data protection and privacy policies that meet the requirements of relevant laws and regulations. Obtain customer consent before collecting and using their data. Ensure that data is protected from unauthorized access and use
Effective marketing and customer engagement	Use various marketing and customer engagement strategies such as search engine optimization (SEO), email marketing, social media marketing, and customer relationship management (CRM) tools

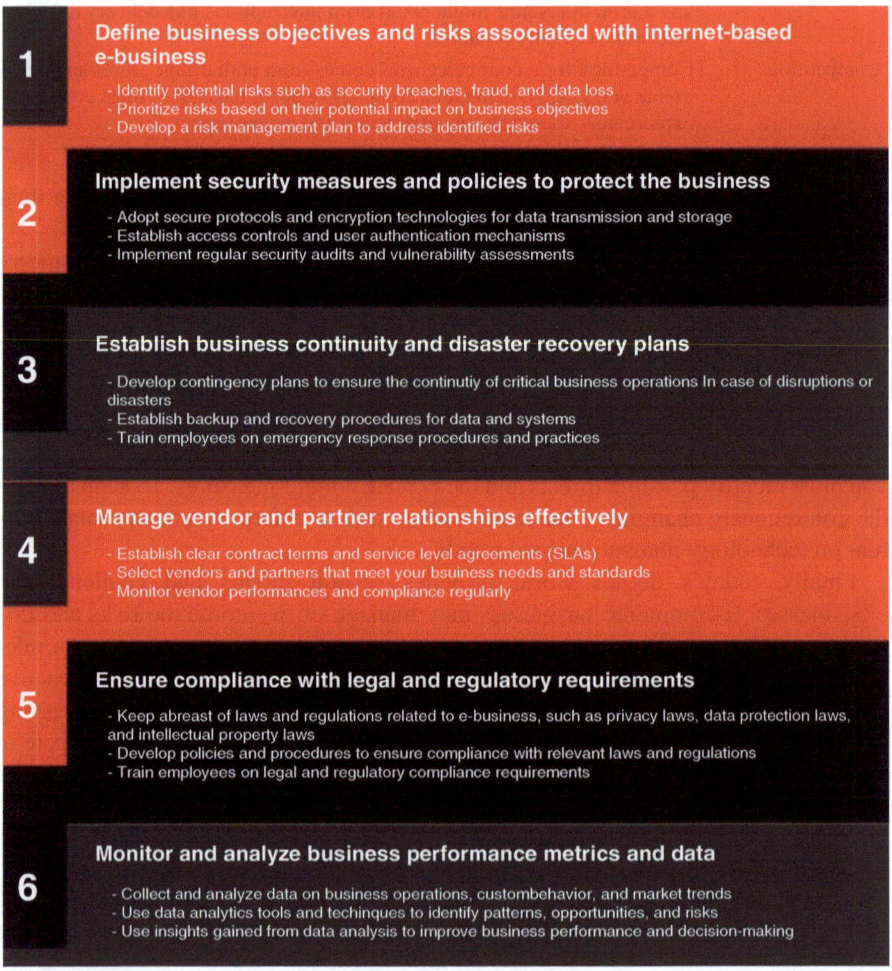

Fig. 4.8 Strategies for mitigating risks and addressing challenges of Internet-based e-business

streaming, and artificial intelligence. Amazon's success may be partly attributed to its dedication to its consumers and its ability to use the massive amounts of customer data it gathers to provide better service and more relevant product suggestions.

Another firm that has profited from the Internet is Airbnb. Airbnb was launched in 2008 as a website where people could advertise their extra rooms and flats for rent, but it has now developed into a worldwide hospitality brand that offers a range of housing alternatives to guests. Airbnb's success may be partly attributed to its ability to compete with traditional hotels by providing tourists with a more unique and customized experience. Its website has made it easier for homeowners to supplement their income by renting out their houses and for budget-conscious travelers to locate low-cost lodging.

Facebook is another firm that has profited from the Internet's pervasiveness. Facebook, which started as a service for college students in 2004, has developed into a worldwide phenomenon with over 2.8 billion monthly active users. Facebook's worldwide reach and robust advertising infrastructure are two of the company's key success drivers. Its comprehensive targeting features enable firms to target particular groups, making it a powerful digital marketing tool.

Netflix is one example of a firm that has thrived by embracing the digital age. Netflix started in 1997 as a DVD rental business, but it has since expanded into a worldwide streaming service with a massive collection of films, TV episodes, and original material. Netflix's success may be attributed partly to its offering customers a more convenient and less costly way to view television shows and films than was previously available. Original content is a crucial distinction for the service that has helped it attract and retain members.

Uber is an excellent illustration of how the Internet can be used to disrupt traditional businesses. Uber began as a taxi service in 2009 and has since grown into a worldwide transportation network that offers not only limousines but also food delivery and freight transport. One reason for Uber's appeal is that it offers clients a more straightforward and economical mode of transportation while allowing drivers to make extra money.

Summary

Introduction

- Definition of e-business: E-business refers to the conduct of business activities, such as buying, selling, and exchanging goods and services, using electronic means, primarily the Internet.
- Importance of the Internet in e-business: The Internet plays a vital role in facilitating e-business by providing a global network infrastructure that enables communication, transactions, and information exchange between businesses and customers.

Evolution of the Internet in E-Business

- Early stages: In the early stages, the Internet was primarily used for essential communication, such as sending emails, and electronic data interchange (EDI), which involved the electronic exchange of business documents between companies.
- Development of the World Wide Web: The creation of the World Wide Web introduced websites and hyperlinks, enabling businesses to establish an online presence and connect information across different web pages, revolutionizing e-business.
- Emergence of e-commerce: E-commerce refers to buying and selling products and services online. With the Internet, businesses could establish online stores and marketplaces, allowing them to reach a wider audience and conduct transactions directly with customers.
- Advancements in technology: Technological advancements, such as faster Internet speeds, mobile devices, and cloud computing, have significantly enhanced the capabilities of e-business, enabling seamless online experiences, efficient data processing, and flexible access to resources.
- Web 2.0 and social media: Web 2.0 represents a shift toward user-generated content and social networking platforms. This evolution has empowered businesses to engage with customers in interactive ways, fostering online communities and leveraging social media platforms for marketing and customer relationship management.

Applications of the Internet in E-Business

- E-commerce platforms: E-commerce platforms are websites or software that enable businesses to showcase and sell their products or services directly to customers. These platforms provide features such as shopping carts, secure payment gateways, and inventory management systems.
- Online marketplaces: Online marketplaces are platforms where multiple sellers can list their products or services, allowing customers to compare options and purchase from various vendors in a single location.
- Online advertising: The Internet offers various advertising channels, such as display ads, search engine marketing, and social media advertising, which businesses utilize to promote their products or services to a targeted audience.
- Online banking and payment systems: Internet-based banking and payment systems enable secure online transactions, allowing businesses and customers to transfer funds, make payments, and manage financial activities electronically.
- Cloud computing and remote collaboration: Cloud computing enables businesses to access and store data, applications, and resources remotely, facilitating collaboration and improving efficiency in e-business operations.
- Big data analytics: The Internet generates vast amounts of data, and businesses can leverage big data analytics to gain insights into customer behavior and market trends and make data-driven decisions to enhance their e-business strategies.

Benefits and Challenges of Internet-Based E-Business

- Benefits: Internet-based e-business offers advantages such as global reach, 24/7 availability, cost-effectiveness, scalability, personalized marketing, enhanced customer engagement, streamlined operations, and improved data analysis capabilities.
- Challenges: Internet-based e-business also presents challenges, including cyber threats, data privacy concerns, technological complexity, intense competition, the digital divide, and the need for continuous adaptation to evolving technology and consumer expectations.

Case Study

Company X is a brick-and-mortar establishment that has been in business for decades. Nevertheless, due to the shifting consumer environment and the expansion of e-commerce, the company needs help. It is considering entering Internet-based e-business in order to remain competitive. They have hired a team of consultants to assess the viability and potential benefits of implementing an e-business model. The consultants are tasked with analyzing the organization's circumstances and providing recommendations based on their findings.

1. What significant changes have influenced the Internet and e-business over time, and how have they impacted the retail sector in particular, based on an overview of the Internet and e-business's development?
2. Given the advantages of e-business over traditional brick-and-mortar businesses, what specific benefits can Company X anticipate from employing an Internet-based e-business model?
3. Discuss the potential risks and issues associated with e-business, including security and privacy concerns. How will Company X manage these challenges and mitigate the associated risks?
4. Provide examples of successful Internet-based retail e-commerce ventures. Analyze the primary factors that contributed to their success and suggest how Company X could employ similar strategies to achieve growth and success in the online marketplace.
5. In e-business, how can Company X effectively use digital marketing and advertising, online purchasing, social media, and user-generated content to engage consumers and increase sales? Provide precise recommendations based on the target market and products of Company X.

Discussion

1. What were the earliest examples of e-business on the Internet?
2. How did the Internet evolve to become a key tool for e-business over time?
3. What were the key developments that have shaped the Internet and e-business?
4. How has the rise of mobile technology and the proliferation of mobile devices impacted e-business?
5. What are the different types of e-business applications that exist on the Internet today?
6. How has online shopping and e-commerce changed the way consumers shop and purchase goods?
7. What are the key features of digital marketing and advertising that make it an effective tool for e-business?
8. How has social media and user-generated content impacted e-business?
9. What are the advantages of cloud computing and software as a service (SaaS) for e-business?
10. What are the advantages of e-business over traditional brick-and-mortar businesses?
11. What are the major risks and challenges associated with e-business, such as security and privacy concerns?
12. What strategies can businesses use to mitigate these risks and address challenges?
13. How can businesses ensure that they are compliant with relevant laws and regulations in the e-business space?
14. What are some examples of companies that have successfully leveraged the Internet to grow and succeed in e-business?
15. What were the key factors that contributed to the success of these companies?
16. What lessons can be learned from these success stories for other businesses looking to succeed in e-business?

Multiple-Choice Questions

1. What year is commonly considered the beginning of e-business on the Internet?
   (a) 1969
   (b) 1989
   (c) 1995
   (d) 2005
2. Which of the following was not a key development that shaped the Internet and e-business over time?
   (a) Introduction of the World Wide Web
   (b) Development of search engines like Google

   (c) Emergence of social media platforms
   (d) Invention of email

3. Which technology is commonly associated with cloud computing and software as a service?
   (a) Bluetooth
   (b) RFID
   (c) VPN
   (d) Virtualization

4. Which of the following is not an application of the Internet in e-business?
   (a) Online shopping and e-commerce
   (b) Digital marketing and advertising
   (c) Supply chain management
   (d) Social media and user-generated content

5. Which term refers to the online buying and selling of goods and services?
   (a) E-commerce
   (b) Social media marketing
   (c) Cloud computing
   (d) Digital advertising

6. What is the primary advantage of e-business over traditional brick-and-mortar businesses?
   (a) Lower operating costs
   (b) Personalized customer experience
   (c) Immediate product delivery
   (d) Physical store presence

7. Which of the following is a risk associated with e-business?
   (a) Limited market reach
   (b) Lack of customer feedback
   (c) High advertising costs
   (d) Security and privacy concerns

8. Which strategy helps mitigate security risks in e-business?
   (a) Implementing strong encryption protocols
   (b) Reducing online advertising efforts
   (c) Avoiding social media platforms
   (d) Offering limited product selection

9. Which of the following is an example of a successful Internet-based e-business?
   (a) Apple Inc.
   (b) Walmart
   (c) Amazon
   (d) Coca-Cola
10. What key factor contributed to the success of Amazon as an Internet-based e-business?
   (a) Wide range of product offerings
   (b) Extensive physical store network
   (c) Minimal online advertising
   (d) Lack of competition in the market
11. Which of the following is a future trend in e-business and the Internet?
   (a) Decreased reliance on mobile devices
   (b) Decline in social media usage
   (c) Increased adoption of virtual reality technology
   (d) Decreased focus on data privacy
12. Which technology enables personalized targeting and advertising in digital marketing?
   (a) Machine learning
   (b) Blockchain
   (c) Internet of Things (IoT)
   (d) Augmented reality
13. Which platform allows users to create, share, and interact with content online?
   (a) Email
   (b) Search engines
   (c) Social media
   (d) Cloud storage
14. What is a challenge associated with online shopping and e-commerce?
   (a) Limited product variety
   (b) Lack of product reviews
   (c) Inability to compare prices
   (d) High shipping costs
15. Which company has leveraged the Internet to grow and succeed in the e-commerce industry?
   (a) Alibaba
   (b) Ford Motor Company
   (c) McDonald's
   (d) General Electric

References

1. Turban, E., et al., *Electronic commerce 2018: a managerial and social networks perspective*. 2018: Springer.
2. Kiang, M.Y., T. Raghu, and K.H.-M. Shang, *Marketing on the Internet—who can benefit from an online marketing approach?* Decision Support Systems, 2000. **27**(4): p. 383–393.
3. Chaffey, D. and F. Ellis-Chadwick, *Digital marketing*. 2019: Pearson uk.

Chapter 5
E-Business Security and Control

Abstract E-business has revolutionized how companies conduct business, bringing numerous benefits such as cost savings, increased efficiency, and global reach. However, with the advantages come challenges, particularly in security and control. This chapter will discuss the various threats faced by e-businesses, as well as strategies for ensuring their security and control. First, an overview of e-business security and control will be provided, examining the different security threats faced by organizations conducting business online. Next, this chapter will delve into the specific threats to e-business security and control, such as hacking, phishing, malware, and denial of service attacks. It will also explore the consequences of these threats, including financial losses, reputational damage, and legal liabilities. To combat these threats, various strategies that e-businesses can adopt to ensure their security and control will be discussed. These strategies include implementing strong passwords, using encryption and firewalls, conducting regular security audits, and providing employee training on security best practices. Additionally, this chapter will examine the role of government regulations and industry standards in promoting e-business security and control. Finally, the legal and ethical considerations in e-business security and control will be addressed. E-businesses need to comply with various laws and regulations, including data privacy and protection laws, intellectual property laws, and consumer protection laws. Ethical considerations also come into play, such as balancing the need for security with customers' privacy rights.

1 Overview of E-Business Security and Control

The phrases "e-business security" and "e-business control" refer to the activities taken by companies to safeguard the integrity of their electronic transactions, data, and information systems against unauthorized access, loss, or corruption brought on by theft. For any business that conducts transactions online, the security and management of an e-enterprise are currently of the utmost importance. Businesses need to invest significantly in comprehensive security and control systems to protect their

H. Taherdoost, *E-Business Essentials*, EAI/Springer Innovations in Communication and Computing, https://doi.org/10.1007/978-3-031-39626-7_5

operations, reputation, and customers' trust in light of the growing threats posed by cybercrime, data breaches, and identity theft.

E-business security and control involve a wide range of technologies, regulations, and procedures that work together to protect the privacy of electronic data while assuring its reliability and accessibility. Encryption is one of the most important technologies used to maintain the security and administration of Internet businesses. Mathematical procedures need to be used during encryption to convert plain text into unreadable coded text by unauthorized parties. Passwords, sensitive information, emails, and other electronic transactions can all be encrypted for increased protection.

Another critical technology utilized for the management and security of online businesses is firewalls. A firewall separates an organization's private network and the Internet. Its main job is to stop harmful traffic from accessing the system and prevent unauthorized network access. For protection against online attacks and other security risks, firewalls are crucial. They can be built with either hardware or software.

Another crucial component of managing and securing e-business is access control. The word "access control" describes the procedures to guarantee that only people with the proper authorization can access electronic resources, data, or information systems. Passwords, biometric identification, and security credentials are just a few of the techniques that can be used to restrict access to a specific environment.

For the security and management of an online business, effective policies and practices need to be implemented in addition to this technology. This entails the creation of a thorough security strategy that outlines the organization's information security policy and offers suggestions for the proper utilization of electronic resources. Employees also frequently receive security training to ensure they are knowledgeable about potential dangers and competent in safely using technology. An online organization's effective governance and security depend on monitoring and auditing. System logs and network traffic monitoring regularly might assist in spotting potential anomalies or security breaches. Then, using this information, security measures may be improved, and future security breaches can be avoided. Senior management must show unwavering dedication to advance e-business security and control. The organization's top executives give information security the attention it deserves and invest enough funds to implement all necessary procedures. This entails investing in the most recent security technologies, hiring competent employees, and regularly reviewing and updating security rules and processes.

Finally, any online business strategy needs to include e-business security and control as essential elements. Due to the rising threats posed by cybercrime and data breaches, businesses need to implement a preventative strategy for securing their sensitive data. By investing in reliable security solutions, implementing effective rules and procedures, and assuring the support of senior management, businesses can protect their operations, reputations, and customers' trust.

Electronic commerce, usually called e-business, has become a crucial component of the global economy. Businesses are becoming increasingly dependent on e-commerce to interact with customers, complete transactions, and manage operations due to the Internet's extensive accessibility. However, as online business needs have grown, so have security and control worries.

The management and security of an e-business's operations are essential for several reasons. First of all, a security breach may be costly in terms of the direct monetary losses and the reputational damage it results in. According to a survey by IBM, the average data breach cost will be $3.86 million by 2020. This sum comprises direct and indirect expenses, such as lost revenue and tarnished brand reputation, as well as legal fees and regulatory fines.

The security of a network can be breached, which could lead to data leaks and financial losses. Competitors might profit from this information, which might be used to conduct more cyberattacks. The loss of sensitive data may also put people at risk since it makes it more likely that their private information will be compromised.

Companies need to implement several steps to guarantee the security and control of e-business operations. These actions consist of the following:

- Authentication: The process of confirming a user's identification and making sure they are who they say they are is known as authentication. Passwords, biometric identification, and two-factor authentication are just a few ways to accomplish this.
- Encryption: Encryption involves scrambling data to prevent unauthorized parties from reading it. Various encryption methods, including SSL and TLS, can be used.
- Access controls limit authorized personnel's access to sensitive data or processes. This can include access controls based on location or device and role-based access controls, which restrict access based on a user's job function.
- Firewalls: A firewall is a piece of hardware or software that prevents unauthorized users from accessing a network. They can be used to monitor traffic and stop bad behavior.
- Intrusion detection and prevention: Systems for detecting malicious activity or unauthorized access to networks are known as intrusion detection and prevention systems (IDPS). They can be set up to prevent threats or notify administrators automatically.
- Regular monitoring and testing: Monitoring and testing e-business systems regularly can help uncover weaknesses and stop security breaches in their tracks.

While putting these protections into place is essential, it is also crucial to consider the human component of e-business security. This involves educating staff members about safe online conduct, such as steering clear of phishing scams and refraining from password sharing.

Companies can also create various rules and processes to help assure the security and control of e-business operations. These may consist of the following:

- Acceptable use policies: These guidelines describe appropriate and improper conduct when utilizing company resources, such as e-business systems.

- Information security policies include instructions for safeguarding confidential data and preventing unwanted access.
- Incident response plans outline the steps to take in case of a security breach or other incident that could jeopardize e-business operations.
- Disaster recovery plans outline steps for recovering from system failures or other catastrophes that could affect e-business operations.

Target Corporation is an example of how important it is for companies to implement sufficient security and control systems for their online activities. A significant data breach at Target in 2013 compromised sensitive personal and financial data belonging to more than 100 million customers. Hackers could obtain private data due to a weakness in the company's payment system, which caused the intrusion.

The Target data breach highlights the potential repercussions of online enterprises' inadequate security and management practices. Along with the immediate financial losses, the breach significantly damaged the company's reputation, leading to many customers abandoning it in favor of rivals.

E-commerce is now a crucial component of how modern businesses operate. E-commerce has advanced concurrently with the security and control measures companies use to protect their operations due to increased Internet usage and other technology. This part aims to give a historical perspective of how e-business security and control have changed from the earliest e-businesses to the present.

Data security and privacy were not top priorities in the early days of online business. E-commerce was initially focused on creating a web presence and making transactions as quick and straightforward as feasible. Because of this, Internet firms either lacked proper security measures or none at all, leaving them open to assault from hackers and other cybercriminals. During this time, hackers were frequently driven by curiosity or a desire to get beyond a system's resistance to intrusion. Security has been a top priority for e-businesses due to hackers using their abilities to steal money as the use of e-business increased in popularity.

In the middle of the 1990s, online firms started using various security measures, such as firewalls and antivirus software, in response to the growing number of security concerns. These safety measures worked well in stopping the more direct attacks but were not enough to stop the trickier ones. Online firms started looking for more cutting-edge protection methods as a direct result.

One of the most critical advancements in online company security was the creation of encryption technology in the late 1990s, which is generally acknowledged. Before sending sensitive data over the Internet, e-commerce businesses were able to secure it using encryption technologies. This stopped the information from getting into the wrong hands. As a result, cybercriminals found it much harder to intercept and steal sensitive data, including credit card numbers and personal information. Encryption software has become essential for the security of Internet enterprises as a direct result of this trend.

Another significant development in the security of online transactions and enterprises came with the advent of digital certificates at the start of the twenty-first century. Digital certificates can validate a website's or online store's validity, which helps avoid phishing scams and other types of online fraud. The great majority of Internet firms today use digital certificates to confirm their identities and safeguard their customers' private information. Digital certificates quickly became the industry standard for e-business security.

As e-businesses expanded, so did the sophistication of the security dangers they had to deal with, necessitating the implementation of increasingly complex security systems. Using artificial intelligence and machine learning to identify and stop cyberattacks has been one of the most important recent advancements. Because of these technologies, it is considerably more challenging for hackers to enter online businesses' networks, recognize threats, and react to them in real time.

Another critical security advancement for Internet businesses is the use of multifactor authentication. Before gaining access to sensitive information or completing a transaction, users of a system using multifactor authentication need to produce two or more kinds of identity, such as a password and a fingerprint. Thanks to this additional degree of security, hackers will find it much more challenging to access critical data or systems.

In conclusion, e-business security and governance development have been a continuous process for the past several decades. E-business security has advanced significantly from the early days of Internet commerce, when businesses did not place a substantial premium on protecting client data and privacy, to the present, when businesses use cutting-edge technology like artificial intelligence and machine learning. Online businesses will continue to invest in cutting-edge security solutions to protect their operations and consumers from more sophisticated cyberattacks. It will be interesting to see how security and control systems develop in response to the constantly evolving security dangers that e-business needs to contend with as it grows. Figure 5.1 depicts the historical development of security and control measures in e-business.

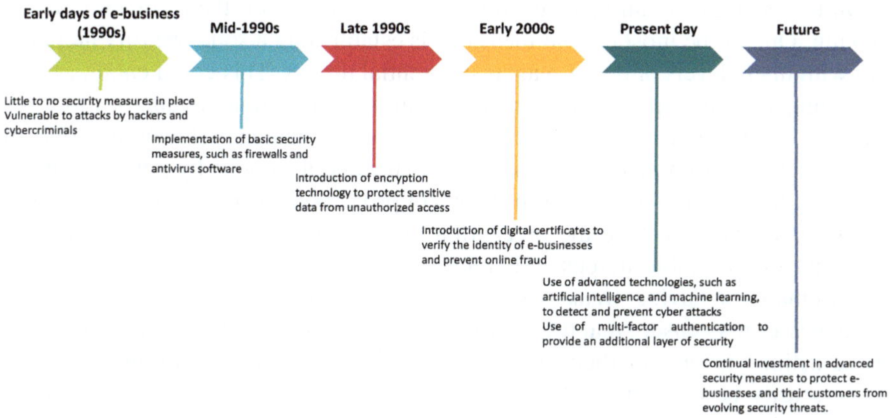

Fig. 5.1 Evolution of e-business security and control over time

2 Threats to E-Business Security and Control

It is impossible to stress the importance of putting in place proper safety procedures given the rising popularity of online business activities like e-commerce and other online commercial interests. Businesses need to be aware of several hazards to safeguard themselves and their customers from harm as more sensitive information is being stored online. An overview of some of the most common risks to online enterprise security is given in this section. These threats include insider threats, phishing scams, cyberattacks, data breaches, and problems with mobile security.

The rise of e-business has fundamentally altered how organizations conduct transactions, opening up new avenues for growth and generating higher total profitability. However, as this digital transformation progressed, new dangers appeared, chief among them cyberattacks. Computer network attacks can do great harm, including monetary losses, harm to a company's reputation, and legal implications. Online firms need to, therefore, fully understand the risks presented by cyberattacks and put precautionary measures in place to lessen these risks.

Phishing assaults and sophisticated malware attacks are two different kinds of cyberattacks. Cyberattacks of the DDoS variety are frequent. In this kind of assault, the attacker attempts to overwhelm and take down a website or server by sending a tremendous volume of traffic. A ransomware attack is deploying malicious software to encrypt a victim's files and demand money in exchange for the decryption key. Data breaches occur when an attacker acquires unauthorized access to confidential information. The victim in each attack has no choice but to comply with the extortion.

One of the biggest problems with cyberattacks is they are constantly evolving. Attackers continually test new tactics and techniques to exploit weaknesses in e-business systems. Therefore, online organizations need to have a preventative strategy for cybersecurity. This requires keeping up with the most recent security issues and taking the necessary preventative measures to combat them.

Online businesses can take several precautions to lessen their vulnerability to hacking. One of the most crucial measures is establishing a solid security architecture with many levels of defense against potential threats. This could entail, among other things, setting up a firewall, using an intrusion detection system, or installing antivirus software. E-commerce companies can dramatically lower the risk of cyberattacks and diminish the impact of any successful assaults by putting in place a security framework with numerous layers of protection and using them all at once. E-businesses can also take the precaution of educating their staff on cyberattack risks and the available defenses. Human error, such as accidentally clicking on bogus emails or having weak passwords, is the primary source of cyberattacks. E-businesses may ensure that their staff are aware of the hazards and take the appropriate actions to protect themselves and the firm by regularly training them on cybersecurity best practices. This training can guarantee that workers effectively protect both themselves and the business [1].

E-commerce businesses can also put strict authentication methods and access restrictions in place to stop unauthorized people from accessing their systems and data. These precautions include enforcing stringent password regulations, utilizing two-factor authentication, and limiting access to sensitive information to only those employees who require it to carry out their job obligations. E-commerce companies can significantly lower their vulnerability to cyberattacks and improve the security of their systems and data by implementing the abovementioned precautions.

Online organizations can employ encryption as an additional essential tool to shield their sensitive data from the risk of being stolen by thieves. Data that has been encrypted is embedded so that it can only be accessed by authorized users who have the correct decryption key. Online organizations can ensure that even if an opponent intercepts sensitive data, it cannot be read or utilized by encrypting it.

Online businesses are significantly at risk from cyber assaults since they could result in significant losses in terms of money and reputation. Therefore, taking precautions to lessen the likelihood of a cyberattack is crucial for Internet organizations. These steps include establishing a solid security structure, offering regular employee training, using strict access controls and authentication procedures, and encrypting sensitive data. By lessening the possibility of being the target of hackers, e-commerce companies can safeguard the security of their systems and data.

In Internet commerce, a "data breach" is any situation in which private or sensitive information belonging to a business or its clients is accessed, acquired, or revealed without authorization. A data breach can cause just as much harm to a company's reputation and financial standing as the following legal duties and fines. This section will examine the typical reasons for data breaches in online businesses and the preventative steps that may be taken.

One of the most frequent reasons for data breaches in Internet businesses is the use of insufficient or stolen login credentials. Among other methods, cybercriminals can get login credentials using phishing, social engineering, and brute-force attacks. Other techniques include password-guessing competitions and keylogging. Once they have successfully gained legitimate credentials, they can use them to access sensitive information or carry out further assaults. Businesses can lower their exposure to this danger by installing strict password laws, multifactor authentication, and regular employee training on cybersecurity best practices. Vulnerabilities in hardware and software are another frequent element that can lead to data breaches in Internet enterprises. Cybercriminals can use these vulnerabilities to install malware that can steal or damage data or get unauthorized access to sensitive data. Cybercriminals may take advantage of these flaws to get unauthorized access to sensitive information. A business can lessen the susceptibility of its software and hardware by conducting regular upgrades, engaging in vulnerability assessments and penetration testing, and implementing efficient security controls like firewalls, intrusion detection systems, and antivirus software.

The security of e-business data is significantly at risk from internal assaults. An "insider threat" occurs when a worker or other reputable organization member misuses or misappropriates essential information. This can happen intentionally or unintentionally due to behaviors like data theft, the sharing of login information, or the unintentional disclosure of sensitive information. These are some instances of this behavior. Businesses can lessen their exposure to this risk by putting strict access restrictions and monitoring systems in place, thoroughly screening potential hires, and regularly training their present workers on data security and cybersecurity [2].

Malware seriously jeopardizes the data security of Internet businesses. Malware is any harmful software created to steal data from a device or network, corrupt that data, or gain control over it. Although malware can take many forms, the most prevalent ones include viruses, worms, Trojan horses, and ransomware. Malware can spread through various channels, including malicious website visits, malicious email attachments, and malicious software downloads. Installing antivirus software, regularly updating their hardware and software systems, and regularly training staff on how to spot and avoid harmful software are all ways that businesses can defend themselves from malware outbreaks.

Data breaches may also originate from unreliable third-party vendors or service providers. Many online firms use third-party vendors for payment processing, data storage, and software development. On the other side, these third-party providers may be a source of data breaches if they have insufficient security procedures or are the target of hackers. Businesses can reduce this risk by conducting thorough vendor assessments, establishing substantial contracts and security guidelines with vendors, and routinely checking the security procedures used by their providers.

In conclusion, e-business data breaches represent a severe risk to the privacy of sensitive data belonging to companies and their clients. By establishing strong security measures and developing a better understanding of the leading causes of data breaches, businesses can lessen the likelihood of having a data breach and thereby preserve their reputations, finances, and legal obligations. Businesses need to, however, maintain ongoing awareness and a proactive approach to cybersecurity due to the constantly evolving nature of the Internet dangers they face.

Phishing scams are a type of cyber assault that entails impersonating a trustworthy company, like a bank, merchant, or government organization, to fool customers into disclosing private information, like usernames and passwords. Phishing scams are an example of social engineering. Online businesses are particularly vulnerable to fraud schemes because they handle much sensitive information regularly, including financial and consumer data. The nature of phishing schemes in online commerce, the techniques used by those who commit these crimes, and the security measures that companies should use to safeguard their clients are all covered in this section.

Phishing methods are frequently used in online business and typically enter through email, instant messaging, and social media platforms. The attacker will send a message that appears to be from a trustworthy company, like a bank or retailer, and will ask the receiver for sensitive data, including login passwords or credit card information. To harm the target, the attacker wants to leverage this information. To urge a fast reaction without the receiver first giving the request significant consideration, these messages typically aim to make the recipient feel a sense of imminence or dread.

Fraudsters who use phishing tactics to target Internet businesses frequently impersonate a legitimate company's website or email address. A website or email address that looks to be from a legitimate source but is actually under the attacker's control is created. An illustration of this would be if an enemy made an email address that appeared to be from a bank but was a slightly altered version of the bank's actual email address. The risk that the recipient may divulge sensitive information rises due to the recipient's potential difficulty in identifying the phishing attempt.

Social engineering, which involves deceit and persuasion to persuade the target to give confidential information, is another tactic used by those who engage in e-business phishing scams. This entails employing psychological strategies to affect the target's behavior, such as creating a sense of doom or imminent disaster. For instance, an attacker might send a message purporting to be from a financial institution notifying the receiver that their account has been compromised and that they need to reveal their login information to stop further harm immediately.

Online businesses and their customers can avoid phishing scams by taking several protective measures. An efficient tactic is to employ two-factor authentication, which requires the user to submit additional identity in addition to their username and password, such as a code texted to their mobile device. The user's mobile device may receive this code. It could be more challenging for an attacker to access sensitive data even if they can access the user's login credentials.

E-businesses may also implement employee training programs that show staff members how to spot and react to phishing schemes. These programs are accessible to employees and can be found online. This can include instructions on how to spot strange emails, check the validity of requests for sensitive information, and report possible phishing attempts. Email filtering and anti-spam software can automatically recognize and block phishing emails, preventing them from ever reaching the recipient's inbox. Online businesses can use this program to stop phishing emails from ever reaching their clients. By decreasing the volume of phishing emails sent to customers and workers, these solutions may lessen the chance that private information may be compromised [2].

Phishing schemes in online commerce seriously threaten the security of sensitive information, and the organization that falls victim to one could suffer significant financial losses and reputational damage. Online firms may safeguard themselves and their customers by implementing security measures like two-factor authentication, employee training programs, email filtering, anti-spam software, etc. E-commerce companies can lessen the possibility that their customers' personal information will be compromised and maintain their customers' trust by using preventative measures against phishing techniques.

An authorized employee, contractor, or vendor accessing a company's network or systems is considered an "insider threat" in electronic commerce. To hurt the company's information assets, including sensitive data, trade secrets, or financial information, these employees may purposely or accidentally abuse their access to the system. An insider threat may originate from several things, including unhappy staff, careless actions, or compromised credentials. In the following paragraphs, we will go over the many kinds of insider threats and the precautions companies can take to protect themselves from these risks.

An illustration of a subtype of the insider threat is a malicious insider. This describes someone who knowingly uses their access to a company's systems or data to gain or harm the business. There is a chance that dishonest personnel will steal sensitive data, falsify documents, or start computer system attacks within the company. These attacks might have been carried out for several reasons, including monetary gain, retaliation, or ideology, to name a few. By setting access controls, regularly monitoring employee behavior, and running background checks on potential hires, businesses can lessen the possibility that malicious insiders would hack their systems.

Insider risks come in different forms, including the careless insider. This describes someone who accidentally damages a company's computer systems or data out of ignorance or carelessness. For example, careless employees may accidentally click on bogus emails or stop updating their software, leaving gaps that outsiders might exploit. Employing security regulations that mandate regular software updates and password changes, along with training and awareness initiatives for staff, can help businesses lower the danger of careless insiders.

The third kind of insider threat comes from a compromised employee. A user whose credentials have been stolen or compromised in some way, whether via phishing, social engineering, or another technique, is referred to by this word. Compromised employees may unintentionally give attackers access to a company's networks or data, which they can exploit for data theft, industrial espionage, or other detrimental activities. Businesses can lower the risk of employee system hacking by implementing multifactor authentication, encrypting critical data, and keeping an eye out for unusual network activity, to name a few strategies.

Businesses need to embrace a thorough, all-encompassing strategy for security to reduce the risk of insider attacks successfully. This entails the development of rules and guidelines for employee behavior as well as the application of technological safeguards like encryption and access controls. Companies should also regularly audit their security procedures and evaluate their risks to spot gaps and develop remedial measures.

Educating employees on their part in maintaining security is crucial to reducing the risks posed by insider attacks. Employees should get training and awareness campaigns on data management procedures, password security, and phishing scams. Companies ought to offer these initiatives. Employees should be encouraged to alert their security teams to any suspicious activity so that the necessary precautions can be taken.

Businesses may also use tools and technology to help identify and avoid threats brought on by employees within the firm, in addition to these security measures. Technologies such as user behavior analytics, for instance, can help in the identification of unusual conduct that may be a sign of an internal threat. Security information and event management, or SIEM, systems can be used to keep an eye on network activity and send security staff early warnings of potential dangers.

In summary, insider threats are a significant concern to online firms and can have dire repercussions like the loss of confidential information or harm to financial assets. Businesses need to have a comprehensive security strategy that incorporates technological controls, policies and procedures, employee education and awareness, and the use of tools and technology to reduce the impact of these threats. By taking precautions against the threat posed by their employees, businesses may safeguard their information assets and uphold the integrity and confidentiality of their data.

How we conduct business has significantly changed due to the widespread use of mobile devices. An unprecedented number of people use their mobile devices to access online business platforms. This has facilitated company operations, increased accessibility, and generated new security flaws. Mobile devices typically have less protection than conventional desktop PCs, which attracts cybercriminals as an appealing target. We will look at some of the most prevalent mobile security flaws that online businesses should be aware of in the following section, followed by a discussion of steps organizations may take to lessen the impact of these threats.

One of the Internet businesses' most pervasive mobile security vulnerabilities is using insecure Wi-Fi networks. Because mobile devices are generally used while the user is moving, users are more likely to connect to public Wi-Fi networks at places like hotels, coffee shops, and airports. However, these networks are typically unencrypted, which makes it simple for hackers to intercept communications and steal private information. E-commerce companies can lessen the impact of this threat by informing their staff about the risks of connecting to unsecured Wi-Fi networks and by promoting the usage of virtual private networks (VPNs) to encrypt critical data.

Another often-used vulnerability on mobile devices is the widespread usage of outdated software. The software on many people's mobile devices needs to be updated, leaving them open to invasions. Cybercriminals are always looking for software and operating system flaws they may exploit. Therefore, online firms should ensure that their employees' devices always have the most recent software updates and security enhancements.

Online businesses need to be mindful of the widespread mobile security flaw known as phishing, commonly referred to as an Internet scam. Phishing is Internet fraud when individuals are duped into giving over private information like login passwords and financial information. This kind of fraud manifests as emails or forum entries that seem to come from reliable organizations, such as banks or online marketplaces. To lessen the risk caused by these frauds, e-businesses should train their staff on how to spot phishing tactics and respond to them.

Another common mobile security weakness in e-business is malware. Malware is any software intended to harm or impair a device's functionality, steal data, or gain unauthorized access to a network. Because they are less secure than traditional desktop computers, mobile devices are frequently the target of malicious malware. E-businesses should mandate that their staff members install antivirus software on their mobile devices and urge them only to download apps from reputable sources to lessen the risk posed by this vulnerability.

Last, lost or stolen devices seriously threaten mobile security in e-business. A data breach may happen if a mobile device that contains sensitive data is misplaced, stolen, or exposed to an unauthorized third party. To reduce this danger, e-businesses should require their staff to use secure passwords, set up two-factor authentication, and encrypt any critical information on mobile devices.

In conclusion, the operation of Internet enterprises needs to be improved by mobile security flaws. Online businesses should be mindful of common risks such as insecure Wi-Fi networks, out-of-date software, phishing scams, malware, and stolen or lost equipment. To reduce the risks brought on by these vulnerabilities, e-businesses should train their personnel on how to spot and defend against these assaults, impose the use of strong passwords and encryption, and encourage the use of virtual private networks (VPNs) and antivirus software. E-commerce companies can safeguard themselves and their clients from the effects of a mobile security breach by taking the safeguards above. Table 5.1 outlines various typical security threats that e-businesses may encounter and provides corresponding prevention measures to mitigate those risks.

E-commerce platforms have swiftly evolved into essential elements of modern life as the global economy moves closer to complete digitalization. These technologies help businesses run more effectively, reach a wider audience, and offer

Table 5.1 Common threats and prevention measures for e-business security

Threat type	Definition	Examples	Impact	Prevention measures
Cyberattacks	Deliberate exploitation of computer systems, networks, or devices	Malware, ransomware, and DDoS attacks	Data loss, system downtime, financial losses	Install security software, use strong passwords, and conduct regular system updates
Data breaches	Unauthorized access to sensitive or confidential information	Hackers, insider threats, lost or stolen devices	Reputation damage, legal penalties, financial losses	Encryption of sensitive data, regular security assessments, implementing access controls
Phishing scams	Deceptive tactics used to trick individuals into divulging sensitive information	Fake emails, websites, social engineering	Data theft, financial losses	Employee training, two-factor authentication, email filtering
Insider threats	Threats from within the organization, such as employees, contractors, or partners	Malicious intent, accidental breaches	Data theft, system damage, reputation damage	Background checks, access controls, employee training
Mobile security vulnerabilities	Security weaknesses in mobile devices or apps	Malware, spyware, unsecured Wi-Fi networks	Data theft, financial losses, reputation damage	Mobile device management, encryption of sensitive data, regular updates, and patches

customers services better suited to their needs. However, the increased reliance on technology also brings new dangers and problems, such as potential weak points that can jeopardize the security of e-business systems. This section looks at some of the most common security issues that can harm e-business systems and shows how to spot and defend against them.

Online business systems are significantly at risk for security breaches because of outdated or unsupported software. A piece of software is more likely to have security issues that could be dangerous the longer it is in use, especially if its makers stop delivering updates and patches as time goes on. Businesses need to take proactive actions to lessen the threat posed by these vulnerabilities since attackers can use them to access sensitive data and systems. Frequent software upgrades and replacing outmoded software with the most recent version are necessary to ensure the security of an e-business system.

Another frequent type of online business system vulnerability is weak passwords. Cybercriminals can utilize automated techniques to anticipate passwords and gain illegal system access, making weak passwords a tempting target. Implementing strict password standards and educating staff members about the value of password security is essential. This should highlight the importance of creating strong passwords, using two-factor authentication, and changing passwords frequently.

Additionally, third-party programs and services can make e-business systems more vulnerable. Businesses are required to only work with trustworthy vendors and carry out extensive quality assurance testing to guarantee the security of their software and services. Regular penetration testing and vulnerability assessments can help find potential security weaknesses and offer advice on fixing them. This can be accomplished by passing the assessments and tests. Inadequate access controls are yet another potential risk associated with computerized business systems. Access controls describe the limitations imposed on who has access to specific data and system resources. Businesses need to ensure that sufficient access controls are in place to prevent unauthorized access to sensitive data and systems. Role-based access controls, careful monitoring of access logs, and routine user privileges audits can all help reduce this risk. Another design problem in online business systems is the need for encryption. Data is transformed into a code that only those with access through encryption can decode. Sensitive information can be intercepted or stolen if it is not encrypted. Companies are required to encrypt sensitive data both in transit and at rest to prevent unwanted access to it.

Weak passwords are another vulnerability that arises in online business systems. Weak passwords are those that an adversary may easily guess or crack. Despite the ease with which an adversary might guess passwords like "password" or "123456," many individuals still use these. Furthermore, many people use the same password across all of their accounts, which might be risky if one of these accounts is compromised. Attackers can use automated techniques to attempt to guess passwords and obtain entrance to e-business systems. Businesses should require their staff and consumers to create complicated passwords that mix alphabetic, numeric, and other special characters to reduce the risk posed by this issue. Two-factor authentication should also boost account security and encourage users to create unique passwords for each account [3].

Social engineering attacks could also compromise e-business systems. Social engineering utilizes psychological manipulation to persuade others to provide sensitive information or take adverse actions. Phishing and resectioning are two examples of social engineering tactics that hackers can use to enter and misuse e-business systems for their malevolent ends. Businesses should constantly teach their personnel and customers to help them recognize and avoid social engineering attempts. This will help to lessen the risk that this vulnerability presents. Additionally, they should create policies and practices that demand identification confirmation before disclosing private information or acting.

E-business systems may be vulnerable if firewalls and network infrastructure are improperly designed. Organizations need to ensure their firewalls are appropriately set to prevent unauthorized access to their systems as the first line of protection against cyberattacks. To find and address any vulnerabilities, it is also critical to routinely monitor network traffic for unusual activities and assess vulnerability. Both of these should be carried out regularly.

To ensure the security and dependability of these systems, it is crucial to identify any potential security weaknesses in e-business systems. Businesses must take proactive steps to find and fix potential security vulnerabilities. The implementation of appropriate access restrictions, the imposition of strict password regulations, the regular conduct of vulnerability assessments and penetration tests, the assurance of the safety of software and third-party services, and the encoding of sensitive data are some examples of these procedures. Businesses may decrease the chance of cyberattacks and data breaches by putting these preventative measures in place, thereby defending not just their customers but also their financial lines. Figure 5.2 depicts the process of identifying potential weaknesses in e-business systems.

Security breaches are too typical in the e-business era and can have severe consequences for businesses and customers. This section will overview some of the most notable security events in recent years. It will look into how these breaches have affected customers and e-commerce companies.

One of the most severe security breaches in recent memory was the 2017 Equifax data leak. The names, birth dates, social security numbers, and addresses of 147 million people—including their complete names—were made public due to the security violation. Customers lost trust in Equifax due to the breach and started to doubt the company's capacity to protect their private information. Regulatory action was taken in reaction to the data breach, and Equifax was required to pay a $700 million settlement to resolve claims arising from the incident.

2018 saw Facebook make public the improper acquisition of up to 87 million users' data by the political consulting firm Cambridge Analytica. This problem was caused by Cambridge Analytica unlawfully accessing Facebook user data. The incident sparked a broader debate on the IT industry's value of data privacy and protection. Concerns about Facebook's data privacy policies were also raised as a result. Following an inquiry into the incident, the Federal Trade Commission penalized Facebook $5 billion.

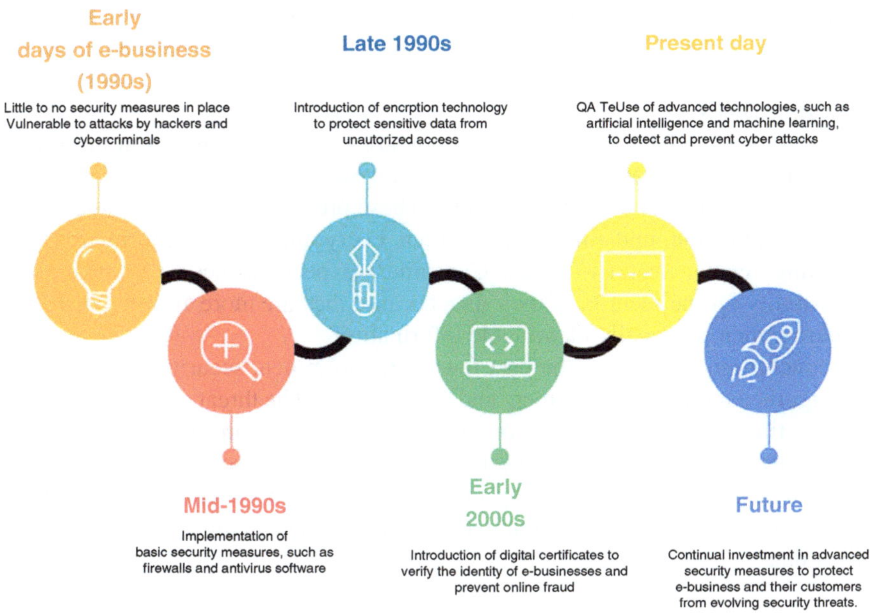

Fig. 5.2 Identification of potential vulnerabilities in e-business systems

When Target experienced a significant security breach in 2013, up to 110 million customers' private and financial information was made public. A third-party vendor with access to Target's systems was the targeted victim of the assault, which used phishing as its approach. A security hole resulted as a result. As a result of the incident, which significantly impacted the corporation, Target's revenues and reputation deteriorated. Additionally, the Federal Trade Commission took enforcement action against the business and penalized Target $18.5 million for failing to secure the privacy of its consumers' information adequately.

Uber revealed in 2017 that 57 million users and 600,000 drivers' personal information had been exposed in a security incident in 2016. The company faced harsh criticism for neglecting to reveal the breach for over a year and paying the hackers $100,000 in ransom to keep it secret. These two techniques were used to cover up the breach. The final results of the incident included the resignation of Uber's CEO and a $148 million settlement with all 50 states and the District of Columbia.

The 2013–2014 security breach, regarded as one of the most significant data breaches in history, exposed the personally identifiable information connected with each of the three billion Yahoo user accounts. The hack resulted from a state-sponsored cyberattack, severely affecting Yahoo and its customers. The announcement led to the CEO of Yahoo resigning, a drop in user confidence, and regulatory action, including a $35 million fine from the SEC. In addition, the breach had a direct negative impact on user confidence in Yahoo.

The effects of these high-profile security breaches on online businesses and their customers cannot be overstated. Companies that suffer an information security breach risk losing their clients' trust, running afoul of the law, and suffering financial losses. Contrarily, consumers bear the danger of directly experiencing severe harm due to a breach. This harm may appear as identity theft, financial fraud, or other cybercrimes.

Businesses have been forced to put more strict security measures in place as a result of these data invasions to safeguard the systems they employ to hold client data as well as the data of their customers. These precautions include using multifactor authentication, updating software often, encrypting critical information, and conducting regular security audits. Consumers expect businesses to take more responsibility for the security of their data because they are more aware of the value of securing their personally identifiable information.

In conclusion, throughout the past few years, prominent security breaches have had a significant effect on customers and e-businesses. The threat of a security compromise still exists today, despite organizations' increased investment in security measures to better safeguard their systems and data. Businesses need to continue to place a high priority on the security of their information systems and the data of their customers. Consumers are required to keep vigilance in preserving the privacy of their data.

3 Strategies for Ensuring E-Business Security and Control

Because of the changing nature of the environment in which they operate, online businesses should maintain the security of their computer systems and network connections. A breach in a company's security can result in reputational damage, financial losses, and legal ramifications. As a result, organizations need to use the best practices for protecting their e-business systems and networks. The following sections will review the most effective methods for safeguarding e-business systems and networks.

One of the most essential core components of e-business system security is the implementation of tight access constraints. Controlling access entails restricting sensitive material to only those granted permission to view it. This includes adopting role-based access control and multifactor authentication to guarantee that sensitive data is only accessible to authorized individuals. Password rules must also be enforced to ensure that passwords are strong, unique, and updated regularly. Another critical part of e-business system security is the implementation of network security measures. This goal necessitates the deployment of firewalls, intrusion detection systems, and intrusion prevention systems. In contrast to firewalls, intrusion detection and prevention systems detect and block illegal network access and hostile network behavior. Another critical part of ensuring the security of e-business systems is encryption. Encrypting data means converting it into an unreadable format for those who are not permitted to access it. This helps to protect critical information in the event of a data breach. Encryption is critical during this data transfer stage since data transmission over the Internet is a particularly vulnerable point where it can be intercepted. Regular security audits are essential to ensure the security of e-business systems and networks. These assessments can assist in identifying potential system vulnerabilities, allowing the system to be improved by applying the required protections. Furthermore, frequently testing various security measures, such as firewalls and intrusion detection systems, is required to ensure their correct operation. Personnel education and training is another critical part of ensuring the security of e-business systems. Employees are usually the weakest link in the security chain; therefore, educating them on the most effective security best practices is critical. This includes teaching people how to recognize phishing and other social engineering attempts and properly maintain sensitive data. In addition to the preceding, it is critical to have a disaster recovery strategy in place in the case of a security breach. This strategy should specify the steps to be taken in the event of a security breach, such as detecting the source of the breach, containing the incident, and restoring normal service levels [1].

Furthermore, it is critical to always use the latest software and operating system versions. Hackers frequently use flaws in computer programs to gain access to computer systems and networks. As a result, it is critical to keep all software and systems up to date with the most recent patches and upgrades. Furthermore, it is critical to have an efficient incident response plan in place. This plan should detail the procedures to be followed in the case of a security breach and the tasks and responsibilities of each incident response team member.

Finally, many harmful actors, including hackers, cybercriminals, and insiders, constantly threaten e-business systems and networks. These risks arise both within and outside of an organization. To protect sensitive data, preserve business continuity, and protect the company's brand, best practices for securing e-business systems and networks need to be implemented. These recommended practices can help organizations reduce the risk of a security breach and ensure the integrity of their e-commerce systems and networks. Among the best practices are substantial access restrictions, network security measures, encryption, frequent security assessments, employee education and training, disaster recovery planning, software and system updates, and incident response. The most successful methods and tactics for protecting e-business networks and systems are included in Table 5.2.

E-commerce is growing in popularity and importance as a direct result of the advent of the digital age in our daily lives. Businesses need to create efficient security policies and procedures to protect their customers' information and business operations by introducing e-commerce and online transactions. This is because thieves are getting more and better at getting through established security protocols. In this essay, we will look at how e-business security standards and practices are implemented. We will specifically examine the many security measures, their advantages and disadvantages, and the best ways to implement them.

Implementing security procedures for online businesses requires encryption technology, which is also one of the main ways to do so. Information is encoded using mathematical techniques during the encryption process to prevent unauthorized access. This can be done in several ways, such as utilizing SSL/TLS protocols to encrypt online traffic or employing encryption software to encrypt files and data

Table 5.2 Best practices for securing e-business systems and networks

Best practice	Description
Regular updates and patches	To avoid vulnerabilities, update all applications and systems with the most recent security patches
Strong passwords and authentication	Make complicated passwords and two-factor authentication mandatory for all user accounts
Employee training	Instruct employees on safe Internet and email usage, including recognizing phishing and other hazards
Access controls	Restriction of access to sensitive data and systems to those who require it
Regular backups	Regularly back up critical information to avoid losing it in a security breach or disaster
Network security	Employ firewalls, intrusion detection systems, and other security measures to prevent external threats
Incident response plan	Create and implement a plan for responding to security incidents, including damage containment and mitigation
Third-party security assessments	Assess the security of third-party vendors and partners with access to your systems and data regularly
Continuous monitoring	Maintain a constant check on your systems and networks for any indications of unusual activity

stored in computer memory. Although it can offer a high level of protection, encryption can also cause speed problems if it needs to be correctly optimized. Implementing encryption can also be challenging [1].

For Internet businesses, firewalls are an additional crucial security protocol. Filtering systems are called firewalls to guard against unwanted network access. These systems could have a hardware or software foundation. They can be used to defend against various cyberattacks, including malware, viruses, and denial-of-service (DoS) attacks. Although setting up and maintaining firewalls is relatively easy, they can make a network vulnerable to attack if they are not configured correctly or maintained up to date.

Another essential part of an online business' security protocol is managing who has access to its network. Access control is a mandatory technique involving managing an organization with access to specific data and systems. Alternative strategies for reaching this goal include password-based authentication, biometric authentication, and multifactor authentication. In large businesses with many users and systems, access control can help limit unwanted access to sensitive data and systems, but it can be challenging to deploy successfully. Particularly in large businesses with access control systems, this is true.

Employing intrusion detection and prevention systems (IDPS) can help online firms efficiently adopt security standards. IDPS monitor network traffic to prevent unauthorized access and potentially dangerous conduct. Numerous techniques, such as signature-based identification, behavioral pattern analysis, and anomaly detection, can accomplish this. However, IDPS may be prone to false positives and negatives if set up and maintained correctly. IDPS have the potential to identify and prevent assaults.

E-commerce companies must create and maintain rigorous security procedures and employ several security rules to apply these protocols effectively. Establishing guidelines for managing security issues, carrying out regular security audits, and training staff on the best security procedures could all be part of this. E-businesses may make sure their security measures are successful and that they are ready to respond to security issues promptly and effectively by having clearly defined security policies and protocols. This can help ensure the success of their security precautions.

Finding a balance between security, usability, and performance is a challenge that must be overcome when adopting security procedures and measures for online organizations. For instance, putting strict access control measures in place could make it harder for users to access data and systems, whereas putting in place overly aggressive intrusion detection algorithms might lead to false positives and a decline in network performance. E-commerce companies need to strike a difficult balance to ensure the security measures they use are practical and effective.

Keep up with the most recent vulnerabilities and threats to execute security policies and procedures successfully. E-commerce companies need to be proactive in recognizing and managing the numerous cyber dangers they face as they change and become more complicated. This may entail staying abreast of the most recent security developments and trends, doing regular security assessments, and implementing new security measures in response to potential risks.

In conclusion, security measures and procedures must be created to protect online businesses and their customers. E-businesses can defend themselves against cyber threats by implementing efficient security procedures and using encryption, firewalls, access control, IDPS, and other comparable technologies. Figure 5.3 shows how a security protocol is implemented for safe computerized business transactions.

Organizations need to conduct more business now online to be competitive in the market, as well as online to stay in business. The demand for security systems that

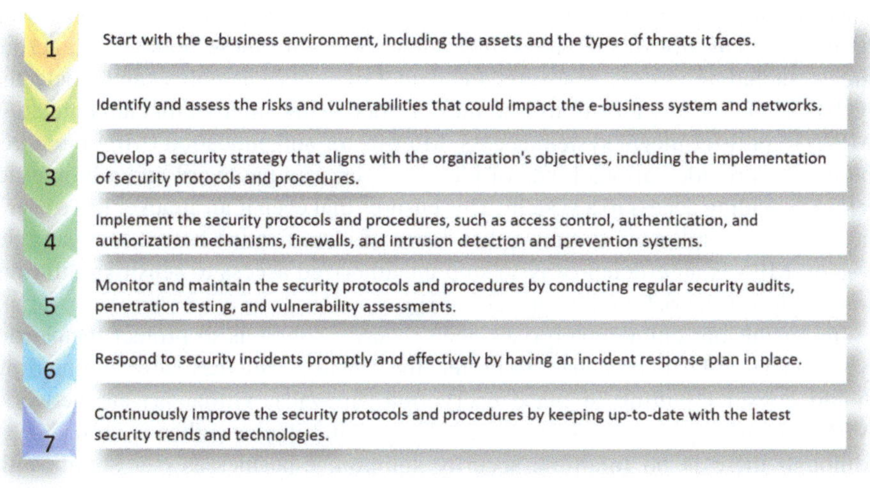

Fig. 5.3 Security protocol implementation for e-business

can ward off cyberattacks, data intrusions, and other forms of threats, on the other hand, has increased as e-commerce has grown. This essay will examine the many security measures that Internet businesses can take and how important it is to protect sensitive information.

One of the most crucial security measures for online businesses at the moment is encryption. Encryption transforms plain text into ciphertext, which someone with the corresponding decryption key can only decrypt. As a result, outside parties cannot access any data sent between the parties. Two of the most used techniques for encrypting data are RSA and AES, both recognized for their dependability and security. Additionally, Internet businesses must use the most recent iterations of the pertinent algorithms to stay one step ahead of possible rivals. For Internet businesses, firewalls are a crucial security tool. An example of a network security device is a firewall, which tracks and limits incoming and outgoing network traffic through established security rules. Unauthorized users cannot access a network without a firewall, and harmful traffic can be stopped before it gets to its destination. Many different kinds of firewalls are available, including packet-filtering firewalls, stateful inspection firewalls, and application-level gateways. An online business needs to select the firewall type that best suits its demands because each form of firewall has a unique set of benefits and drawbacks.

Online businesses should carefully consider using virtual private networks, or VPNs, to secure their data, encrypting it, and using firewalls. Virtual private networks, or VPNs, are secure connections that give remote users access to networks. Regardless of their or the organization's geographical location, they allow users to send and receive data as if directly linked to the network. Users can send and receive data just like they would if physically linked to the network. Virtual private networks (VPNs) offer safe access to sensitive data from anywhere globally and are especially helpful for online enterprises with remote staff or business partners. Intrusion detection and prevention systems, or IDPS, are a further crucial security tool for Internet businesses. Software programs called IDPS scan network traffic for signs of malicious activity and notified administrators when a potential threat is found. Limiting traffic from dubious sources and separating infected devices from the network can thwart the attack. They have access to these options. Numerous threats, including viruses, malware, and denial-of-service assaults, can be detected by IDPS. Online organizations should consider using multifactor authentication (MFA) to prevent illegal system access. Multifactor authentication (MFA) requires the user to provide various kinds of identity, such as a password and a biometric, before providing access to a system. This ensures that even if an attacker has a user's password, they still need to show other forms of identification before accessing the system. MFA is an efficient way to stop unauthorized access to sensitive information and lower the risk of data breaches [2].

In conclusion, online enterprises need security solutions to protect them from cyberattacks and data breaches. To protect the sensitive data they store, online firms should consider putting in place encryption, firewalls, virtual private networks (VPNs), identity protection systems (IDPS), and multifactor authentication (MFA). Although there is no infallible security solution, e-businesses can significantly

Table 5.3 Comparison of security technologies for e-businesses

Security technology	Function	Advantages	Disadvantages
Encryption	Converts plain section into the intersection to protect data transmission from unauthorized access	Provides a high level of security, widely recognized as secure and reliable	It can be computationally expensive and may impact network performance
Firewalls	Monitors and controls the incoming and outgoing network traffic based on predetermined security rules to prevent unauthorized access and block malicious traffic	It can prevent unauthorized access to a network and offers network visibility and control	It can be complex to configure and may impact network performance
Virtual private networks (VPNs)	Enables remote users to access a network using encrypted connections securely	It provides secure access to sensitive information from anywhere worldwide and offers network visibility and control	It can be complex to set up and maintain and may impact network performance
Intrusion detection and prevention systems (IDPS)	Monitors network traffic for signs of malicious activity and alerts administrators when a potential threat is detected	Can detect various types of attacks, including viruses, malware, and denial-of-service attacks and can take action to prevent attacks	It may generate false positives and may require ongoing updates and maintenance
Multifactor authentication (MFA)	Requires users to provide multiple forms of identification before granting access to a system	It provides an extra layer of security, even if a user's password is compromised	It can be inconvenient for users and may require additional setup and maintenance

lower the chance of security issues by combining these technologies. Thanks to this, they can function safely in the digital age. A comparison of the security technologies accessible to e-businesses is given in Table 5.3.

4 Legal and Ethical Considerations in E-Business Security and Control

The requirement for proper safety and management practices is more critical than ever due to the growing popularity of doing business online today. Several rules and regulations in Internet enterprises protect sensitive information about customers and clients. By putting these laws into effect, it is made sure that Internet firms need to adhere to specific criteria for the protection of this data. Depending on where an online business is based, these laws and regulations may vary from one location to another. However, they all aim to foster reliable and safe online business operations.

The European Union passed the General Data Protection Regulation (GDPR) in 2018. Since then, it has become one of the most well-known and significant pieces of online company security law. A project called the General Data Protection Regulation (GDPR) aims to protect people's privacy and give them more choices over how their data is collected, processed, and stored. Any organization that collects or processes the personal data of EU citizens is subject to the regulation, regardless of where the company's headquarters are. Online firms must first seek people's express agreement before collecting their data to adhere to the General Data Protection Regulation (GDPR) and then put any necessary security measures to safeguard it. The General Data Protection Regulation (GDPR) carries stiff fines and violation penalties.

Another significant law relating to the security of Internet businesses is the Health Insurance Portability and Accountability Act (HIPAA), passed by the US Department of Health and Human Services. A person's medical history kept private and safe in compliance with the Health Insurance Portability and Accountability Act (HIPAA) is called protected health information, or PHI. It applies to business partners who handle protected health information (PHI) on behalf of covered entities, including healthcare providers, health plans, and healthcare clearinghouses. HIPAA requires e-businesses to provide their staff with HIPAA compliance training and security measures to protect protected health information (PHI). These precautions' fundamental components may be administrative, technical, or physical. Substantial fines and other sanctions may be applied if HIPAA regulations are broken [1].

Many nations have more broad regulations regarding the control and security of e-business in addition to these specific pieces of legislation. For instance, the Computer Fraud and Abuse Act (CFAA) in the United States forbids unauthorized access to a computer and the use of a computer to commit fraud or any other illegal action. The CFAA likewise prohibits the trading of passwords and other access credentials. E-businesses are expected to put in place the necessary security measures to secure sensitive data and notify people of a data breach under the laws governing data protection and privacy in many nations worldwide.

Although laws and regulations are an essential part of creating safe online business practices, there are other things online firms need to consider. Online businesses need to fulfill not only legal commitments when it comes to protecting the information of their consumers and clients but also ethical and moral ones. Even though laws and regulations do not mention these needs expressly, their significance for creating credibility and upholding a positive reputation remains unaffected.

Several industry standards and best practices have been developed to help online enterprises uphold their moral and ethical commitments. To ensure that online firms that manage credit card information do so securely and competently, major credit card companies developed the Payment Card Industry Data Security Standard (PCI DSS) as an example. The Payment Card Industry Security Standards Council set this requirement. Similarly, the International Organization for Standardization (ISO) has created several information security standards. An information security management system can be created, implemented, maintained, and continually improved using the ISO 27001 framework [2]. In conclusion, the problem of e-business security and control is complicated and varied, requiring consideration of the relevant legal and ethical issues. E-businesses must consider the expectations of their customers and clients, strive to achieve industry standards and best practices, and comply with laws and regulations, which are crucial for defining minimum standards and fostering safe business practices.

Putting the proper security measures in place is more crucial than ever since Internet commerce will only grow and become more intertwined in our daily lives. "E-business security" covers many issues, such as safeguarding private information from online attacks, ensuring legal and regulatory compliance, and upholding ethical standards. The ethical issues surrounding the security of Internet businesses will be the main topic of this section.

One of the most significant ethical issues relating to the security of Internet businesses is the obligation to protect the personal information of one's customers and clients. Online businesses gather and store financial and personal data like names, addresses, and phone numbers. Frequently, this information is quite sensitive and can be exploited for fraud, identity theft, and other financial crimes. E-commerce businesses have to protect this information from being seen by unauthorized individuals or revealed without authorization and ensure that it is only used for proper activities.

Transparency is another crucial moral consideration in the context of online company security. Customers and clients should be given clear and explicit information about how Internet firms gather, use, and store their personal information. Online firms should be open and honest about their security guidelines and practices. An essential part of this process is informing customers of any data breaches or other security incidents that may affect their personally identifiable information. Transparency makes building trust with consumers and clients more accessible, crucial for maintaining one's reputation and creating lasting professional connections [4].

E-commerce companies must also weigh the moral implications of their alliances with outside suppliers and service providers. E-commerce companies frequently use third-party vendors to deliver payment processing and website hosting services. However, there is no guarantee that these suppliers will always follow the same moral principles as the online business. E-commerce businesses must thoroughly examine, screen, and monitor third-party providers to ensure they are trustworthy and dependable. These companies also need to hold their third-party providers responsible for any security lapses or other ethical transgressions.

E-commerce companies need to consider the more significant ethical ramifications of their business practices in addition to these ethical concerns. For instance, Internet enterprises need to consider how their activities may affect the environment, including the energy used and carbon emissions produced by their data centers and other IT equipment. When making business decisions, e-commerce enterprises need to consider not just the financial but also the social and political repercussions, including how their activities will affect the economy as a whole and the communities in which they operate [3].

To conduct themselves ethically, e-businesses can abide by several industry-specific standards and best practices. For instance, the Payment Card Industry Data Security Standard (PCI DSS) is a set of guidelines for online firms that manage customers' credit card information. The International Organization for Standardization (ISO) has also created several information security and privacy-related standards. E-businesses can also join trade organizations and industry associations, which opens up networking opportunities and information-sharing with other industry professionals.

Finally, an online organization's security architecture needs to include ethical issues. E-commerce businesses have to safeguard the private information of their customers and clients, be open and truthful about the security measures they take, and consider the broader ethical ramifications of their business activities. E-businesses can ensure they follow ethical standards and uphold the trust of their customers and clients by joining industry associations and trade groups, adhering to standards and best practices pertinent to their industry, and putting into practice standards and guidelines that are industry-specific. A comparison of the ethical issues related to e-business security methods is shown in Table 5.4.

Due to the expansion of the digital economy, e-commerce companies are becoming increasingly susceptible to assaults, including data breaches, hacking, and other threats. Online firms risk severe legal and moral repercussions when they neglect to protect sensitive data by putting proper security measures in place. The following section will examine various situations when firms have suffered moral or legal repercussions due to lax Internet company security procedures.

One of the most well-known instances in recent memory of an online corporation facing legal ramifications for lax security measures was the 2017 Equifax data

Table 5.4 Comparison of ethical considerations in e-business security practices

Ethical considerations	Description
Transparency	E-businesses should be transparent regarding data collection and usage policies and provide customers with plain and concise privacy notices
Respect for privacy	Online business enterprises should protect customers' personal information from unauthorized access and disclosure
Fairness	E-businesses should ensure that their security measures do not create barriers to access for marginalized communities or engage in unjust discrimination against certain groups of people
Responsibility	E-businesses should be accountable for security breaches or data leaks involving their consumers' information
Proactive approach	E-businesses should take a proactive approach to security by instituting regular risk assessments, performing security audits, and keeping abreast of the most recent security threats and best practices
Collaboration	E-businesses should establish common standards and best practices for e-business security in collaboration with their consumers, industry partners, and regulatory bodies

breach. More than 145 million people's personal information, including their Social Security numbers and birth dates, were compromised due to a data breach at Equifax, one of the biggest credit reporting companies in the world. Because Equifax neglected to patch a known vulnerability in its system, hackers could access the private information they were after. The business agreed to reimburse the impacted clients $700 million due to the consumer class action lawsuit and government agency probes.

Yahoo is yet another business that may face ethical and legal consequences due to lax security measures. Yahoo experienced not one but two significant data thefts in 2013 and 2014 that compromised three billion user accounts in total. Sensitive data, including email addresses, dates of birth, and answers to security questions, were exposed due to the breaches. Yahoo was subjected to several class action lawsuits and criticized for its slow notification of security vulnerabilities. Following a settlement in 2018, the company was obliged to pay the impacted clients $50,000,000. The SEC and the EU also threatened the company with extra fines.

In addition to facing legal implications, businesses that fail to put in place proper security measures for their online activities run the possibility of suffering severe ethical penalties. For instance, Target experienced a data breach in 2013 that exposed more than 40 million consumers' credit card information. Due to the breach brought on by weakness in Target's payment system, the retailer came under heavy fire for inadequately protecting its consumers' data. Target's CEO resigned due to the data breach, which also caused the company to face a significant drop in revenue and brand harm.

Due to the Cambridge Analytica scandal in 2018, which involved a political consulting business inappropriately accessing the private data of millions of Facebook users, Facebook attracted a lot of media attention. Due to the uproar, Facebook's data protection and privacy policies came under intense scrutiny, and the business received harsh criticism for its inadequate data security measures. Numerous investigations and fines were levied against Facebook, and the company reached a $5 billion settlement with the Federal Trade Commission.

Marriott International is another business dealing with unethical fallout from lax security protocols. The personal information of up to 500 million customers may have been exposed due to a data compromise Marriott announced in 2018. The reservation system for the hotel chain was infiltrated by malevolent parties who got unauthorized access to it. A class action lawsuit was brought against Marriott because the business did not effectively protect the personal information of its clients. The security breach seriously hurt the company's reputation, and some guests decided not to book rooms at the chain of hotels as a result.

In conclusion, online businesses need to implement essential security measures to protect critical information and thwart cyberattacks. Companies that violate these norms risk severe legal and moral repercussions, such as being sued, paying a fine, and tarnishing their reputations. Therefore, online organizations need to prioritize data privacy and follow industry norms when creating and implementing new policies. They will be able to achieve this and establish themselves as reliable, trustworthy businesses while also avoiding the moral and legal issues that have historically beset corporations.

Summary

Introduction to E-Business Security and Control:

- Definition: E-business security refers to the measures and practices implemented to protect electronic transactions, information, and systems from unauthorized access, use, disclosure, disruption, modification, or destruction.
- Importance: With the increasing reliance on digital platforms for conducting business, ensuring security and control is crucial to safeguard sensitive data, maintain trust, and prevent financial losses.

Threats to E-Business Security and Control:

(A) External Threats

1. Malware: Malicious software designed to infiltrate, damage, or gain unauthorized access to computer systems.
2. Phishing: Deceptive tactics, often through emails or websites, trick users into revealing sensitive information.
3. Denial of Service (DoS) Attacks: Overwhelming a system or network with excessive traffic, rendering it inaccessible to legitimate users.
4. Hacking: Unauthorized intrusion into computer systems or networks to gain access, steal information, or disrupt operations.

(B) Internal Threats

1. Insider Threats: Employees or insiders with privileged access who misuse or intentionally harm the organization's resources or data
2. Social Engineering: Manipulating individuals into divulging confidential information or performing actions compromising security
3. Data Leakage: Unintentional or intentional unauthorized disclosure of sensitive information

Strategies for Ensuring E-Business Security and Control:

(A) Strong Authentication and Access Controls

- Implementing multifactor authentication, secure passwords, and role-based access controls to prevent unauthorized access

(B) Secure Communication and Data Encryption

- Using secure protocols (e.g., SSL/TLS) and encryption to protect data during transmission and storage, ensuring confidentiality and integrity

(C) Regular Security Audits and Updates

- Conducting regular assessments of security measures, identifying vulnerabilities, and promptly applying patches and updates to software and systems

(D) Employee Training and Awareness

- Educating employees about security best practices, identifying potential threats, and promoting a security-conscious culture

Legal and Ethical Considerations in E-Business Security and Control:

(A) Compliance with Data Protection Regulations

- Adhering to applicable laws and regulations, such as the General Data Protection Regulation (GDPR), to protect customer data and privacy

(B) Ethical Handling of Data

- Respecting customer privacy, obtaining informed consent for data collection, and ensuring secure storage and proper usage of personal information

(C) Intellectual Property Rights Protection

- Safeguarding digital assets, trade secrets, trademarks, and copyrights from unauthorized use, reproduction, or theft

Case Study

The XYZ Corporation is an online retailer that carries an assortment of products. Over the past several years, the company's customer base and sales have expanded substantially. Due to its expansion, however, XYZ Corporation has become a prominent target for cyberattacks and data intrusions. Concerned about the security and control of their e-commerce activities, the management team desires to analyze extant practices and develop plans to ensure the security and protection of consumer data. They have hired a consulting team to thoroughly evaluate their e-business security and control procedures.

1. Determine potential vulnerabilities in the e-business systems and networks of XYZ Corporation, considering the company's unique business model and activities. How might cyberattackers exploit these vulnerabilities?
2. Evaluate the importance of security and control in e-commerce for the XYZ Company. How would a significant security breach impact the company's reputation, consumer trust, and financial performance?
3. Examine the evolution of e-business security and control over time and how they have adapted to new threats. What emerging technologies and trends can XYZ Corporation use to enhance the security and control of its e-business?
4. Based on best practices, recommend specific solutions for XYZ Corporation to protect its e-business systems and networks. How can businesses effectively integrate security measures and a consistent user experience?
5. Evaluate the legal and ethical considerations that XYZ Corporation should have regarding e-business security and control. What are the potential legal and ethical repercussions if they fail to protect consumer data or use transparent security measures adequately?

Discussion

1. What is e-business security and control, and how does it differ from traditional security and control measures?
2. What are some examples of e-business security threats that businesses should be aware of?
3. Why is it essential for businesses to prioritize security and control in their e-business operations?
4. How has the evolution of e-business impacted security and control measures over time?
5. What are some common threats to e-business security, and how do they differ from traditional security threats?
6. How can businesses identify potential vulnerabilities in their e-business systems?
7. What are some high-profile security breaches that have occurred in recent years, and what impact have they had on businesses and consumers?
8. How can businesses protect themselves from cyberattacks and other e-business security threats?
9. What are some best practices for securing e-business systems and networks?
10. How can businesses implement security protocols and procedures to protect their e-business operations?
11. What role does encryption and other security technologies play in ensuring e-business security and control?
12. How can businesses stay up-to-date with the latest security and control measures for e-business?
13. What laws and regulations are relevant to e-business security and control, and how do they impact businesses?
14. What ethical considerations should businesses take into account when implementing security and control measures for e-business?
15. What are some examples of companies that have faced legal or ethical consequences for inadequate e-business security measures?
16. How can businesses balance the need for security and control with user privacy and transparency in their e-business operations?

Multiple-Choice Questions

1. What is e-business security and control?

   (a) Ensuring physical security of e-commerce websites
   (b) Protecting electronic transactions and data from unauthorized access
   (c) Monitoring employee productivity in online businesses
   (d) Preventing online advertising fraud

2. Why is security and control important in e-business?
   (a) To prevent competitors from accessing sensitive information
   (b) To comply with legal and ethical standards
   (c) To increase website traffic and conversion rates
   (d) To minimize the risk of financial loss and reputation damage
3. How has e-business security and control evolved over time?
   (a) It has become less important due to advancements in technology.
   (b) It has shifted focus from physical security to virtual security.
   (c) It has become more expensive and complex to implement.
   (d) It has remained unchanged since the inception of e-business.
4. What are common threats to e-business security?
   (a) Competition from other online businesses
   (b) Loss of customer trust and confidence
   (c) Server downtime and website errors
   (d) Cyberattacks and data breaches
5. Which of the following is a potential vulnerability in e-business systems?
   (a) Strong encryption methods
   (b) Regular software updates and patches
   (c) Weak passwords and authentication mechanisms
   (d) Multifactor authentication
6. Which of the following is an example of a high-profile security breach?
   (a) Facebook's Cambridge Analytica scandal
   (b) Amazon's acquisition of Whole Foods
   (c) Apple's introduction of Face ID
   (d) Google's search algorithm update
7. What are best practices for securing e-business systems and networks?
   (a) Sharing passwords among employees to simplify access
   (b) Regularly backing up data and storing it off-site
   (c) Allowing unrestricted access to sensitive information
   (d) Disabling firewalls and antivirus software
8. What is the purpose of implementing security protocols and procedures?
   (a) To complicate the user experience and deter potential customers
   (b) To ensure compliance with industry regulations
   (c) To increase operational costs and reduce profitability
   (d) To protect against unauthorized access and data breaches

9. Which security technology is commonly used to protect data during transmission?
   (a) Firewall
   (b) VPN (virtual private network)
   (c) Intrusion detection system (IDS)
   (d) Antivirus software
10. What are relevant laws and regulations in e-business security and control?
   (a) Intellectual property laws
   (b) Environmental protection laws
   (c) Employment and labor laws
   (d) Data protection and privacy laws
11. What ethical consideration is important in e-business security and control?
   (a) Maximizing profits at any cost
   (b) Ensuring transparency and user consent
   (c) Exploiting customer data for targeted advertising
   (d) Obstructing access to competitor websites
12. Inadequate security measures can lead to:
   (a) Increased customer loyalty and trust
   (b) Enhanced brand reputation and public image
   (c) Legal and ethical consequences
   (d) Higher conversion rates and sales
13. Which law or regulation aims to protect personal data and privacy in the European Union?
   (a) HIPAA
   (b) CCPA
   (c) GDPR
   (d) FERPA
14. Which of the following is an example of a case where a company faced legal consequences for inadequate security measures?
   (a) Equifax data breach
   (b) Amazon's acquisition of Whole Foods
   (c) Google's search algorithm update
   (d) Apple's introduction of Face ID
15. Which ethical consideration is important when collecting customer data for e-business?
   (a) Sharing customer data with third parties without consent
   (b) Providing customers with clear and accessible privacy policies
   (c) Exploiting customer data for targeted advertising without disclosure
   (d) Using deceptive practices to obtain customer data

References

1. Whitman, M.E. and H.J. Mattord, *Management of information security*. 2013: Cengage Learning.
2. Ladan, M.I. *E-Business Security Challenges*. in *The Second International Conference on Digital Enterprise and Information System (DEIS2013)-Malaysia*. 2013. Citeseer.
3. Zhou, M., et al. *Security and privacy in cloud computing: A survey*. in *2010 Sixth International Conference on Semantics, Knowledge and Grids*. 2010. IEEE.
4. Kshetri, N., *1 Blockchain's roles in meeting key supply chain management objectives*. International Journal of information management, 2018. **39**: p. 80–89.

Chapter 6
Billing and Payment Systems

Abstract The importance of e-business and its associated invoicing and payment systems cannot be understated as the world becomes more digital. These programs are required for online transactions since they let clients swiftly and conveniently pay for goods and services. E-businesses can also benefit from the efficiency and cost reductions of electronic payments. To ensure that these systems are secure, reliable, and accessible to all users, various challenges need to be overcome. This chapter will primarily concentrate on the invoicing and payment methods used in the e-business context. This chapter will go through the many invoicing and payment methods typically used in this field and the precautions that need to be taken to avoid fraud. The chapter will also look at the frequent obstacles people face while implementing these systems and offer solutions to those problems. By the end of this chapter, readers will have a better knowledge of the importance of effective invoicing and payment systems in e-business and the best ways to develop these systems.

1 Billing and Payment Systems in E-Business

The way businesses send out invoices and collect payments has significantly changed due to the e-business industry's spectacular growth. The billing and payment techniques of the last several decades have yet to be improved to meet the unique needs of online enterprises. Online businesses, as a result, depend on invoicing and payment systems created mainly to handle the complexity of online financial transactions. With a robust infrastructure for invoicing and payment, e-commerce can run. Invoice generation, payment tracking, and customer data management are all responsibilities of billing systems. Conversely, payment systems make trading money easier for customers and companies.

Customers' experience is enhanced by integrating billing and payment systems into online businesses. Customers can make payments via credit cards, electronic wallets, and Internet banking, among other methods. Since the billing system is

H. Taherdoost, *E-Business Essentials*, EAI/Springer Innovations in Communication and Computing, https://doi.org/10.1007/978-3-031-39626-7_6

integrated with all these payment options, the system will instantly update the customer's account and record the payment. Online businesses often put their billing and payment processing systems on the cloud, making them available from anywhere. Businesses can do this to manage payments and invoicing from a single platform, saving time and minimizing errors. E-commerce companies can also alter their billing and payment procedures to suit their requirements [1].

One of the significant advantages of using a billing and payment system in an online business is that it increases security. These systems make online business transactions safer for both the consumer and the store by encrypting sensitive data, such as credit card information. Two further benefits of using it are its versatility and convenience. Thanks to billing and payment systems, customers can make payments whenever, wherever, and using whichever manner they desire. The ability to service customers worldwide and accept payments in various currencies and languages makes this vital to these companies.

Finally, billing and payment processes are essential elements of an online business. They increase the security, efficiency, and user-friendliness of billing clients and collecting payments for enterprises. These technologies will become more crucial as e-commerce develops, and companies will need to invest in them to keep a competitive edge in the online market.

It is impossible to exaggerate the significance of effective invoicing and payment systems, particularly in light of the rise in online commerce. E-commerce has had a considerable impact on companies, and as a result, more than standard billing and payment methods are required to meet the unique needs of online transactions. Businesses may boost income, improve client satisfaction, and streamline operations with the help of an effective billing and payment system. Increased productivity is among the most essential advantages of a solid invoice and payment system. Time is vital in online commerce, and billing and payment delays can harm the entire consumer experience. The creation of invoices, processing of payments, and account reconciliation are all tasks that can be automated with an effective invoicing and payment system. In turn, this lessens the workload on the staff and gives them more time to concentrate on other parts of the business operations.

Practical procedures for invoicing and payments can also increase client satisfaction. Customers expect a hassle-free experience when they make purchases online, and any invoicing or payment problems may result in irate customers and missed sales. Customer satisfaction, retention rates, and customer loyalty may all be raised with the aid of a billing and payment system that is simple to use and provides a range of payment alternatives.

Solutions for managing invoices and payments can also aid companies in more efficient money management. With real-time reporting and analysis, businesses can learn more about their cash flow, revenue, and expenses. As a result, they are more equipped to manage the business's finances, spot opportunities for growth, and streamline their processes to maximize profits. Businesses can cut costs by using an efficient billing and payment collection system. Businesses may boost efficiency and reduce the risk of mistakes by automating procedures like billing and payment processing. As a result, this can lead to lower operating costs and more profitability.

Table 6.1 Benefits of effective billing and payment systems for e-businesses

Benefit	Description
Increased efficiency	Automating invoicing and payment processes frees up time for other business-related responsibilities
Improved customer satisfaction	Multiple payment options and an easy-to-use billing system will increase customer loyalty
Better financial management	Real-time reporting and analysis to aid in making wise decisions and maximizing profitability
Reduced costs	Automation lowers operating expenses and error risk; cost reduction is furthered by negotiating better terms with payment providers
Expanded customer reach	Accepting payments in many languages and currencies, expanding one's market, and improving profits

To further cut costs, e-commerce companies can also bargain better rates with payment processors.

Lastly, a successful system for billing and collecting payments can help firms grow their clientele. E-commerce websites can draw customers worldwide and break into new areas when they can accept payments in various currencies and languages. This could lead to increased earnings, more excellent prospects for growth, and a broader consumer base.

In conclusion, the growth and profitability of online enterprises depend on a reliable billing and payment system. It can promote productivity, raise client satisfaction, improve financial management, lower expenses, and broaden the customer base for the company. Businesses will need to invest in effective billing and payment systems as the volume of online commerce increases to keep their competitive advantage and satisfy their consumers' constantly changing expectations. Table 6.1 illustrates the benefits of implementing effective billing and payment systems for e-businesses.

2 Types of Billing and Payment Systems

Payment systems have become an integral element of our daily lives. These technologies enable the rapid and effective transmission of funds between accounts. There are numerous methods for exchanging currency, each with advantages and disadvantages. This section will briefly discuss the primary payment methods, including credit and debit cards, e-wallets, bank transfers, electronic checks, cryptocurrencies, cash on delivery, mobile payments, prepaid cards, bill-to-phone, loyalty points and rewards, and gift cards.

Credit and debit cards are among the most common online payment forms that merchants accept. Customers appreciate their convenience, while businesses value the accelerated processing times and reduced hazards associated with cash handling that these methods provide. When using credit and debit cards for transactions in an online business, a company needs to be aware of and prepared to address several potential risks and issues. The widespread acceptance and familiarity of credit and debit cards as payment methods are among their most significant advantages. Most individuals have at least one credit or debit card, making them a convenient and straightforward payment method. This may be advantageous for businesses as it may help them attract more customers and increase sales, which is crucial in an age where online purchasing has become the norm.

Additionally, businesses can process credit and debit card transactions much more rapidly, reducing the risks of handling cash. This strategy significantly reduces the time and resources required for the manual processing and management of cash because card payments include electronic funds transfer. Additionally, businesses are less likely to experience theft, cash loss, or cash injury, and fraud or the use of counterfeit currency is less likely to occur. However, there are some potential drawbacks to paying for an online business with a credit or debit card. The risk of chargebacks, which occurs when a customer disputes a transaction with the business that issued their card and then requests a refund, is among the greatest. Chargebacks can be costly for businesses because they frequently incur fees and penalties, resulting in revenue loss, and harm the company's reputation.

Fraud is an additional risk associated with using credit and debit cards. When cybercriminals and hackers use stolen card information to make unauthorized purchases or pilfer sensitive data, customers and businesses are harmed. To secure the integrity of their payment processing systems and protect themselves from the risk of data theft, businesses need to employ robust encryption and continuously monitor for suspicious behavior. Regarding the use of credit and debit cards for transactions in online businesses, companies should adhere to the industry's established norms and standards. This requires compliance with the Payment Card Industry Data Security Standard (PCI DSS). This standard outlines the security requirements that cardholder data-handling businesses need to meet. A company that disregards these regulations risks incurring fines, confronting legal action, and suffering reputational harm.

It is impossible to exaggerate the significance of having trustworthy payment options in today's age of expanding e-commerce. E-wallets are one such payment option that has become prominent over the past few years. E-wallets, digital wallets, allow users to store and manage all their payment information in one location. This enables online financial transactions to be completed quickly and with less effort.

The user-friendliness of e-wallets is one of their primary advantages. Users need to only register with a service provider and link their payment information to their account to utilize an electronic wallet. When the system is configured for users, they can make payments with a few clicks and do not need to input their payment information each time they purchase. Users save time and increase their level of security when making online purchases because they do not need to share their payment information with multiple merchants.

Another advantage of using e-wallets is the variety of payment options they provide. Customers of e-wallet providers can frequently link various payment methods to their accounts, including credit and debit cards, bank transfers, and even digital currencies like Bitcoin and Ethereum. Customers who shop online typically use e-wallets because they may select the best payment solution for their needs.

Using an electronic wallet when making online purchases offers additional security to the transaction. Most electronic wallets encrypt and tokenize user data to prevent hackers from accessing payment and other sensitive information. Furthermore, many electronic wallet providers offer their customers fraud prevention and dispute resolution services, which boosts users' confidence when doing business online. E-wallets can also help businesses reduce the transaction costs associated with traditional payment systems. Businesses that accept payments via electronic wallets can avoid the high operating costs of traditional point-of-sale systems and the high fees often connected with credit and debit card payments. This could result in significant cost reductions for small- and medium-sized enterprises.

Electronic wallets also allow businesses to communicate with a more extensive client base. Businesses that accept e-wallets as payment have a better chance of attracting more customers, especially younger generations who favor digital payment methods. This is especially true for millennial-targeted firms. As a result, businesses that employ e-wallets may see increased income and growth opportunities.

Transfers between bank accounts have long been a popular means of payment, and they continue to play an essential role in the payment infrastructure for e-businesses. It is a safe and dependable payment method because it directly transfers funds from one bank account to another. The following paragraphs will discuss the benefits and drawbacks of using bank transfers for online business transactions. One of the most significant benefits of using this method is the security of bank transfers. Bank transfers are one of the safest ways to conduct financial transactions since they directly transfer funds between two bank accounts. The danger of fraud or identity theft is reduced by eliminating the obligation for the buyer and seller to divulge sensitive payment information.

The dependability of bank transfers is another advantage of this technique. While other payment methods may experience technical difficulties or payment processing delays, bank remittances are typically completed within a few business days. Payment disputes can be reduced by handling transactions quickly and efficiently. Transfers between bank accounts are another successful payment method for online businesses. Because many institutions charge nominal or nonexistent transaction fees for online bank transfers, this strategy is economically feasible for businesses of all sizes. Businesses that engage in e-commerce may find that it decreases operating costs and increases profitability.

One of the numerous issues with the practice is the inconvenience of bank transfers. Unlike other payment alternatives, such as those that may be completed instantaneously, bank transfers can take several days to process. Customers that require quick payment confirmation may have a delay in acquiring their products or services, which could be costly. Another disadvantage of using bank transfers is the process's intricacy. Some clients, particularly those unfamiliar with Internet banking, may need help initiating a bank transfer. This could lead to payment issues or delays, aggravating the customer and the e-business.

Electronic checks are commonly utilized as a backup payment option in online company transactions. Businesses can receive payments from the buyer's bank account promptly and digitally using this payment method, also known as an electronic check or an electronic funds transfer (EFT). Before the vendor can begin accepting electronic checks, the buyer needs to provide the seller with information about their bank account and routing number.

Customers benefit significantly from the increased convenience that electronic checks provide. This is one of the most significant advantages of using electronic payments. Customers no longer need to write a handwritten check and mail it in, saving time and delaying processing. The speed with which electronic checks are processed reduces the time it takes for the vendor to get paid for their goods or services. When using electronic checks, the risk of fraud is lower than when using traditional paper checks. This is an additional benefit of using electronic payments. Electronic tests are processed in private using encryption to protect sensitive information. Electronic checks also include built-in verification methods, which reduce the likelihood of a fraudulent transaction. Companies can save money by accepting payments via electronic checks. Electronic checks are less expensive to process than alternative payment methods such as credit cards over time, which helps firms save money. In addition to credit cards, cash and debit cards can be used to make payments. Furthermore, the processing of electronic tests can be done automatically, eliminating the need for human processing and the associated costs.

Customers need to have a bank account to use electronic checks. As a result, marketing to prospective clients who do not have bank accounts or prefer alternative payment methods may need to be revised. Concerns over transmitting sensitive data, such as bank account information, via the Internet may impede the widespread adoption of automated checks.

In recent years, cryptocurrencies have emerged as a creative and cutting-edge method of conducting online financial transactions. Banks and other financial institutions regulate traditional payment systems, but cryptocurrencies are decentralized and operate on a peer-to-peer network. As a result, their safety, productivity, and cost-effectiveness improve. The usage of cryptocurrencies as a form of payment for online purchases of goods and services is rapidly approaching the top of the list.

One of the most significant advantages cryptocurrencies provide for online transactions is security. Because cryptocurrency transactions use advanced encryption techniques, they cannot be compromised by fraud or hackers. As a result, both merchants and customers benefit from a high level of security. A transaction cannot be modified once visible and recorded on the blockchain. Another advantage of cryptocurrencies is their lightning-fast transaction speeds. Transactions involving cryptocurrency can be completed in seconds compared to traditional payment methods, which might take several hours to several days. This transaction speed is essential for online businesses, which must complete transactions quickly to provide customers with a flawless experience [2].

Furthermore, cryptocurrencies have lower transaction fees when compared to more traditional payment methods. When no middlemen are present, transaction fees are dramatically reduced, allowing businesses to save significant money. As a result, cryptocurrencies are an appealing option for start-ups and small businesses with limited resources. Virtual currencies are also incredibly easy to use and accessible to almost everyone with an Internet connection. Because developing countries frequently lack or have a weak traditional banking infrastructure, they are an especially appealing option for businesses in those countries. Because of this, they are a very appealing alternative for enterprises. Businesses in these sectors may discover that utilizing cryptocurrencies to conduct business with consumers worldwide is safe, dependable, and less expensive.

However, several challenges come with using cryptocurrencies in online business transactions. Volatility is among the most challenging problems. Cryptocurrency prices could be more stable, making it challenging for businesses to successfully manage their finances and establish long-term strategies. Additionally, the absence of regulation and oversight may leave cryptocurrencies open to scams and fraud, which can worry cryptocurrency-using businesses and their clients.

Many online retailers accept cash on delivery (COD), a payment method that enables customers to pay for their orders after the item has been delivered to their homes. Customers can use cash on delivery (commonly known as COD) if they would rather pay with cash or do not have access to electronic payment options. Although COD may be more practical for customers, it presents numerous risks and challenges for online businesses.

The vast improvement in accessibility is one of the main advantages of COD. If a customer does not have access to or does not want to use a credit or debit card, they can still make an online purchase using cash on delivery (COD). Broadening their consumer base may help Internet firms generate more revenue.

However, managing COD can take much work for online businesses. Because cash transactions are inherently risky, online businesses need to take steps to minimize the possibility of fraud or nonpayment. This may entail confirming the customer's identity and the specified delivery address. Businesses that accept COD face an additional challenge in the form of rising delivery and cash handling costs. E-commerce companies must consider the expenses related to handling currency payments and scheduling deliveries, which can raise overall operating costs. Furthermore, COD may cause delays in payment processing and longer delivery times. E-commerce enterprises need to alter their business operations and cash flow estimates to consider that payment is not received until after the goods have been dispatched.

Many online businesses continue to accept COD payments despite their challenges because it is so popular with customers. E-commerce companies can successfully provide COD as a payment option and respond to their consumers' particular needs and preferences if they take the appropriate safeguards to reduce risk and account for any additional costs.

The emergence of mobile payment options has fundamentally changed how transactions in e-commerce are carried out. Mobile payments have become a viable and dependable transaction method in recent years, partly due to the widespread use of smartphones and other portable electronics. Customers can use mobile payments to make purchases using their mobile devices without carrying cash or credit cards. Customers no longer need to carry cash or credit cards because of this.

One of the most significant advantages of mobile payments is their convenience. Consumers can make purchases whenever and wherever they want with the help of their mobile devices and payment solutions. People may now shop without worrying about carrying cash or credit cards, which is incredibly convenient for those traveling or making impulsive purchases. Mobile payments' streamlined and uncomplicated purchasing process decreases the time and effort needed to execute a transaction. One advantage of using these techniques is the security of mobile payments. Modern authentication and encryption techniques are used in mobile payments to safeguard customer information and stop fraud. Most companies that support mobile payments demand that customers set up password-protected accounts. This adds an extra layer of security.

Mobile payments make spotting and reporting fraudulent activity easier because users can view their transactions in real time. Mobile payments can also improve the shopping experience for consumers. Mobile payments provide a streamlined and accelerated shopping experience with features like automatic payment processing and one-click completion. The possibility of repeat business can thus rise, as can client satisfaction. Customers have a range of payment alternatives, including bank wire transfers, credit and debit cards, electronic wallets, and mobile payments, in addition to these other methods.

Additionally, mobile payments offer opportunities and advantages to businesses. Businesses can grow their consumer base and enter new markets by providing mobile payment solutions. Thanks to mobile payments, businesses can offer their goods and services to customers needing access to more traditional payment methods like credit cards and checks. Sales and revenue can go up as a result of this. Mobile payments can also assist businesses in cutting expenses by streamlining the payment process and eliminating the need for manual processing. Automation of the payment process and a decrease in the requirement for manual processing are two ways to accomplish this.

Mobile payments have a variety of drawbacks in addition to their advantages. Market fragmentation is one of the main challenges to be solved. Several companies offer mobile payments, each with payment options, costs, and security measures. Because of this, it could be challenging for businesses to select the service provider and payment method that best suit their requirements. Additionally, some customers might be reluctant to accept mobile payments due to worries about the security and privacy of their financial information.

An increasing number of people use pre-loaded credit cards to make online payments. Customers using their platform can shop online conveniently and securely without disclosing personal financial information or worrying about going over their spending limits. Prepaid cards are available offline or online and can be filled with a predetermined amount of money. The card can be used to make purchases at online merchants that accept it after being loaded with money.

One of the most important advantages of prepaid cards is that they allow those who do not have access to traditional banking services to do online business still.

Prepaid cards can only be bought with cash; neither a bank account nor a credit history is necessary. This makes them a desirable alternative for people who do not have bank accounts or who do not have enough money in their bank accounts, enabling them to take advantage of the advantages of online shopping.

Prepaid cards also provide confidentiality and privacy that is unattainable with traditional financial transaction techniques. Prepaid card users are not obliged to divulge private financial data, such as their bank account or credit card number, to complete the transaction. This guarantee reduces the possibility of identity theft and other types of fraud when customers make online purchases. Users of prepaid cards also benefit from the ability to track their spending habits and budgets better. Customers' prepaid cards are pre-loaded with a set amount, so they can only spend the money now on them. They may be able to avoid common credit card hazards like excessive spending and debt building.

A prepaid card might be helpful as a tool for budgeting as well. Customers can manage their spending better and stick to their budgets by using the budgeting and expenditure-tracking features that come with some types of prepaid cards. Those trying to control their finances or save money may find this to be of particular use. For companies who want to provide their clients with an alternate payment method, prepaid cards are a potential alternative payment option. Businesses may increase their customer base and make it simpler for customers to make transactions when they accept prepaid cards. Using prepaid cards also helps businesses lower their risk of chargebacks and fraudulent transactions, which can be expensive and time-consuming.

Customers can make purchases through an online storefront that accepts "bill-to-phone" payments. Using this technique, the purchase price will increase the customer's monthly phone bill. This payment method has rapidly gained popularity due to its portability and ease, especially among mobile phone users who prefer to keep their credit card information private when purchasing.

Customers only need to provide their mobile phone number to access the bill-to-phone feature throughout the buying process. After that, the purchase cost is either charged to the customer's phone account or removed from their available prepaid credit. Many mobile network service providers provide this payment option. It can be used for a wide range of transactions, including subscriptions, purchasing physical goods, and purchasing digital content. The ease of usage of the bill-to-phone service is just one of its numerous benefits. Customers do not need to register for an account or enter financial information. Thus, the checkout process is straightforward and quick. This can boost the frequency of impulse purchases while decreasing the number of shoppers who leave their shopping carts unattended. Utilizing bill-to-phone has the added benefit of being accessible. If they have a mobile phone, customers can use this payment method regardless of their credit history or current financial status. Especially in areas with low credit card usage rates, this can assist e-commerce firms in growing their consumer base. Rural places are where this is especially true.

However, there are some limitations to using bill-to-phone. Safety and security are among the main worries with this payment option because it only needs the customer's phone number as additional proof. Due to this, it is susceptible to fraud, particularly if a customer's phone is taken or lost. Charge-to-phone transactions are usually restricted to lower spending due to the necessary invoicing and payment processes. This makes it unacceptable for regular payments and high-value transactions.

The growth of e-commerce has compelled companies to create innovative strategies to draw customers and increase their brand loyalty. One of these approaches is signing up for loyalty points or rewards programs that accept payments from Internet businesses. Every time a customer makes a purchase or interacts with a particular firm, they can earn loyalty points, which are a type of cash. These points can be exchanged for advantages, such as VIP event invitations, exclusive discounts, and gifts.

One of the main advantages of loyalty points for payments made through online retailers is that they aid in client retention and promote repeat business. If customers feel they are getting something in return for their ongoing support, they are likelier to stick with a business. By offering incentives to customers, such as price breaks or freebies, a firm might persuade them to return. Incorporating loyalty points into an online store's payment process can benefit businesses by increasing existing client loyalty and new client acquisition. Businesses can set themselves apart from rivals and attract customers interested in buying their goods or services by providing a rewards program. Customers are likelier to patronize a company that values and rewards their loyalty. Another benefit of employing loyalty points as a means of payment in online stores is that they can give those stores essential consumer data. Businesses that monitor their customers' purchases and commitment to the rewards program might learn more about their preferences and behaviors. The creation of targeted marketing campaigns and the improvement of the overall customer experience can both benefit from the use of this data. Another efficient way to reward customers online is to employ loyalty points throughout the payment process.

On the other hand, loyalty points are a more adaptable type of money that may be used to purchase a range of benefits. This is because giving out discounts and freebies to customers may be expensive for firms. This enables companies to give considerable client incentives while also assisting them in cost management.

To successfully deploy a loyalty points program in an online store that accepts payments, careful design and execution are necessary. Businesses need to provide clients with enticing benefits, and the program itself needs to be simple, easy to use, and understandable. To make sure that customer loyalty point programs are working efficiently and adding value for both the consumer and the business, businesses also need to be able to monitor and manage them.

Gift cards are now used more frequently for payment by Internet businesses. Similar to gift cards, they can only be used at a particular store or collection of retailers and need to be purchased in advance. Gift cards provide several advantages for both consumers and businesses.

Gift cards in online retail offer customers more ease and flexibility when purchasing. This is one of their main advantages. Customers do not have to utilize traditional payment options like credit cards or bank transfers to make purchases from the business if gift cards are offered. Gift certificates can be used to purchase any good or service the retailer offers. As a result, customers might save time and effort, and their shopping experience would be more streamlined. Gift cards can assist businesses in increasing their revenue and sales, which is just one of their numerous benefits. The business is paid upfront when customers buy gift cards, regardless of when the card is ultimately used to buy goods. This might increase a business's cash flow and bring in more money.

The use of gift cards can also increase client retention and loyalty. Customers are more likely to return to a retailer's store after receiving gift cards from that store. Customers may also be given gift cards as an incentive or reward for making multiple purchases at a store or referring their friends and family as new customers. Additionally, customers can offer businesses helpful information by using gift cards. Retailers can gather information about customers, including their email addresses and past purchases, when they buy gift cards from consumers. These details can be used to tailor advertising campaigns and provide customers with more valuable goods and services.

However, a few drawbacks to using gift cards in online shopping need to be considered. The possibility of gift card fraud, which could involve using stolen or counterfeit cards, is one of the difficulties. Retailers need to put security measures in place to guard against these risks, such as assigning special codes or serial numbers to each gift card. Additionally, there is a chance that gift cards will only be used partially, which is a potential problem. A customer's gift card may not have enough money to pay for their purchase. Customers may be entitled to refunds or shop credits in certain circumstances, which can be costly for retailers. Table 6.2 compares the benefits and drawbacks of various e-business payment options.

3 Security and Fraud Prevention in Billing and Payment Systems

The prevalence of fraudulent and illegal online activity is increasing while online commerce is expanding. Since it is used in online transactions, sensitive information is susceptible to hacking by cybercriminals. Therefore, best practices implementation is crucial if one wants to conduct secure transactions and lower the risk of online fraud.

One of the most crucial ways to guarantee financial transaction security and reduce fraud risk is to use secure payment systems. By encrypting the data and putting extra security measures in place, these systems guard critical information like credit card numbers, passwords, and personal details. Well-known examples of secure payment systems that guard customers against fraud and illegal access include payment gateways like PayPal, Stripe, and Square.

Another suggested best practice is safeguarding your website from predators. This entails using HTTPS encryption for your website, secure passwords, and regularly updating your website's software and security features. Additionally, verifying

Table 6.2 E-business payment methods: advantages and disadvantages

Payment method	Advantages	Disadvantages
Credit/debit cards	Widely accepted, quick, and convenient transactions, protection against fraud	Risk of overspending, high interest rates, and fees, the potential for identity theft
E-wallets	Secure, convenient, and easy to use; some offer cashback and rewards	Limited acceptance, the potential for account hacking or fraud, transaction fees
Bank transfers	Direct transfer from a bank account, low transaction fees, secure	Slow processing times, the potential for errors, not widely accepted
Electronic checks	Easy to use, low transaction fees, secure	Not widely accepted, the potential for fraud, limited processing times
Cryptocurrencies	Decentralized, low transaction fees, secure	Limited acceptance, high volatility, the potential for hacking or fraud
Cash on delivery	No need for electronic payment, secure	Potential for fraud, delayed payments, not widely accepted
Mobile payments	Convenient and easy to use; some offer rewards or cashback	Limited acceptance, the potential for fraud, transaction fees
Prepaid cards	Easy to use and budget-friendly; some offer rewards or cashback	Limited acceptance, the potential for fees, risk of loss or theft
Bill-to-phone	Convenient, quick, and easy transactions	Limited acceptance, the potential for billing errors or disputes, high transaction fees
Loyalty points/ rewards	Offers incentives for repeat purchases, which can be used as payment	Limited acceptance, the potential for reward program changes or cancellations, may not have monetary value
Gift cards	Easy to use, can be used as payment	Limited acceptance, risk of loss or theft, the potential for expiration dates or fees

the clients' identities before beginning the transaction processing is crucial. This could entail requesting two-factor authentication, checking billing addresses and other personally identifying information, and watching for unusual activity.

Additionally, you and your staff must gain the knowledge to recognize and stop fraud. This includes having the capacity to spot potentially suspect financial activities, such as those involving significant sums of money, repeated purchases of the same thing, or orders placed from odd places. It also necessitates keeping up with the newest fraud-related trends and technology, like phishing scams and malware assaults. Another essential step in the fight against fraud is to examine your transactions and accounts routinely. Performing routine audits of transaction records, checking for unusual behavior, and using fraud detection tools to find suspicious financial transactions are all required to achieve this. It is crucial to act quickly in case of any questionable activity, such as getting in touch with the customer or payment processor to confirm the transaction as soon as possible [3].

It is also crucial to have a strategy in place for dealing with fraud incidents. This can involve alerting the appropriate authorities, working with your payment processor to reverse any fraudulent transactions, and updating your security procedures to

Fig. 6.1 Best practices in ensuring secure transactions and preventing fraud in e-commerce

prevent such incidents. If you have a plan, you can lessen the harm that fraud causes and rapidly resume normal business activities.

In conclusion, preventing fraud in online commerce and ensuring the security of online transactions requires a diverse approach. You can defend your business and your customers from the risks of online fraud and cybercrime by utilizing secure payment methods, maintaining a secure website, verifying customer identities, educating yourself and your staff, monitoring transactions and accounts, and having a plan for responding to fraud incidents. The best practices for protecting online transactions and reducing the possibility of fraud are shown in Fig. 6.1.

The proliferation of online shopping has increased the likelihood of falling victim to online deception or cybercrime. As the number of online business transactions increases, fraudsters are constantly developing new techniques to exploit security flaws in Internet businesses' invoicing and payment systems. To safeguard themselves and their customers, businesses must implement effective fraud detection and prevention procedures.

Implementing stringent identity verification measures is one of the most important methods for preventing fraud in online businesses' invoicing and payment operations. Additional forms of identification, such as a driver's license or passport, may be required to confirm that a consumer is who they claim to be. In addition, businesses can use biometric authentication techniques such as fingerprint or facial recognition to confirm that the individual completing the transaction is who they claim to be. Due to these preventative measures, identity thieves are prevented from engaging in fraudulent activity using stolen identities.

Using machine learning and artificial intelligence (AI) algorithms is another successful method for detecting and preventing fraud. These algorithms can examine vast quantities of transaction data and flag potentially suspicious behavior for further investigation if they detect it. Because they constantly learn and adapt to new types of fraudulent behavior, these algorithms can prevent fraudulent transactions and remain one step ahead of fraudsters. For instance, AI systems can analyze transaction patterns and identify anomalies that may indicate fraudulent activity. Anomalies can include transactions that deviate from the typical behavior of a customer or involve significant sums of money.

In addition to deploying cutting-edge technologies, businesses need to develop open and comprehensive fraud prevention strategies. These guidelines should delineate the steps employees should take to detect and prevent fraud, such as verifying clients' identities, searching for unusual transaction patterns, and reporting suspicious activity to management. In addition, it is essential to train personnel frequently on these policies and keep them abreast of the most recent fraud trends and techniques.

Consistently monitoring transaction activity is another excellent method for combating fraud in an online business's invoicing and payment operations. Identifying suspicious activity may involve analyzing transaction records, researching payment trends, and employing technology that detects fraudulent behavior. Businesses can avoid additional losses and safeguard their customers' sensitive information if they promptly identify and address fraudulent transactions. For example, companies can use software that monitors transactions in real time and alerts them to the possibility of fraudulent activity before the transaction is finalized.

In addition, the billing and payment systems utilized by online enterprises need to be secure. This includes using encryption technologies to protect sensitive client data and regularly updating security software to combat evolving threats. In addition, businesses need to implement PCI DSS-compliant payment gateways. Businesses need to have a fraud prevention strategy in place if they encounter fraud. This may involve notifying law enforcement, collaborating with payment processors to reverse fraudulent transactions, and enhancing security procedures to prevent future occurrences. With a comprehensive response plan, businesses can reduce the damage caused by fraudulent behavior and swiftly resume commercial operations.

Lastly, e-commerce companies need to implement fraud detection and prevention methods. Businesses can protect themselves and their customers from online fraud and cybercrime by implementing stringent identity verification procedures, machine learning, artificial intelligence algorithms, clear fraud prevention policies, transaction monitoring, secure systems, and extensive response plans. Due to the constant evolution of fraudsters' methods, legitimate businesses need to maintain high vigilance and proactivity in preventing fraud. Businesses can develop their reputations in the e-commerce industry and maintain their customers' trust if they

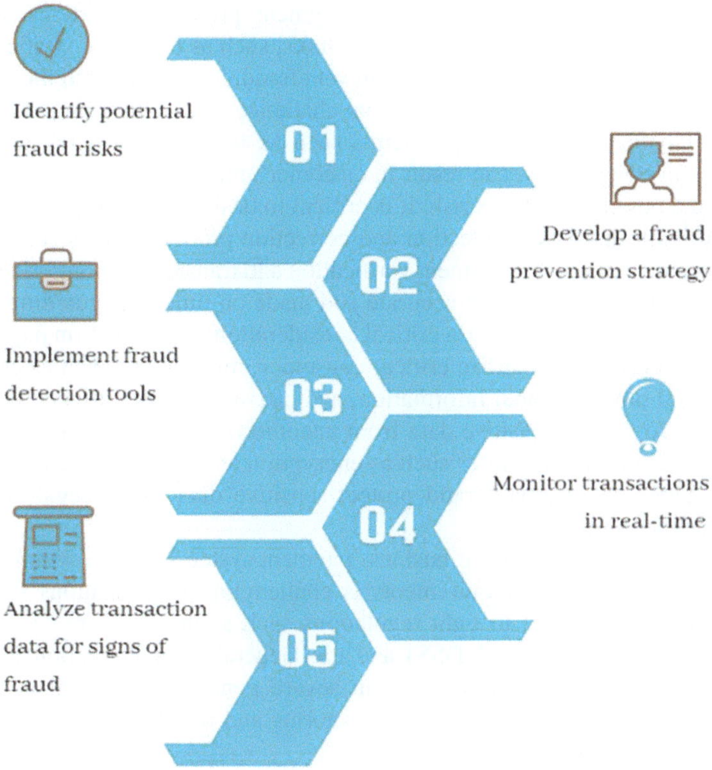

Fig. 6.2 Fraud detection and prevention measures in e-business billing and payment

adhere to these rules and take the necessary precautions. The many methods used to identify and stop fraud in electronic business billing and payment systems are shown in Fig. 6.2.

4 Challenges and Solutions in Billing and Payment Systems

As the number of people conducting business online grows, so will the demand for trustworthy and secure invoices and payment solutions. However, these systems present several challenges and concerns that need to be addressed to be effective. This part discusses some of the most significant obstacles and challenges in an online business's invoicing and payment systems and some tips for resolving such issues and difficulties.

The danger of fraud is one of the most severe challenges that payment systems represent for Internet businesses. Both the company and the consumer are at risk of severe monetary losses due to fraudulent activity, such as the theft of credit card information or unauthorized access to payment processing systems. Adopting strong identification and verification procedures, such as biometric verification and two-factor authentication, is critical to prevent fraudulent actions and illegal access.

Transaction faults create yet additional challenges for online company payment systems. Payment transaction problems, such as incorrect invoicing amounts or unsuccessful transactions, can result in customer unhappiness and loss of income. To limit the possibility of this risk, it is critical to design trustworthy payment systems with built-in mistake detection and correction processes. This entails putting processes and methods for payment verification and transaction monitoring in place to guarantee that payments are accurate and made on time. Safeguarding sensitive consumer information is an extra critical consideration for online firm payment systems. Data breaches jeopardize critical consumer information, such as credit card numbers and other personal information, and can significantly affect a company's reputation. To protect sensitive data from unauthorized access and potential cyber threats, strict security measures such as encryption and firewalls need to be implemented. These safeguards should protect sensitive data from illegal access and potential cyber threats [3].

Furthermore, ensuring that e-business payment systems comply with the numerous legal and regulatory requirements is challenging. Organizations that handle payment data are subject to tight regulations such as the Payment Card Industry Data Security Standard (PCI DSS) and the General Data Protection Regulation (GDPR). Noncompliance may result in severe penalties and legal punishment. Businesses need to follow all applicable norms and standards to avoid legal and financial consequences.

Payment system integration with other e-business operations can be complex and provides its issues. Connecting payment and order fulfillment systems, for example, necessitates regular communication between the two systems, which can be challenging. Furthermore, using multiple payment systems can create complications that need to be managed and coordinated effectively. Furthermore, scalability remains a challenge for online business payment systems. When a company expands, its payment system needs to handle an increase in the number of transactions processed and new payment methods. Failure to scale the payment system can result in system failures and delays, leading to dissatisfied customers and revenue loss. Customers' trust is critical to the success of online merchant payment systems. Customers want information about their payment methods to be kept private and safe. As a result, it is critical to create and keep client trust by establishing strict security measures, offering clear and transparent payment procedures, and delivering dependable customer support services [4].

Finally, e-business payment systems provide a variety of challenges and concerns that need to be addressed before the systems can be considered functional. Among these are fraud prevention, error detection and correction, data protection, compliance with legal and regulatory requirements, integration with other e-business operations, scalability, and consumer trust. Businesses that design and implement appropriate policies and strategies can mitigate these challenges and provide consumers with secure and dependable payment solutions. The issues and barriers that

Table 6.3 Summary of issues and challenges in billing and payment systems in e-business

Issue/ challenge	Description	Impact	Possible solutions
Security	The requirement for secure invoicing and payment systems to safeguard sensitive data and financial transactions	Unauthorized access to sensitive data, data breaches, and financial losses	Encryption, firewalls, authentication, access control, IDPS, SSL, and backup and recovery systems
Fraud	Possibility of fraudulent activities, including identity theft, credit card fraud, and chargeback fraud	Financial losses, damage to reputation, and legal consequences	Implementing fraud detection tools like real-time transaction monitoring, machine learning algorithms, and chargeback prevention mechanisms
Technical errors	Technical issues, including system malfunctions, payment processing errors, and payment gateway problems	Customer dissatisfaction, financial losses, and damage to reputation	Regular system maintenance, having a backup and recovery plan in place, and using reliable payment gateway providers
Regulatory compliance	Compliance with regulations, including PCI DSS, GDPR, and CCPA	Legal consequences, financial penalties, and damage to reputation	Conducting regular compliance audits, ensuring all stakeholders know their responsibilities, and implementing the necessary controls to ensure compliance
Payment fraud	The possibility of payment fraud, such as credit card information theft, phony chargebacks, and account takeovers	Financial losses, damage to reputation, and legal consequences	Implementing fraud detection tools like real-time transaction monitoring, machine learning algorithms, and chargeback prevention mechanisms
Chargebacks	The method by which a consumer contests a transaction and requests a refund from the merchant	Financial losses, increase in administrative costs, and damage to reputation	Providing clear and concise billing statements, having a customer-friendly refund policy, and addressing customer concerns promptly
Payment processing fees	The fees that payment gateway providers charge for payment processing	Increased business expenses and reduced profit margins	Negotiating fees with payment gateway providers, using cost-effective payment processing solutions, and passing on some or all fees to customers

exist in billing and payment systems in the context of e-commerce are highlighted in Table 6.3. The problems and difficulties e-business payment systems confront are shown in Fig. 6.3 and related solutions.

With the rise and popularity of online businesses, there is a greater need for dependable and secure payment and billing solutions. Along with the benefits of

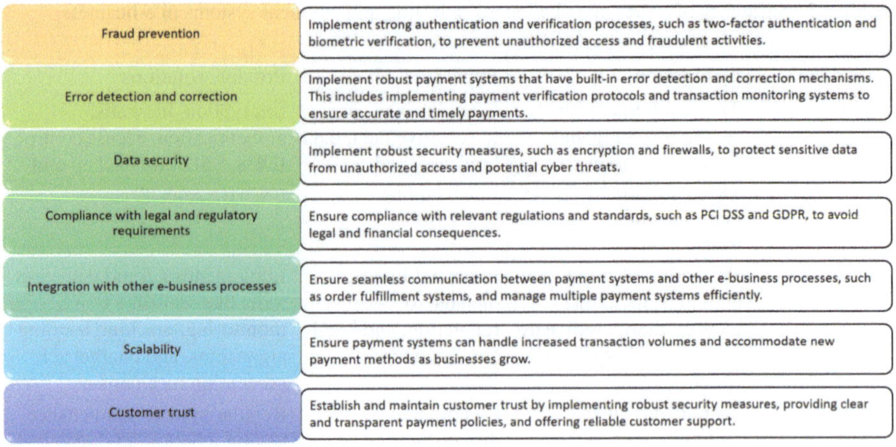

Fig. 6.3 Challenges and issues in e-business payment systems and their solutions

doing business online, several problems and obstacles arise to ensure the efficacy of these systems. This piece will look at some of the best practices for dealing with arguments and issues related to an online business's payment and invoicing systems.

The possibility of fraud is one of the most significant concerns confronting online company payment systems. Because of the increasing incidence of cybercrime, businesses are responsible for protecting their customer's sensitive financial information. Strong authentication and verification processes, such as biometric verification and two-factor authentication, may be required to prevent fraudulent activity and unauthorized access.

Transaction errors pose a new issue for online corporate payment systems. Customers dissatisfied with their bills due to incorrect billing amounts or unsuccessful transactions may create a revenue loss. To mitigate the potential impact of this risk, organizations should prioritize deploying trustworthy payment systems with built-in error detection and correction methods. To achieve this purpose, developing systems for transaction monitoring and payment verification may be necessary. This ensures that payments are made appropriately and on time.

The privacy of user information is another critical factor for online payment systems. Data breaches jeopardize customer-sensitive information, such as credit card numbers and other personal information, and can severely harm a company's reputation. As a result, it is critical to implement tight security measures, such as firewalls and encryption, to protect sensitive data from unauthorized access and future cyberattacks.

One of the most difficult challenges for e-business payment systems is ensuring that all rules and regulations are obeyed. Payment data handling organizations are subject to stringent laws such as the General Data Protection Regulation (GDPR) and the Payment Card Industry Data Security Standard (PCI DSS). You may face severe penalties and legal consequences if you do not comply. To avoid financial and legal consequences, businesses need to follow all applicable rules and standards.

Integration of payment systems with other e-business operations might need to be revised. For example, connecting order fulfillment and payment systems requires regular communication, which is only sometimes straightforward to establish. Furthermore, using multiple payment methods can cause complications that need to be carefully coordinated and managed.

Scalability is another essential consideration when developing payment systems for online enterprises. To stay up with the business expansion and the related increase in transaction volume, payment processing systems need to be able to handle both new payment methods and higher transaction volumes. Failure to scale the payment system may result in system problems and delays, causing customers to be dissatisfied and businesses to lose money [5].

To summarize, payment processing systems for online firms need to earn and maintain the trust of their customers. Businesses need to take steps to win and maintain their customers' trust. Customers have the right to expect that information about their financial transactions will be kept private and secure. This could include implementing strict security measures, providing clear payment terms, and providing dependable customer support.

Finally, e-business payment systems face various challenges and problems that need to be addressed for the systems to function correctly. Implementing strong authentication and verification processes, prioritizing reliable payment systems with integrated error detection and correction mechanisms, ensuring data security through encryption and firewalls, ensuring compliance with legal and regulatory requirements, efficiently integrating with other e-business processes, scalable payment systems, and establishing an audit trail are all strategies for resolving disputes and addressing challenges in e-business. Companies can reduce their risk exposure while providing safe payment solutions to clients using these processes. Figure 6.4 depicts numerous techniques for overcoming payment and billing challenges in e-commerce businesses.

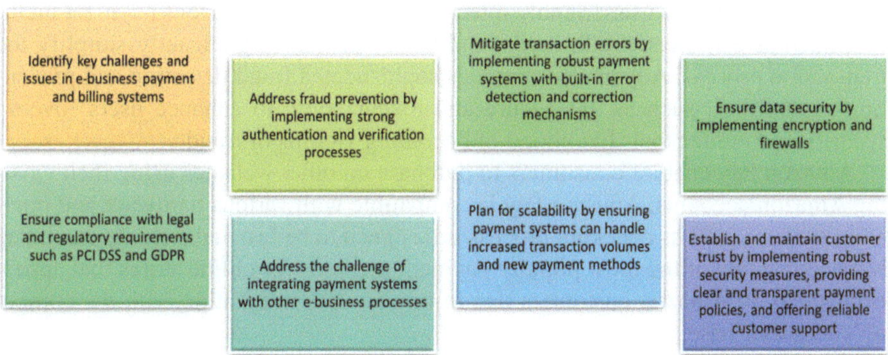

Fig. 6.4 Strategies for addressing challenges in e-business payment and billing

In the ever-changing world of e-commerce, it is crucial to have efficient and trustworthy invoicing and payment systems. A well-designed payment system can enhance the user experience, reduce transaction costs, and increase profits. However, designing and implementing such a system can be challenging because it requires knowledge of numerous factors, such as customer preferences, security concerns, and regulatory agency requirements. This section examines several successful invoicing and payment systems used in e-business and the factors that contribute to their success.

PayPal is currently one of the most popular methods for accepting payments via the Internet. The usability, security, and user-friendliness of the platform contribute to PayPal's success. Users can send and receive payments via email without using a credit card. Additionally, PayPal protects both the consumer and the seller against fraudulent transactions. In addition, the fact that PayPal can be integrated with major e-commerce platforms such as Shopify and eBay has contributed to its prominence as the payment method of choice for many businesses.

Stripe is another prominent and successful online payment provider. A payment gateway named Stripe enables businesses to accept online payments quickly and securely. Stripe's key features include the ease of integration, the speed of payment processing, and the support for various payment methods, including credit cards, Apple Pay, and Google Pay. Another vital selling factor for Stripe is its sophisticated fraud detection and prevention techniques. Stripe places a high priority on security.

Square is a well-known payment option that has recently gained significant popularity. Square offers numerous payment processing-related services. Included in these services are billing, in-person payments, and Internet transactions. It is notable for its ease of use, swift payment processing, and transparent pricing. Square's pricing structure is reasonable and transparent; no hidden fees or protracted contract obligations exist. Square's user-friendly interface and excellent customer support have also contributed to the company's rise to prominence among small- and medium-sized businesses.

Adyen is a successful electronic payment system. Adyen is a payment gateway that enables businesses to accept online payments in numerous monetary systems and dialects. Adyen's user-friendly interface and support for various payment methods, such as e-wallets, credit cards, and bank transfers, significantly simplify the transaction process. Adyen also provides sophisticated fraud detection and prevention capabilities, such as 3D Secure and tokenization, to enhance users' overall security and reduce their likelihood of becoming victims of fraud.

Amazon Pay enables consumers to purchase on other websites using their existing Amazon accounts. Amazon Pay is compatible with various hardware and operating systems, and its checkout process is designed to be fast and easy. Amazon Pay has become one of the most popular payment methods for online stores that appeal exclusively to Amazon customers due to its seamless integration with the company's vast user base.

Developing and deploying a billing and payment system for a web-based business can be challenging and time-consuming. However, businesses comprehensively comprehend the essential components of effective payment systems. In that

Table 6.4 Examples of successful billing and payment systems and their key features in e-business

Payment system	Key features	Company
PayPal	Fast and easy payments Ability to accept payments from over 200 countries Integration with many e-commerce platforms Fraud detection and prevention measures Buyer and seller protection policies	PayPal Holdings, Inc.
Stripe	Customizable checkout forms Support for multiple payment methods Built-in fraud prevention and detection Real-time transaction tracking Automated invoice generation and payment reminders	Stripe, Inc.
Square	User-friendly point-of-sale system Support for online and in-person payments Customizable invoicing and payment options Advanced reporting and analytics Fraud prevention tools	Square, Inc.
Amazon Pay	Easy checkout experience for Amazon customers Ability to pay with Amazon account information Support for multiple currencies Automatic refunds and chargebacks Fraud protection measures	Amazon.com, Inc.
Google Pay	Ability to store multiple payment methods Seamless integration with Google products In-app and online payments Fraud protection and detection Easy refunds and dispute resolution	Google LLC

case, they can design and implement a payment system that meets their specific needs and enhances the customer experience. Case studies presented in this part illustrate the necessity of simplicity, security, user-friendliness, and support for various payment methods for effective payment systems. By incorporating these components into their payment processing systems, businesses can increase revenue, reduce transaction costs, and provide clients with a fast and secure payment method. Table 6.4 presents examples of effective billing and payment systems used in e-business and their essential characteristics.

Summary

Types of Billing and Payment Systems

- Billing and payment systems are crucial components of e-business that enable transactions and financial interactions between businesses and customers in online environments.
- Online Invoicing: This system allows businesses to generate and send invoices electronically, streamlining the billing process and reducing paperwork.

- Recurring Billing: This system automates regular payments by charging customers regularly, such as monthly subscriptions or membership fees.
- Mobile Payments: Mobile payment systems facilitate transactions through smartphones and other mobile devices, providing convenience and flexibility for customers.
- E-Wallets: E-wallets securely store customers' payment information and allow for quick and easy transactions across various platforms and merchants.

Security and Fraud Prevention in Billing and Payment Systems

- Security measures are essential to protect sensitive financial information and prevent fraudulent activities in billing and payment systems.
- Encryption: Encryption ensures that data is securely transmitted between parties, preventing unauthorized access and protecting customer information.
- Tokenization: Tokenization replaces sensitive data, such as credit card numbers, with unique tokens, reducing the risk of data breaches and enhancing security.
- Two-Factor Authentication (2FA): 2FA adds an extra layer of security by requiring users to provide two forms of identification, typically a password and a unique code sent to their mobile device.
- Fraud Detection Systems: Advanced algorithms and machine learning techniques detect and prevent fraudulent transactions, minimizing business financial losses.

Challenges and Solutions in Billing and Payment Systems

- Billing and payment systems face various challenges, such as system integration, customer disputes, and payment failures. Implementing effective solutions is crucial to ensure smooth operations.
- System Integration: Integrating billing and payment systems with existing e-commerce platforms or accounting software can be complex. Streamlining the integration process and ensuring seamless data transfer is vital.
- Customer Disputes: Resolving billing disputes promptly and efficiently is essential to maintain customer satisfaction. Implementing clear and transparent billing policies and excellent customer support are robust solutions.
- Payment Failures: Addressing payment failures, such as declined transactions or technical issues, is crucial to prevent revenue loss. Employing automated retries, alternative payment methods, and proactive customer communication can help mitigate these challenges.
- Regulatory Compliance: Adhering to legal and industry regulations, such as PCI DSS (Payment Card Industry Data Security Standard), is essential to protect customer data and maintain trust. Regular audits and compliance monitoring are vital solutions.

Case Study

Online retailer XYZ is rapidly expanding and offers a wide variety of products. The company's clientele and sales volume have grown significantly over the past year. XYZ is implementing a new accounting and payment system to streamline its operations and enhance the customer experience. To serve its extensive customer base, the company must ensure its system is effective, secure, and capable of administering multiple payment methods.

1. At XYZ, payment disputes and chargebacks have increased recently. How can the organization mitigate these obstacles while maintaining a positive customer relationship?
2. XYZ desires to offer its customers various payment options, including credit card payments, PayPal, and digital wallets. What are the advantages and disadvantages of each payment method in terms of security, usability, and customer confidence?
3. As XYZ expands internationally, it must accommodate the various national payment preferences. How can the organization modify its billing and payment system to accommodate regional payment methods and currencies while preserving a consistent client experience?
4. XYZ is concerned with security, especially when storing and transmitting sensitive payment information from consumers. What encryption and security technologies should the organization employ to protect client data and prevent unauthorized access?
5. XYZ desires to earn a reputation for providing a reliable and quick invoice and payment system. What essential features and best practices could be implemented into the organization's system to provide a seamless payment process for clients while mitigating errors and delays?

Discussion

1. What are billing and payment systems in e-business, and how do they differ from traditional payment systems?
2. Why are effective billing and payment systems important for e-businesses?
3. How have billing and payment systems evolved over time to meet the needs of e-businesses?
4. What are some key features of effective billing and payment systems for e-businesses?
5. What are the most common types of payment systems used in e-business, and how do they work?
6. What are the advantages and disadvantages of different payment systems for e-businesses and their customers?
7. How do payment systems differ in terms of fees and processing times?

8. What factors should businesses consider when selecting a payment system for their e-business operations?
9. What are some best practices for ensuring secure transactions and preventing fraud in billing and payment systems?
10. How can businesses implement fraud detection and prevention measures in their payment systems?
11. What role does encryption and other security technologies play in billing and payment system security?
12. How can businesses balance the need for security with the need to provide a seamless payment experience for customers?
13. What are some common issues and challenges that can arise in billing and payment systems, such as payment disputes and chargebacks?
14. How can businesses effectively address these issues and challenges to maintain customer satisfaction and minimize losses?
15. What are some successful billing and payment systems, and what key features contribute to their success?
16. How can businesses adapt to changing trends and technologies in the billing and payment system landscape?

Multiple-Choice Question

1. What is the purpose of billing and payment systems in e-business?
   (a) To track customer orders
   (b) To process payments for goods and services
   (c) To manage inventory
   (d) To generate sales reports
2. Which of the following is not a common type of payment system in e-business?
   (a) Credit card payments
   (b) PayPal
   (c) Bank transfers
   (d) Cash on delivery
3. What is a disadvantage of credit card payments?
   (a) High transaction fees
   (b) Slow processing time
   (c) Limited global acceptance
   (d) Lack of security measures
4. Which of the following is an example of a digital wallet?
   (a) Visa
   (b) Mastercard
   (c) Venmo
   (d) American Express

5. What is an important consideration for ensuring secure transactions in billing and payment systems?
   (a) Publicly sharing customer payment information
   (b) Storing payment data in plain text
   (c) Implementing encryption and other security technologies
   (d) Using weak passwords for payment accounts
6. How can fraud be prevented in billing and payment systems?
   (a) Allowing anonymous payments
   (b) Implementing fraud detection measures
   (c) Sharing customer payment information with third parties
   (d) Using unsecured payment gateways
7. Which security technology is commonly used to protect sensitive payment data during transmission?
   (a) SSL (Secure Sockets Layer)
   (b) HTTP (Hypertext Transfer Protocol)
   (c) SMTP (Simple Mail Transfer Protocol)
   (d) IP (Internet Protocol)
8. What is a potential challenge in billing and payment systems that can lead to disputes?
   (a) Delayed order fulfillment
   (b) Offering multiple payment options
   (c) Providing accurate billing information
   (d) Offering discounts and promotions
9. What is a chargeback in the context of billing and payment systems?
   (a) A fraudulent transaction
   (b) A refund requested by a customer
   (c) A fee charged by a payment gateway
   (d) A delayed payment processing time
10. How can payment disputes be resolved in billing and payment systems?
    (a) Ignoring customer complaints
    (b) Offering partial refunds
    (c) Avoiding direct communication with customers
    (d) Implementing a clear and transparent dispute resolution process
11. What is a key feature of a successful billing and payment system?
    (a) Limited payment options
    (b) Lack of customer support
    (c) Quick and seamless payment processing
    (d) High transaction fees

12. Which of the following payment systems is known for its global acceptance and user-friendly interface?
    (a) Credit card payments
    (b) Bank transfers
    (c) Digital wallets
    (d) Cash on delivery
13. What is an advantage of using PayPal as a payment system?
    (a) High transaction fees
    (b) Limited merchant support
    (c) Quick and secure payment processing
    (d) Lack of buyer protection
14. How does encryption contribute to the security of billing and payment systems?
    (a) It increases transaction fees.
    (b) It slows down payment processing time.
    (c) It protects sensitive payment data from unauthorized access.
    (d) It limits the number of payment options available.
15. Which of the following is an example of a successful billing and payment system?
    (a) A system that frequently experiences payment disputes
    (b) A system with slow and unreliable payment processing
    (c) A system that provides 24/7 customer support and resolves issues promptly
    (d) A system that offers limited payment options and high transaction fees

References

1. Kshetri, N., *1 Blockchain's roles in meeting key supply chain management objectives.* International Journal of information management, 2018. **39**: p. 80–89.
2. Rittinghouse, J.W. and J.F. Ransome, *Cloud computing: implementation, management, and security*. 2016: CRC Press.
3. Turban, E., et al., *Electronic commerce 2018: a managerial and social networks perspective.* 2018: Springer.
4. Alzoubi, H., et al., *The effect of e-payment and online shopping on sales growth: Evidence from banking industry.* International Journal of Data and Network Science, 2022. **6**(4): p. 1369–1380.
5. Goel, R., *E-commerce*. 2007: New Age International.

Chapter 7
Supply Chain Management in E-Business

Abstract Supply chain management (SCM) controls the flow of goods and services from the start of manufacturing until the finished product is delivered to the consumer. In today's fast-paced corporate climate, the necessity of SCM for the success of e-business has expanded. The rise of e-commerce has disrupted the old supply chain paradigm, bringing new opportunities and difficulties. Understanding the key SCM components and how technology may be utilized to improve the process is critical for every e-business's success. This chapter is primarily concerned with SCM in e-business contexts. Because of the growing importance of e-commerce, the integration and coordination of various SCM components have gained prominence in this environment. This chapter focuses on how technology can improve the efficacy and efficiency of SCM operations. Furthermore, the chapter discusses SCM challenges in e-business and potential solutions. By the end of this chapter, the reader will have a better knowledge of the function of SCM in e-business and the essential elements impacting its performance.

1 Introduction to SCM in E-Business

SCM manages the flow of goods and services from the point of origin to the consumption site. In e-business, SCM refers to integrating information technology, logistics, and business processes to manage the flow of products and services in an online marketplace. SCM aims to create a streamlined and efficient supply chain that can deliver products quickly and affordably to consumers.

Inventory management is an integral component of SCM in e-commerce. Inventory management entails monitoring and regulating merchandise flow throughout the supply chain, from raw materials to finished products. In e-commerce, inventory management is crucial because it enables businesses to ensure they have the right amount of inventory at the right moment to meet customer demand. Inventory management can also help businesses save money by reducing excess stock and preventing stockouts.

H. Taherdoost, *E-Business Essentials*, EAI/Springer Innovations in Communication and Computing, https://doi.org/10.1007/978-3-031-39626-7_7

In e-business, logistics administration is another essential component of SCM. Logistics management is the planning, execution, and administration of the movement of goods and services along the supply chain. The importance of logistics management in e-business stems from the fact that it enables enterprises to deliver products to consumers promptly and efficiently. This is particularly important in the online economy, where customers expect prompt and reliable delivery.

In e-business, SCM also encompasses the application of information technology to manage the flow of products and services. E-businesses manage their supply chains utilizing several tools and technologies, such as enterprise resource planning (ERP) systems, customer relationship management (CRM) systems, and SCM software. These tools allow businesses to track inventory levels, monitor client orders, and manage the flow of products throughout the supply chain.

Controlling supply chain risk is one of the most significant SCM challenges in e-business. Interruptions in the supply chain, cyberattacks, and natural calamities are just a few of the risks that e-commerce businesses face. These risks have the potential to significantly affect the flow of products and services throughout the supply chain, as well as cause significant financial losses. To mitigate these risks, businesses need to develop supply chain risk management techniques, such as diversifying suppliers, instituting cybersecurity safeguards, and developing contingency plans [1].

SCM in e-business entails managing relationships with suppliers and consumers. Effective supplier and customer relationship management is necessary for a seamless and efficient supply chain. This requires establishing clear communication channels, building trust, and collaborating to identify and resolve supply chain challenges.

In conclusion, SCM is a fundamental aspect of e-commerce that requires integrating information technology, logistics, and business processes to regulate the flow of products and services in the online marketplace. Effective SCM enables businesses to quickly and affordably deliver products to customers, reduce inventory costs, and mitigate supply chain risk. However, managing supply chain risk and establishing solid relationships with suppliers and consumers are significant obstacles in e-business SCM. E-commerce companies need to devise solutions to these problems and build a supply chain capable of reaching the requirements of the online marketplace.

Successful e-commerce depends on efficient SCM. As the world becomes more digital, businesses turn to e-commerce to reach customers, expand their markets, and reduce costs. Managing the flow of products and services in an online environment presents unique challenges that demand meticulous planning, organization, and execution. This part will discuss the importance of outstanding SCM for e-businesses.

Good SCM is a prerequisite for satisfying customer demand. Customers anticipate rapid delivery times and superior products from e-businesses operating in a fast-paced, competitive environment. To meet these demands, e-businesses need efficient and dependable supply chains that deliver products and services on time and within budget. Suppliers, logistics providers, and other supply chain participants should work closely to ensure that products are available when and where they are needed.

Excellent SCM is crucial for e-businesses for the additional reason that it can aid in cost reduction. By streamlining supply chain processes, businesses can streamline operations, reduce inventory carrying costs, and enhance efficiency. This may result in reduced costs for both the business and its customers. Using data analytics and predictive modeling, for instance, businesses can better predict demand and manage inventory levels accordingly, reducing the need for excess inventory and the risk of stockouts.

In addition to ensuring product quality and safety, effective SCM is crucial for ensuring product quality and safety. Before reaching the consumer, e-commerce products frequently undergo several processes, including production, packaging, shipping, and handling. Any disruption in the supply chain could result in quality or safety issues, damaging the company's reputation and leading to dissatisfied customers. By implementing effective SCM strategies, businesses can ensure that their products are handled and transported safely and meet the quality standards demanded by their customers.

Moreover, effective SCM can help e-businesses respond rapidly to market shifts. Businesses that rapidly adapt to shifting consumer preferences and trends will likely thrive in the digital economy. Having a flexible and adaptable supply chain enables e-businesses to react to shifts in demand, the introduction of new products, and the competitive landscape. This can aid their efforts to remain competitive and maintain a good market position.

Effective SCM can also help online businesses reduce their environmental impact. As more consumers become concerned with sustainability and environmental responsibility, businesses demonstrating their dedication to these values are more likely to attract and retain customers. Using eco-friendly supply chain techniques, such as reducing packaging waste, optimizing delivery routes, and utilizing renewable energy sources, e-businesses can reduce their carbon footprint and differentiate themselves from competitors.

Effective SCM is crucial to the success of an e-business. By ensuring that products are delivered on time, at a reasonable price, and to the required quality standards, e-businesses can satisfy client demand and acquire a competitive edge. By streamlining supply chain procedures, e-businesses can reduce costs, boost efficiency, and quickly adapt to market shifts. By employing eco-friendly supply chain strategies, e-businesses can reduce their environmental impact and demonstrate their commitment to sustainability. Figure 7.1 demonstrates how important efficient SCM is to the success of e-businesses.

Fig. 7.1 Importance of effective SCM for e-businesses

2 Components of SCM in E-Business

SCM is an essential business process that includes the design, implementation, and management of the movement of products and services from their origin to their point of consumption. Effective SCM requires a comprehensive comprehension of the process's essential components. This essay will examine the most crucial aspects of SCM.

Planning is crucial to SCM because it entails developing strategies and methods to meet customer demand while optimizing resources. The significance of planning as a component of e-business SCM will be examined in this paper. Effective planning is necessary for e-businesses to satisfy their customers' demands while minimizing costs and maximizing profits. Planning entails recognizing client requirements and devising methods to meet them effectively. Demand planning, which entails forecasting future product and service demand, is frequently the initial step in this procedure. By accurately estimating demand, reducing costs, and ensuring on-time customer delivery, e-businesses can better manage their inventory and production operations. Planning for capacity is another essential aspect of SCM planning. Capacity planning involves analyzing the production capabilities of an e-business and identifying areas where capacity can be increased or enhanced. This may necessitate the purchase of new equipment, the hiring of additional personnel, or the outsourcing of particular processes. By effectively managing capacity, e-businesses can ensure they can meet demand and avoid bottlenecks or delays in the supply chain.

Transportation planning is an additional essential aspect of SCM planning. This procedure includes identifying the most efficient and cost-effective means of transporting products from the manufacturer to the customer. This may involve determining the best mode of conveyance, such as air, sea, or land, and the optimal delivery routes. Logistics management, such as customs clearance, storage, and tracking, is included in transportation planning to ensure that goods are delivered on time and in excellent condition to customers. Effective SCM planning includes risk and contingency management. This may involve preparing for unforeseen events such as natural disasters or supply chain disruptions. By anticipating and preparing for potential risks, e-businesses can lessen the impact of such disasters on operations and maintain supply chain continuity.

Good SCM planning has numerous benefits, including increased efficiency and cost-effectiveness. By anticipating demand, managing capacity, and optimizing transportation routes, e-businesses can reduce operational expenses and enhance profitability. Effective planning can also benefit e-businesses by reducing lead times, increasing inventory turnover, and enhancing customer satisfaction, contributing to increased sales and repeat business. In addition to these benefits, effective SCM planning can help e-businesses achieve their sustainability objectives. By optimizing conveyance routes and reducing unnecessary traffic, e-businesses can lower their carbon footprint and contribute to a more sustainable future. E-commerce organizations can also benefit from effective planning to reduce waste, minimize environmental impact, and foster ethical sourcing practices.

Purchasing is a crucial component of SCM for online enterprises. It identifies, analyzes, and selects suppliers for the firm's required products or services. Effective procurement can benefit e-businesses by lowering costs, enhancing quality, and ensuring on-time delivery of products to customers. This part examines procurement as a component of SCM for e-businesses and analyzes the key strategies involved.

The first step in e-business sourcing is identifying the products and services required to operate the business. This involves analyzing consumer demand for the products or services and any special requirements or preferences. After identifying the requirements, the e-business can investigate potential suppliers to meet them. Potential suppliers are evaluated based on their capabilities and suitability to meet e-business requirements. This includes evaluating the supplier's financial stability, quality control procedures, and on-time delivery capacity. E-commerce businesses also need to consider the supplier's location and ability to provide the requisite support and service. After evaluating potential suppliers, e-businesses can begin contract negotiations with the selected vendors. Important terms and conditions, including pricing, delivery, and payment terms, should be agreed upon. The objective is to reach a mutually beneficial agreement that meets the requirements of both parties. After negotiated and finalized contracts, e-businesses can begin managing their relationships with suppliers. This includes monitoring supplier performance and addressing any problems that arise. E-businesses need also to establish strong connections with their suppliers to meet the ever-changing demands of the company.

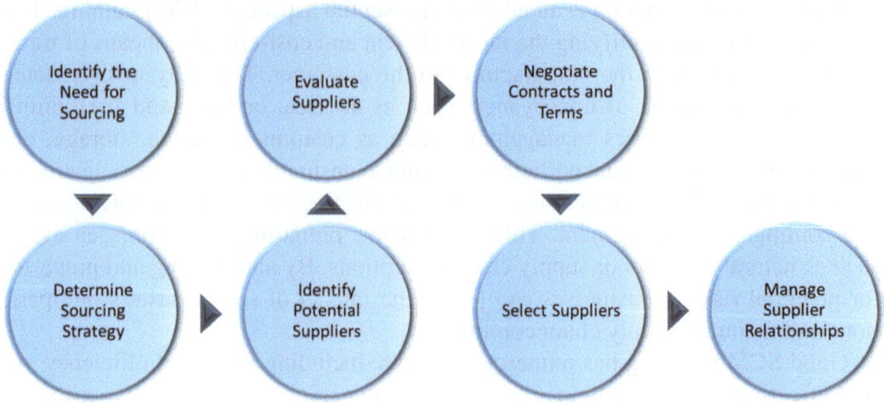

Fig. 7.2 Sourcing process in SCM for e-businesses

Successful e-commerce depends on efficient procurement. It can assist businesses in reducing costs, enhancing product quality, and ensuring on-time customer delivery. However, sourcing can be a time-consuming and complex process. E-commerce companies should thoroughly evaluate prospective suppliers and negotiate contracts that meet their needs. They also need to manage their supplier relationships to ensure that they can meet the evolving business requirements. Figure 7.2 demonstrates the sourcing process involved in electronic SCM.

Inventory management plays a critical role in the context of SCM for online businesses. It requires keeping a check on commodity flow as it moves from suppliers to warehouses and, finally, to end consumers. To correctly manage their inventory, a company needs to find a means to strike a balance between having enough product on hand to meet their customers' needs and eliminating surplus inventory, which can lead to higher prices. Keeping inventory quantities in real time is one of Internet business's most challenging parts of inventory management. Because of the volatility of online sales, businesses need to be able to adjust their inventory levels to meet changing client demands quickly. You need a solid inventory management system that provides real-time visibility into inventory levels to accomplish this. Another critical component of good inventory management is forecasting future demand. Businesses need to have an accurate understanding of future demand to ensure that they have enough inventory on hand to meet the needs of their customers. This includes analyzing past sales data, market trends, and other client demand factors.

Inventory management necessitates keeping track of the expenditures associated with the inventory. This covers procuring, storing, and paying workers to handle inventory. Online businesses need to exercise considerable caution in managing these expenses to keep them from eating into their profits. Inventory turnover management is an essential aspect of inventory management. This relates to how quickly a product is drained and replenished after being sold. A high inventory turnover suggests that items are selling quickly and efficiently, whereas a low inventory turnover may indicate that products are not selling as well as expected. E-commerce

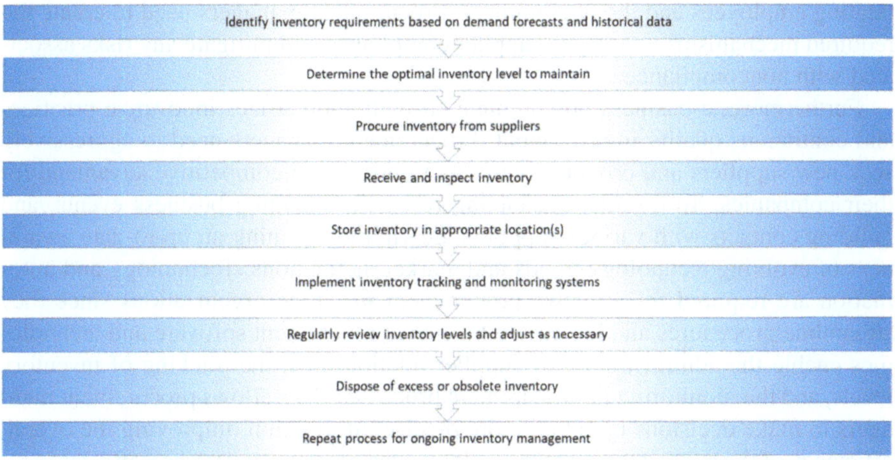

Fig. 7.3 Inventory management process in e-business SCM

businesses need to evaluate how inventory management may affect their supply chain partners. Strong relationships with suppliers and logistical partners are required for efficient inventory management. These collaborations are essential to ensuring inventory is delivered on schedule and in good shape. Figure 7.3 depicts the inventory management process in the context of e-business SCM.

Procurement is a critical component of SCM in e-business that should not be disregarded. It comprises purchasing commodities and services from third-party vendors to support producing and distribution of goods or services. Procurement is a crucial function in e-business because it ensures that the company has access to the resources it requires to meet the needs of its consumers. To guarantee that the company remains competitive in the market, the procurement process in an e-business needs to strike a careful balance between cost, quality, and speed. One of the most essential purposes of procurement in e-business is to ensure that the company has a consistent supply of high-quality materials and services. To ensure that suppliers can consistently offer the goods and services the firm requires, procurement managers need to build strong relationships with individuals who provide the company's goods and services. Negotiating contracts and rates, monitoring supplier performance, and managing any challenges or conflicts that may arise along the route are all part of the job.

Managing the costs of acquiring goods and services is another critical component of e-business procurement. To ensure that the company remains competitive in the market, procurement managers need to balance the cost of goods and services and the quality of the products. To do this, procurement strategies should focus on cost reduction, value creation, and risk mitigation. Procurement is critical in an e-business scenario for controlling spending and ensuring that the organization complies with all applicable legislation and requirements. This includes ensuring suppliers comply with legislative obligations and uphold ethical standards regarding treating employees and the environment. Procurement managers need to create the required mechanisms to analyze supplier compliance and mitigate any risks associated with noncompliance.

Furthermore, e-business procurement strongly emphasizes innovative practices and continuous quality improvement. Procurement managers need to aggressively seek new suppliers and products regularly to preserve a competitive advantage for their companies. To do this, conducting research, attending business events, and building contacts with various suppliers are vital to retaining an up-to-date awareness of evolving technology trends and market inclinations. Technology and automation are required for effective procurement in online organizations since they streamline procedures and increase efficiency. Procurement software and technologies enable the administration of supplier relationships, the tracking of inventory levels, and the monitoring of procurement processes. This allows procurement managers to make decisions based on more precise information, improving the overall efficiency of the procurement process. Finally, the procurement process in e-business necessitates a collaborative effort by procurement managers, suppliers, and other relevant stakeholders. To ensure that the purchasing procedure aligns with the company's overarching aims, procurement managers need to work closely with the heads of other departments, such as those in charge of production and logistics. They also need to establish excellent relationships with the firms that supply them and collaborate to identify areas that require development and implement improvements that benefit both the company and the supplier. Figure 7.4 illustrates the procurement process within the SCM of e-business, highlighting the various steps involved in the procurement process.

E-business SCM relies heavily on distribution and storage. Warehouses store and handle things until they are needed. Transporting items from the warehouse to their final destination is what distribution includes. Warehousing and distribution are critical in ensuring the timely and effective delivery of products to clients in e-business. One of the primary benefits of warehousing in e-commerce is the ability to store things closer to the consumer. Because products can be sent directly from the warehouse to the consumer, shipping costs and delivery times can be reduced.

Furthermore, warehousing can assist in ensuring that products are available when needed. This is especially important in e-commerce, where customers expect on-time delivery and reliable service. Another critical aspect of e-business SCM is distribution. The distribution aims to carry items from the warehouse to their final destination, such as retail outlets or individual clients. Effective distribution systems can assist businesses in cutting delivery times and costs while guaranteeing that product arrive in good condition.

One of the critical challenges of e-business distribution is managing a massive volume of minor orders. A high level of coordination and organization is required to ensure that things are delivered on time and at the correct location. To provide precise delivery information to customers, e-commerce sites also need to be able to track the status of each purchase in real time.

Adapting to changing customer expectations is another critical distribution difficulty in e-business. This demands an adaptive distribution infrastructure that

Fig. 7.4 Procurement in e-business SCM

quickly responds to changes in demand. E-commerce companies need to be able to scale up or down their distribution operations in response to changes in consumer demand. Effective warehouse and distribution management is critical for the success of e-businesses. E-commerce companies can cut costs, enhance delivery times, and improve the overall consumer experience by improving their supply chain processes. E-businesses need to invest in contemporary warehouse and distribution systems and staff with the expertise to handle these complex processes to achieve this.

Transportation is an integral part of e-business SCM because it guarantees that things are moved from one site to another promptly and efficiently. It entails the movement of raw materials, completed commodities, and products from suppliers to manufacturers and distributors and finally to the end user. Effective transportation management is necessary for the supply chain to run smoothly and for products to arrive on time and in excellent condition. One of the critical benefits of transportation in SCM is that it allows organizations to control inventory levels more effectively. Transportation enables the flow of items from suppliers to manufacturers and distributors, allowing ideal inventory levels to be maintained. Businesses can reduce the inventory they need to store by using transportation to transfer things as needed, lowering costs and enhancing efficiency.

Another significant advantage of transportation in SCM is that it allows businesses to respond more quickly to consumer demand. Businesses can respond quickly to fluctuations in demand, such as an increase in demand for a particular product or a change in consumer preferences, by maintaining a dependable transportation network. This lets companies change their manufacturing and distribution plans in real time, improving their market competitiveness. Transportation is also crucial in e-business to ensure that products are delivered to customers quickly and efficiently. E-commerce enterprises need to rely on transportation networks to deliver products to customers frequently. Effective transportation management is essential to ensure that products are delivered on time and in outstanding shape, which can boost customer loyalty and overall satisfaction.

However, transportation can cause problems in e-commerce. E-commerce businesses, for example, need to deal with a range of transportation choices, including air, sea, and land transportation. Choosing the best form of transportation for each cargo may be a complicated and time-consuming process, particularly for companies that send items globally. Furthermore, e-commerce businesses should deal with customs clearance, tariffs, and other regulations affecting cross-border product transportation.

Customer service is an essential component of SCM in the e-business industry. As more people shop online, it is more important than ever to deliver a great customer experience. Superior customer service is more likely to beat competitors, becoming a crucial point of differentiation in the e-commerce business. E-business customer care requires a coordinated supply chain-wide effort. Each interaction needs to be methodically controlled from when a consumer puts in an order until the product is delivered to ensure a good experience. This comprises the speed and accuracy with which orders are processed, the timeliness with which they are delivered, and the clarity with which they are communicated along the process.

One of the critical advantages of customer service in SCM is increased client loyalty. Customers who believe they have been handled well are more inclined to make additional purchases. This can lead to higher income for the organization and good referrals to their peers from delighted clients. Another critical advantage of customer service in SCM is increased operational efficiency. Businesses may limit the number of queries and complaints by providing consumers with clear and accurate information, saving time and resources.

Furthermore, efficient customer service can reduce the likelihood of supply chain errors or blunders, resulting in costly returns or replacements. In e-commerce, customer service is critical in managing product returns and exchanges. E-commerce businesses need to have clearly defined rules and procedures to handle returns and exchanges efficiently and effectively. Businesses may reduce the friction associated with returns and exchanges by offering clear information and user-friendly procedures, enhancing customer satisfaction and saving expenses.

However, providing excellent customer service in an e-business environment can take time and effort. E-commerce businesses need to manage customer care across numerous channels, including email, phone, and social media. To manage well, significant resources and specialist skills may be required. E-commerce companies also need to manage client delivery time expectations, which can be hampered by foreign shipping and supply chain delays.

In the e-business era, data analytics has evolved as a critical component of SCM. Businesses that can adequately evaluate the massive amounts of data generated by online transactions are better positioned to make educated decisions and optimize supply chain operations. E-commerce enterprises can improve productivity, cut costs, and improve the customer experience by leveraging data analytics. One of the critical advantages of data analytics in SCM is demand forecasting. E-commerce enterprises can develop more accurate demand projections by analyzing previous sales data and other relevant variables, lowering inventory costs, and improving order fulfillment.

Furthermore, data analytics can assist businesses in identifying consumer behavior trends and patterns, which can then be used to inform product development and marketing strategies. Another critical benefit of data analytics in SCM is improved operational efficacy. E-commerce enterprises can detect bottlenecks and inefficiencies in their supply chains and design plans to optimize operations by studying supply chain data. This could include improving transportation routes, warehouse layout, and shortening lead times. E-commerce enterprises can lower costs while increasing customer happiness by increasing their efficiency.

In e-business, data analytics is critical for managing supply chain risks. E-commerce enterprises can detect potential hazards and design mitigation plans by evaluating data from suppliers, logistics providers, and other partners. This may entail identifying alternative suppliers, creating contingency plans, and developing risk management frameworks. However, implementing effective data analytics in SCM can be difficult. E-commerce companies need the technical expertise and infrastructure to gather, process, and analyze vast volumes of data in real time. Furthermore, data privacy and security should be carefully managed to ensure that consumer data is protected and complies with applicable regulations.

Effective SCM is critical for success in the e-business world. With the increasing complexity of supply chain networks and rising consumer expectations, e-commerce enterprises need to improve their supply chain operations by using best practices. This section will look at the best practices for e-commerce SCM. Real-time inventory management is one of the most effective best practices for optimizing SCM in e-commerce. E-commerce enterprises may ensure that products are available when customers need them by implementing systems that enable real-time visibility into inventory levels. Real-time inventory management can also help organizations cut inventory expenditures and avoid stockouts. Effective order management is another critical best practice for optimizing SCM in e-businesses. E-commerce enterprises can ensure that orders are completed quickly and correctly by using systems that automate order processing and tracking.

Furthermore, effective order management can lower order processing costs while increasing order fulfillment rates. Transportation management is an integral part of SCM in e-companies, and using simplified transportation management strategies can help organizations improve delivery times while lowering costs. This includes optimizing shipping routes, choosing the best carriers, and utilizing technology for real-time cargo tracking. Adequate warehousing and distribution are critical for ensuring product availability for order fulfillment and shipment. E-commerce businesses should apply best practices for warehousing and distribution, such as optimizing warehouse layouts, implementing automation technologies for material handling and picking, and establishing an efficient distribution network to save transportation costs. Collaboration and communication with suppliers and logistical partners are also required for e-commerce companies to optimize their SCM. Establishing good connections with suppliers and logistical partners and employing collaborative technological solutions can help e-commerce enterprises improve supply chain visibility and responsiveness. Data analytics is a critical component of SCM in e-businesses. Companies can use data analytics tools and methodologies to get insights into supply chain performance, identify areas for improvement, and optimize supply chain operations. This includes assessing key performance indicators (KPIs) like order fulfillment rates, inventory turnover, and transportation costs and using the results to drive continuous improvement. Table 7.1 presents a set of recommended strategies for maximizing SCM efficiency in electronic businesses.

Table 7.1 Best practices for optimizing SCM in e-businesses

Challenge	Best practices
Inventory management	Implement real-time inventory management systems, prioritize demand forecasting, and establish safety stock levels to avoid stockouts
Order fulfillment	Automate order processing and tracking, streamline returns and exchanges, and establish clear customer communication
Transportation management	Optimize transportation routes, select reliable carriers, use technology to track shipments, and consider alternative shipping methods
Warehouse and distribution	Optimize warehouse layout and equipment, implement automation technologies for material handling and picking, and leverage distribution networks to reduce costs
Supplier management	Establish strong relationships with suppliers, set clear performance expectations, and use collaborative technology to improve communication and coordination
Data analytics and insights	Use data analytics tools to gain insights into supply chain performance, identify inefficiencies and areas for improvement, and make data-driven decisions

3 Use of Technology to Improve SCM

The usage of technology has changed numerous industries, including SCM. Technology integration has substantially increased the efficiency and efficacy of supply chain operations. By leveraging technology, businesses can save money, improve procedures, and increase supply chain visibility. This part looks at how technology can be used to increase the efficacy and efficiency of SCM. Automated systems are one way that technology is being utilized to improve SCM. With the help of automated solutions, the quantity of manual labor necessary for SCM operations can be reduced. These systems can automate various processes, including order processing, inventory management, and delivery. This could increase accuracy, lower labor costs, and significantly save time. Data analytics is another way that technology is being utilized to improve SCM. Businesses can utilize data analytics to improve their supply chains' efficiency and identify improvement areas. Businesses may enhance the effectiveness of their supply chains, cut costs, and increase customer satisfaction by analyzing data from several sources, such as sales data, logistics data, and customer data.

The application of technology has the potential to increase supply chain visibility. Businesses can gain real-time access to the status of their inventory, orders, and shipments by utilizing tracking systems and other digital solutions. This insight can assist firms in better understanding possible problems and taking proactive efforts to address them, enhancing supply chain responsiveness and minimizing the chance of interruptions. Technology can help improve supply chain communication and visibility. Cloud-based technologies and electronic data interchange (EDI) enable suppliers, manufacturers, distributors, and customers to cooperate quickly. Businesses may manage their supply chain relationships more efficiently, boost supply chain efficiency, and save costs through enhancing collaboration and communication. Technology allows supply chain data to be more accurate and dependable. Companies can record precise and trustworthy data about their inventory, orders, and shipments using barcoding, radio-frequency identification (RFID), and automated data capture. Using this information to make better supply chain decisions can reduce the chance of errors and enhance overall supply chain performance [2].

Finally, by utilizing technology, supply networks may be made more sustainable. Businesses can lower their carbon footprint and boost the environmental sustainability of their operations by adopting eco-friendly technology such as electric vehicles and renewable energy sources into their supply chains. Furthermore, with the help of technology solutions such as predictive analytics, businesses can optimize supply chain routes and reduce transportation-related emissions. Finally, the use of technology has significantly increased the efficiency and efficacy of SCM. Table 7.2 lists some technological solutions that can be used to improve the effectiveness of SCM. Modern SCM has become more reliant on technology, from automating procedures to improving data quality and visibility to encouraging collaboration and sustainability. As organizations investigate and employ new technology to become more productive, adaptive, and sustainable, the SCM sector will continue to grow. Figure 7.5 depicts numerous technological methods for improving the efficiency and sustainability of SCM.

Table 7.2 Technology solutions for improving SCM

Technology solution	Benefits
Automated systems	Reduce manual workload, improve accuracy, save time, and reduce labor costs
Data analytics	Gain insights, optimize operations, reduce costs, and improve customer satisfaction
Tracking systems	Improve visibility, identify potential issues early, improve responsiveness, and reduce the risk of disruptions
Collaboration tools (e.g., cloud-based platforms, EDI)	Improve communication, manage relationships, improve efficiency, reduce costs
Barcoding, RFID, and automated data capture	Improve data accuracy and reliability, make more informed decisions, and reduce the risk of errors
Sustainable technologies (e.g., electric vehicles, renewable energy sources)	Reduce carbon footprint, improve environmental sustainability
Predictive analytics	Optimize supply chain routes, reduce transportation-related emissions

Businesses are increasingly employing SCM solutions to streamline processes and increase productivity. This section will examine several real-world examples of e-businesses utilizing SCM technologies to streamline operations and obtain a competitive advantage.

Amazon is a well-known example of a company that has revolutionized the e-commerce industry by employing cutting-edge SCM systems. Automation and robotics technology support Amazon's expansive network of depots and distribution centers. This involves using automated guided vehicles (AGVs) and conveyor systems to transfer products throughout warehouses and fulfillment centers. In addition, Amazon has optimized its inventory management and order fulfillment procedures by employing cutting-edge algorithms and analytics tools. Using these technologies, Amazon can provide rapid and efficient delivery times to customers around the globe.

A Spanish clothing retailer, Zara is another online business adapting SCM systems effectively. Zara monitors demand and adjusts production accordingly using cutting-edge data analytics and real-time inventory monitoring systems. By utilizing these technologies, Zara can rapidly adapt to altering market trends and deliver new products on schedule. Zara has implemented RFID technology to track its products throughout the supply chain and quickly locate and replenish inventory as required.

Walmart is another online company that has implemented SCM tools to streamline operations. Walmart optimizes its supply chain operations using a sophisticated inventory management system that integrates real-time data analytics with cutting-edge algorithms. The company also monitors its inventory and improves supply chain visibility using RFID technology. To improve its delivery capabilities, Walmart has implemented cutting-edge logistics technologies, such as its private fleet of trucks and an extensive network of distribution centers.

Alibaba, a Chinese e-commerce colossus, is another e-business that has successfully implemented SCM systems. Alibaba streamlines its supply chain processes

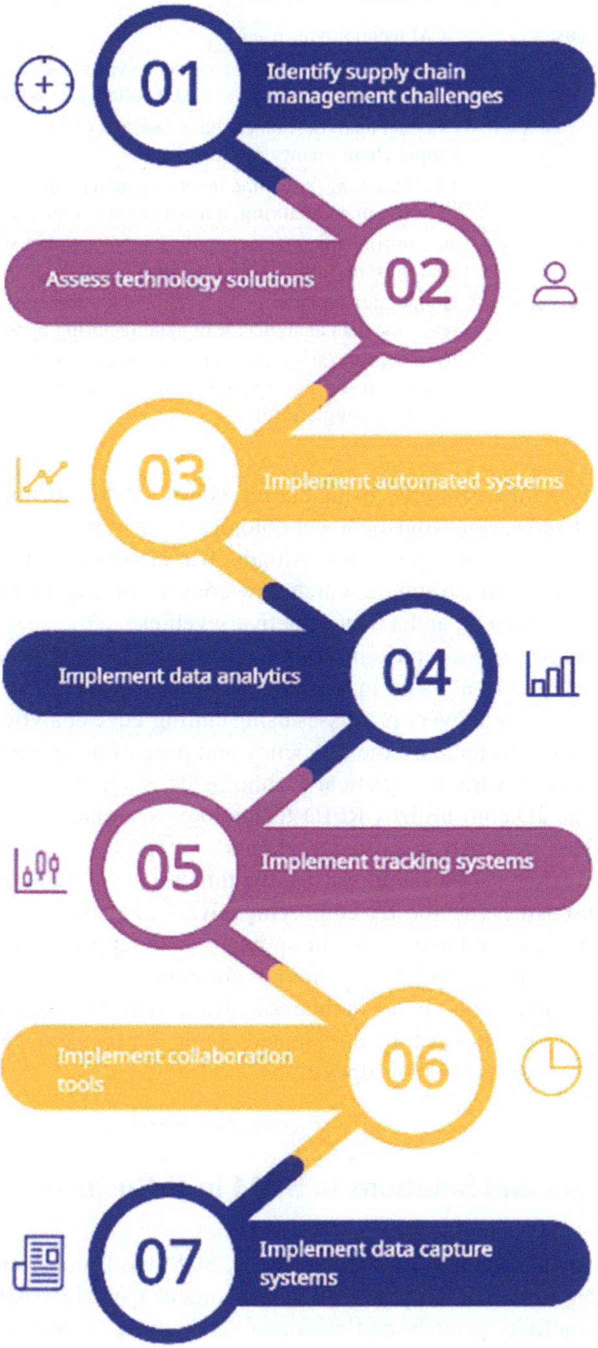

Fig. 7.5 Technology solutions for improving SCM efficiency and sustainability

Table 7.3 E-businesses that utilize various SCM technologies in their operations

Company	Industry	SCM technologies used
Amazon	Retail	Automated inventory management systems, RFID tracking, predictive analytics, cloud-based collaboration tools
Alibaba	E-commerce	Big data analytics, blockchain, Internet of Things (IoT) devices, supply chain finance platform
Zara	Fashion	RFID tracking, real-time inventory management systems, in-house manufacturing, quick-response supply chain
Dell	Technology	Just-in-time manufacturing, direct-to-consumer sales model, data analytics, real-time order tracking
Procter & Gamble	Consumer goods	Collaborative planning, forecasting, and replenishment (CPFR), real-time data analytics, lean manufacturing principles
UPS	Logistics	Route optimization algorithms, telematics, GPS tracking, automated sorting and packaging systems, cloud-based logistics management platform

and expedites consumer deliveries using cutting-edge analytics techniques. The corporation has utilized various logistical technologies, such as robotics and automated warehouses, to expedite its operations. Alibaba has also made substantial investments in its logistics infrastructure, which now consists of a vast network of warehouses, fulfillment centers, and a fleet of delivery vehicles.

JD.com could be used as an example of a last resort. This Chinese e-commerce colossus has emerged as a leader in applying technologies for SCM. JD.com optimizes the efficacy of its delivery processes using cutting-edge analytics and machine learning techniques. To increase the efficiency and pace of its operations, the company has invested in various logistical technologies, such as drone delivery and robotics systems. JD.com utilizes RFID technology to monitor its inventory and ensure that orders are timely delivered to clients.

These real-world examples demonstrate the importance of SCM solutions to the growth of the e-business sector. By employing advanced analytics, robotics, RFID, and other technologies, e-businesses can optimize their supply chain processes and gain a competitive advantage. As the market continues to develop and new technologies emerge, an increase in the number of advanced SCM systems implemented by e-businesses worldwide can be anticipated. Table 7.3 lists e-businesses and the SCM technology they use in their operations.

4 Challenges and Solutions in SCM in E-Business

In the quick-paced digital environment of today, SCM has become more significant in e-commerce. Effective supply chain management can aid businesses in cost reduction, productivity growth, and customer satisfaction enhancement. However, e-commerce SCM can come with some difficulties. This section looks at some issues and challenges that e-businesses encounter with SCM.

Lack of insight into the supply chain is one of the most critical issues facing e-businesses in SCM. Many e-businesses have extensive supply chains that involve numerous suppliers, manufacturers, and distributors. As a result, it could be challenging to have a complete picture of the supply chain, which could lead to inefficiencies and delays. E-commerce businesses need to invest in technology that offers real-time visibility into their supply chains to overcome this difficulty.

E-businesses also need to manage their supply chains to survive the intense competition in the e-commerce sector. E-businesses need to constantly innovate and set themselves apart from the competition in a market with so many market participants. This cannot be easy because e-businesses need to continuously enhance their supply chain processes, incorporate new technologies, and forge solid relationships with their suppliers and logistics providers.

The intricacy of global supply networks is another obstacle that e-commerce needs to get beyond. Customs brokers, freight forwarders, and carriers are only a few partners involved in international supply chains, which are also governed by some rules and compliance standards. To avoid expensive delays and fines, e-businesses should comprehend the regulatory and compliance hurdles in managing an international supply chain [3].

For e-businesses, SCM is more challenging due to the possibility of supply chain disruptions. Various events, including transportation problems, labor disputes, and natural disasters, can bring on supply chain hiccups. E-businesses need to have backup suppliers, backup inventory, and contingency transportation providers in place to minimize the effects of these disruptions.

The management of the supply chain for e-businesses needs to include data management. As SCM technology grows more widespread, e-businesses are producing and collecting massive volumes of data from numerous sources, including sales, logistics, and customer data. Managing and comprehending this data can be challenging, especially for small- and medium-sized e-businesses who need more funds to invest in sophisticated data management tools.

Not least, e-commerce businesses need to be mindful of the ethical and social responsibility issues with SCM. E-commerce businesses need to ensure that their supply chain procedures adhere to these standards since consumers are becoming more concerned about labor exploitation, environmental sustainability, and ethical sourcing. This can be challenging, particularly for e-businesses that depend on several suppliers and logistics companies, but preserving a positive brand reputation is necessary.

Finally, SCM in e-businesses can be challenging due to the complexity and rivalry between supply chains and the e-commerce industry. To effectively manage their supply chains, e-businesses need to make technology investments, cultivate positive working relationships with suppliers and logistics providers, and be aware of legal, ethical, and compliance requirements. E-commerce businesses may raise the effectiveness and efficiency of their supply chains, reduce costs, and boost consumer satisfaction by overcoming these obstacles. Table 7.4 shows the numerous challenges and impediments that SCM faces in e-business.

Table 7.4 Challenges and issues in SCM for e-businesses

Challenge	Description	Solution
Lack of visibility	Difficulty in gaining a complete picture of the supply chain due to multiple parties involved	Invest in technology that provides real-time visibility into the supply chain
High competition	They need to innovate and differentiate themselves to stay ahead constantly	Continually improve supply chain processes, adopt new technologies, and build strong relationships with suppliers and logistics providers
International supply chains	Involves multiple parties and is subject to various regulations and compliance requirements	Have a clear understanding of the regulatory and compliance issues involved in international SCM
Supply chain disruptions	This can occur due to various factors and lead to delays	Have contingency plans in place, such as alternative suppliers, backup inventory, and contingency logistics providers
Data management	Large volumes of data are generated and collected from various sources, but managing and analyzing this data can be challenging	Invest in sophisticated data management tools

An essential part of e-business operations, SCM calls for careful planning, execution, and monitoring. When it comes to managing their supply chains, e-businesses, however, confront some difficulties and problems, including a lack of visibility, competitiveness, international legislation, supply chain interruptions, data management, and ethical and social responsibility issues. This section examines several strategies e-commerce businesses can employ to overcome these challenges and enhance their SCM.

One of the essential strategies for improving supply chain management in e-businesses is to invest in technology that allows real-time visibility into the supply chain. E-businesses can use various technology solutions, such as cloud-based platforms, artificial intelligence, and the Internet of Things (IoT), to gain more visibility into their supply chains. These solutions enable e-commerce businesses to swiftly identify and address supply chain issues by providing real-time inventory levels, order status, and logistical movement data. Another crucial strategy for improving SCM in e-businesses is to forge great connections with suppliers and logistics service providers. By closely collaborating with their partners, e-businesses can uncover opportunities for improvement, share information and data, and plan and execute the supply chain. E-commerce can encourage suppliers and logistics providers to meet performance goals and quality standards, which can enhance the overall effectiveness and efficiency of the supply chain.

By staying current with regulatory and compliance standards, e-commerce enterprises can adopt a proactive strategy for managing international supply chains. This entails conducting routine risk assessments, keeping track of regulatory changes, and creating backup plans during a disruption. By working closely with customs brokers, freight forwarders, and carriers, e-businesses can manage the complexities of international supply chains and ensure compliance with local rules and laws [2].

E-commerce businesses can use some strategies, such as adding redundancy to their supply chain networks, expanding their supplier base, and investing in alternative logistics providers, to lower the risk of supply chain interruptions. E-businesses can also employ data analytics and predictive modeling to spot impending disruptions and take proactive measures to lessen or mitigate their effects.

Effective data management is also necessary for enhancing SCM in e-businesses. E-commerce businesses can gain insights into their supply chains' performance, identify improvement areas, and make data-driven decisions using business intelligence and data analytics solutions. By investing in data management systems and procedures, e-businesses can boost the effectiveness of their supply chains, reduce costs, and gain a competitive advantage.

Finally, e-businesses need to prioritize ethical and social responsibility concerns in their SCM strategies. This includes developing sustainable and ethical sourcing practices, ensuring workers are treated fairly and with dignity, and working with suppliers and logistical providers who share their values. By setting and upholding moral and social standards, e-businesses can enhance their brand's reputation, draw in and maintain customers, and help create a more just and sustainable global supply chain.

To sum up, managing the supply chain is a complex and challenging task for e-businesses, but some solutions can be put into place to get around these challenges and increase the efficiency and effectiveness of the supply chain. By investing in technology, developing strong partnerships, staying current with regulatory and compliance requirements, adopting proactive risk management strategies, prioritizing data management, and putting ethical and social responsibility issues first, e-businesses can increase their competitiveness, cut costs, and improve their brand image and reputation. Table 7.5 lists SCM enhancement strategies in the framework of e-businesses.

Table 7.5 Strategies for improving SCM in e-businesses

Strategies for improving SCM in e-businesses	Description
Investing in technology	Gain real-time supply chain visibility using cloud-based platforms, AI, and IoT
Establishing solid relationships with suppliers and logistics providers	Share information and data, work together to design and implement the supply chain, and look for areas for improvement
Staying up-to-date with regulatory and compliance requirements	Conduct regular risk assessments, track regulatory changes, and develop contingency plans
Adopting risk management strategies	Increase supply chain network redundancy, diversify supplier base, invest in alternative logistics providers, and employ data analytics and predictive modeling to identify potential disruptions
Prioritizing data management	To get insights into supply chain performance, find areas for improvement, and make data-driven decisions, use data analytics and business intelligence solutions

Summary

Definition of Supply Chain Management in E-Business

- Supply chain management (SCM) in e-business refers to coordinating and integrating activities involved in the flow of products, services, information, and finances within a digital business environment.
- It encompasses the entire lifecycle of a product or service, from sourcing raw materials to delivering the final product to customers using electronic technologies and online platforms.

Components of Supply Chain Management in E-Business

- Procurement: The process of sourcing and purchasing raw materials, components, and finished goods from suppliers, which involves online procurement platforms and digital payment systems
- Inventory Management: Efficiently managing and tracking stock levels and ensuring the availability of products through online inventory management systems
- Warehousing and Distribution: Storing, organizing, and distributing products through online fulfillment centers and logistics networks
- Transportation and Logistics: Optimizing the movement of goods through digital tracking, real-time visibility, and efficient routing systems
- Demand Planning and Forecasting: Utilizing data analytics and online tools to predict customer demand and plan inventory levels accordingly
- Order Fulfillment and Customer Service: Ensuring timely order processing, accurate order fulfillment, and excellent customer service through online platforms and automated systems
- Returns Management: Managing product returns and exchanges efficiently through online return portals and reverse logistics processes

Integration of Technology in Supply Chain Management in E-Business

- E-commerce Platforms: Integration of online marketplaces and e-commerce websites to facilitate seamless transactions, enable product catalog management, and provide customer support
- Enterprise Resource Planning (ERP) Systems: Utilizing ERP software to integrate and manage various business processes, including procurement, inventory management, and order fulfillment
- Supply Chain Visibility Tools: Implementing online tools and systems that provide real-time visibility into the supply chain, allowing stakeholders to track and monitor inventory, shipments, and delivery status
- Internet of Things (IoT): Using IoT devices and sensors to collect and share data throughout the supply chain, enabling better monitoring of inventory levels, product quality, and shipment conditions
- Artificial Intelligence (AI) and Machine Learning (ML): Applying AI and ML algorithms to analyze data, optimize supply chain operations, and make accurate demand forecasts

- Blockchain Technology: Leveraging blockchain for secure and transparent supply chain transactions, traceability, and verification of product authenticity

Challenges and Solutions in Supply Chain Management in E-Business

- Integration Complexity: Overcoming challenges related to integrating different software systems, platforms, and stakeholders in the digital supply chain
- Cybersecurity Risks: Addressing security threats and vulnerabilities in e-business supply chains to protect sensitive data, prevent data breaches, and ensure secure transactions
- Scalability and Flexibility: Adapting the supply chain to accommodate fluctuations in demand, changing customer preferences, and scaling operations to meet growth requirements
- Data Management and Analytics: Establishing robust data management practices and utilizing advanced analytics tools to extract meaningful insights for informed decision-making
- Collaboration and Communication: Enhancing collaboration and communication among supply chain partners through online platforms, shared data, and real-time information exchange
- Sustainability and Ethical Considerations: Addressing environmental concerns, promoting responsible sourcing and production, and ensuring ethical practices throughout the e-business supply chain

Case Study

Company XYZ is an online retailer that carries a wide selection of products. They collaborate with multiple suppliers and logistical partners to fulfill requests for a global clientele. The company aims to establish a streamlined and efficient supply chain management system to ensure on-time delivery of high-quality products. However, they are experiencing supply chain management issues and are seeking to improve their operations.

1. Which factors should Company XYZ prioritize when selecting e-business suppliers to maintain a stable and effective supply chain?
2. How can Company XYZ effectively manage inventory levels in its e-commerce operation to avoid stockouts and transport costs while meeting customer demand?
3. How can Company XYZ improve its e-commerce logistics operations, focusing on increasing order fulfillment and delivery speed?
4. How can company XYZ use technology, such as RFID or blockchain, to enhance supply chain visibility and traceability, resulting in greater customer satisfaction?
5. What challenges might Company XYZ face in maintaining product quality control in their e-commerce, and what steps can they take to ensure quality consistency throughout their supply chain?

Discussion

1. How is supply chain management different in the context of e-business compared to traditional businesses?
2. What are some key benefits of effective supply chain management for e-businesses?
3. How can e-businesses leverage supply chain management to gain a competitive advantage?
4. How does sourcing impact the overall supply chain management process for e-businesses?
5. What role does inventory management play in supply chain management for e-businesses?
6. What are some best practices for optimizing logistics in the context of e-business supply chain management?
7. How can technology be used to enhance supply chain management in e-businesses?
8. What are some examples of supply chain management technologies and how do they improve efficiency and effectiveness?
9. How can e-businesses ensure that the technology they use in supply chain management is secure and reliable?
10. What are some common challenges that e-businesses face in supply chain management and how can they be addressed?
11. How can e-businesses ensure quality control throughout the supply chain management process?
12. What strategies can e-businesses use to select the best suppliers and vendors for their supply chain management needs?

Multiple-Choice Question

1. Supply chain management in e-business refers to:

   (a) Managing the flow of goods and services in physical stores
   (b) Managing the flow of goods and services in online businesses
   (c) Managing the flow of information within a company
   (d) Managing customer relationships in online businesses

2. Effective supply chain management is important for e-businesses because it:

   (a) Increases customer satisfaction
   (b) Reduces shipping costs
   (c) Improves product design
   (d) Enhances employee morale

3. Which of the following is a key component of supply chain management in e-business?

   (a) Marketing strategy
   (b) Financial planning

   (c) Sourcing
   (d) Human resources management

4. Inventory management is an essential component of supply chain management in e-business. It involves:
   (a) Tracking and monitoring customer behavior
   (b) Determining the best pricing strategies
   (c) Managing stock levels and replenishment
   (d) Analyzing competitor performance
5. Logistics in supply chain management for e-businesses refers to:
   (a) Managing the physical flow of goods and services
   (b) Implementing effective advertising campaigns
   (c) Conducting market research for new product development
   (d) Analyzing financial statements for investment decisions
6. Which of the following is an example of supply chain management technology?
   (a) Social media marketing
   (b) Virtual reality simulations
   (c) Radio-frequency identification (RFID)
   (d) Customer relationship management (CRM) software
7. The use of technology in supply chain management can improve efficiency by:
   (a) Reducing the need for human labor
   (b) Increasing product prices
   (c) Creating additional supply chain complexities
   (d) Limiting communication with suppliers
8. Blockchain technology can be utilized in supply chain management to:
   (a) Ensure product quality control
   (b) Enhance employee training programs
   (c) Increase market research capabilities
   (d) Improve customer relationship management
9. Challenges in supply chain management in e-business can include:
   (a) Lack of Internet connectivity
   (b) Excessive product demand
   (c) Inefficient advertising campaigns
   (d) Supplier selection difficulties
10. One strategy for resolving supply chain management challenges in e-businesses is:
   (a) Implementing regular quality control checks
   (b) Reducing product variety and options
   (c) Increasing shipping times
   (d) Ignoring customer feedback

11. Which of the following is a solution for improving supply chain management in e-businesses?
    (a) Limiting supplier partnerships
    (b) Centralizing all inventory in one location
    (c) Automating order fulfillment processes
    (d) Avoiding technology integration
12. Effective sourcing in supply chain management involves:
    (a) Reducing customer support services
    (b) Implementing strict return policies
    (c) Selecting reliable suppliers
    (d) Increasing product prices
13. The integration of technology in supply chain management can lead to:
    (a) Increased shipping delays
    (b) Higher transportation costs
    (c) Improved inventory accuracy
    (d) Reduced customer satisfaction
14. The main goal of supply chain management in e-business is to:
    (a) Maximize profit margins
    (b) Minimize customer loyalty
    (c) Minimize supplier partnerships
    (d) Optimize the flow of goods and services
15. Which of the following is *not* a benefit of effective supply chain management in e-businesses?
    (a) Improved customer satisfaction
    (b) Reduced operational costs
    (c) Enhanced product quality
    (d) Increased employee turnover

References

1. Chen, Y., et al., *Recent advancements on three-dimensional electrospun nanofiber scaffolds for tissue engineering*. Advanced Fiber Materials, 2022. **4**(5): p. 959–986.
2. Hofmann, E., *Supply Chain Management: Strategy, Planning and Operation, S. Chopra, P. Meindl*. 2013, Elsevier Science.
3. Harland, C.M., et al., *A taxonomy of supply networks*. Journal of Supply Chain Management, 2001. **37**(3): p. 21–27.

Chapter 8
E-Procurement

Abstract E-commerce is completed by incorporating e-procurement, which manages the purchasing process through electronic systems and technology. E-procurement has gained prominence due to Internet use and technological advancements, offering benefits such as lower transaction costs, improved operational effectiveness, and simplicity of use. This chapter begins by defining e-procurement in relation to e-business. It explores various e-procurement methods like online auctions, request for proposals (RFPs), and reverse auctions. The advantages and disadvantages of each method are discussed, providing a comprehensive understanding. Additionally, the chapter delves into implementing e-procurement systems and processes, including successful business models and strategies to achieve goals. It also explores linking e-procurement with other systems like inventory management and accounting to maximize productivity and effectiveness. Benefits of electronic procurement, such as reduced transaction costs, improved supplier relationships, and increased openness, are examined. The chapter addresses challenges like supplier selection, legal compliance, and security concerns. By the end of this chapter, readers will have a comprehensive understanding of e-procurement, its implementation strategies, benefits, and drawbacks. Whether they own a small company, are involved in e-commerce, or are procurement experts, this chapter equips them with the knowledge and skills needed to effectively utilize electronic procurement for online business success.

1 Definition and Explanation of E-Procurement in the Context of E-Business

E-procurement is the process of purchasing products and services online, which is an essential component of e-business. It encompasses all aspects of a company's digital acquisition of products and services, from discovering potential suppliers to executing transactions and receiving items. One of the critical benefits of e-procurement is cost reduction. Automating the procurement process can save

H. Taherdoost, *E-Business Essentials*, EAI/Springer Innovations in Communication and Computing, https://doi.org/10.1007/978-3-031-39626-7_8

money by removing the need for human participation and paper-based processes. Additionally, e-procurement enables better supplier pricing negotiations and more effective inventory management, saving time and effort for faster procurement procedures. Another advantage is increased openness, as all procurement operations are digitized, making it easier to analyze buying patterns, monitor supplier performance, and identify issues. E-procurement can help companies comply with regulatory standards, as all transactions can be readily logged and audited.

There are two types of electronic buying: direct procurement and indirect procurement. Direct procurement refers to how a corporation obtains the materials and services required for manufacturing its products. In contrast, indirect procurement includes commodities and services necessary for a company's operation but not directly engaged in manufacturing its products or services, such as computer components or paper fasteners. E-procurement can take several forms, such as electronic catalogs, auctions, and sourcing.

If effectively implemented, e-procurement can significantly impact a company's bottom line, making it a crucial element of e-commerce. By utilizing e-procurement platforms, businesses can speed up the supplier selection and payment processing stages of the procurement cycle, thus boosting efficiency. Effective e-procurement gives companies a more significant say in the purchase process, as standardization can help develop rules that ensure compliance with regulatory requirements and internal procurement policies. Furthermore, supplier relationships may benefit from standardization, as suppliers will know what to anticipate and how to do business with the company. E-procurement can also save money for e-commerce sites by reducing staff time, paper expenses, and input mistakes. By increasing the number of suppliers, e-procurement can lead to reduced pricing and more negotiating power. Effective use of e-procurement can also enhance the quality of items and services bought by online enterprises by allowing them to search among a larger pool of possible suppliers and define criteria for their suppliers to fulfill. This can help ensure the quality of goods and services, making e-procurement a valuable tool for e-commerce companies. Transparency and openness are critical to the success of e-procurement, as businesses can use them to log their purchases, monitor supplier development, and identify potential risks and opportunities. The availability of this information can help firms make better decisions and improve their overall competitiveness. By simplifying the procurement process and enhancing the quality of the goods and services bought, businesses can focus on their core capabilities and innovate, better satisfying the demands of their consumers [1].

Finally, e-businesses can only exist with effective e-procurement. Its advantages include greater productivity, management, cost savings, quality, accessibility, and transparency. E-commerce companies may profit from effective e-procurement since it frees up resources to concentrate on core strengths and innovate. The importance of e-procurement to the sustainability of online enterprises will only grow as e-commerce advances. Figure 8.1 illustrates the importance of effective e-procurement for e-businesses.

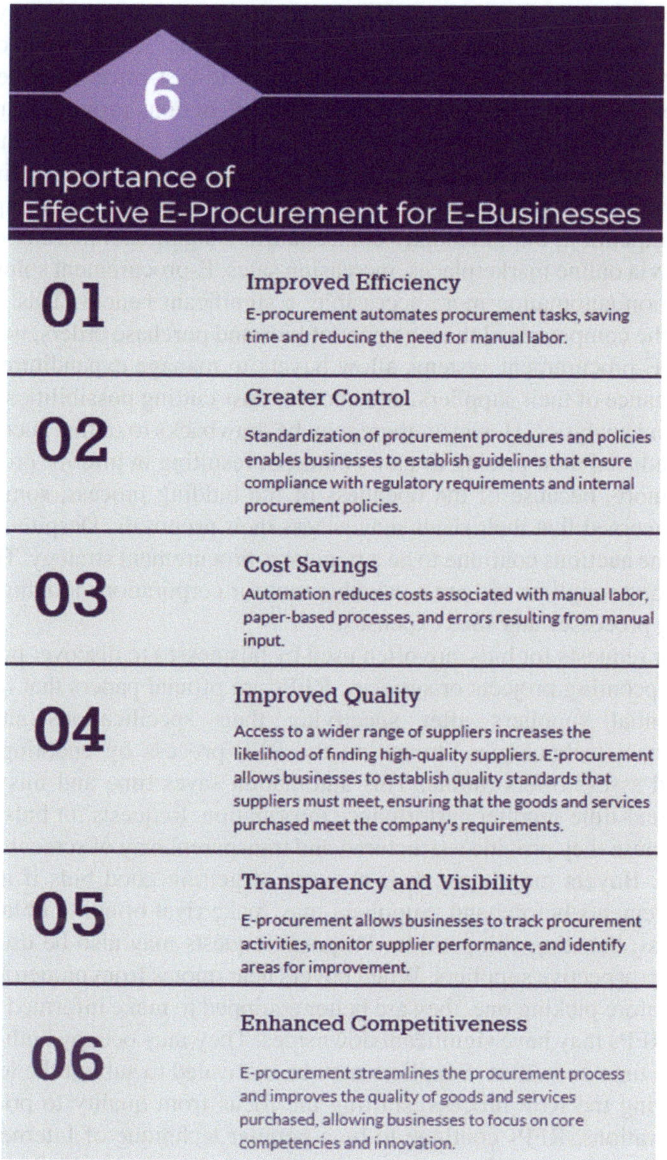

Fig. 8.1 The Importance of effective e-procurement for e-businesses

2 Overview of Common Types of E-Procurement

Using digital tools and systems to improve the effectiveness and efficiency of the traditional buying process is known as e-procurement. E-procurement comes in many forms, including:

Many businesses utilize online auctions as an e-procurement strategy to obtain products and services. Under this technique, vendors must participate in online auctions for specific items and services. Typically, the lowest-priced supplier wins the auction, performing the transaction online. Online auctions provide many benefits over more conventional means of purchasing goods and services, including more competition and transparency, shorter processing times, and cheaper pricing. Buyers may receive the best possible rates for products and services by participating in a bidding competition. Suppliers may contact a more significant number of possible consumers via online marketplaces, increasing sales. E-procurement solutions make online auction automation more accessible, a significant benefit. This automation allows for the computerized management of bids and purchase orders, which boosts efficiency. E-procurement systems allow buyers to manage expenditures, evaluate the performance of their suppliers, and uncover cost-cutting possibilities using real-time data and analytics. However, there may be drawbacks to online auctions. Some suppliers reduced their pricing to gain contracts, resulting in inferior products.

Furthermore, because of the openness of the bidding process, some suppliers may be concerned that their rivals may access their proposals. Despite these drawbacks, online auctions continue to be a popular e-procurement strategy. They benefit customers and suppliers and are a valuable tool for corporations looking to streamline buying processes and save expenses.

RFPs, or requests for bids, are often used by businesses to discover possible suppliers for upcoming projects or services. RFPs are official papers that request bids from potential suppliers after specifying their specifications and criteria. E-procurement technologies streamline the RFP process by enabling buyers to receive and assess offers online. This automation saves time and mistakes while providing real-time supplier performance information. Requests for bids (RFPs) are helpful because they provide a structured and transparent way of screening potential contractors. Buyers may boost their chances of getting good bids if they outline their requirements beforehand. Suppliers may make rival offers to obtain the buyer's business, boosting competition. Proposal requests may also be used to locate and assess prospective suppliers. When buyers hear quotes from numerous potential suppliers before picking one, they are better equipped to make informed judgments. However, RFPs may have significant downsides. They may occupy both the buyer's and seller's time and effort. Suppliers may be motivated to submit the lowest feasible bid during the RFP process, shifting the focus from quality to price. Despite these reservations, RFPs continue to be a popular technique of Internet purchase. They make it easier to evaluate several possible suppliers by establishing a uniform approach to choosing service providers. Using e-procurement tools to automate RFPs saves time and money while providing valuable data and analytics on supplier performance [2].

Many organizations utilize reverse auctions as innovative electronic procurement to purchase the items and services they demand. In traditional auctions, buyers try to outbid one another by proposing ever more excellent prices to the seller; in reverse auctions, sellers seek to undercut one another by lowering their offers. Electronic procurement systems are widely used to conduct reverse auctions, in which buyers send requests for proposals to prospective suppliers. Suppliers must submit their ideas within a defined deadline as part of the standard bidding process. The contract is usually handed to the lowest bidder at the end of an auction. Reverse auctions can cut costs by creating healthy competition. Consumers may ask bids from numerous suppliers to get the best purchase deal. E-procurement systems allow for the electronic management of bids and purchase orders and can further boost efficiency. However, there may be complications if reverse auctions are used. Some suppliers reduced their pricing to gain contracts, resulting in defective products.

Consequently, less important factors, such as the supplier's quality or competence, may need to be considered in favor of pricing. Despite these disadvantages, reverse auctions are a popular means of Internet buying. They provide various advantages to customers, including lower pricing and a more comprehensive range. E-procurement technologies that automate reverse auctions may make them more efficient and give more meaningful data and analytics on supplier performance.

Many organizations depend on electronic cataloging for commonly bought products and services. The technique creates a database of pre-screened products and services that customers can purchase without inquiry or haggling. Electronic catalogs can boost production while lowering expenses. Buyers save time and effort by not having to haggle rates with suppliers for each transaction when they purchase a pre-approved inventory of items and services. E-cataloging also aids in buying standards, allowing purchasers to manage better spending and discover opportunities for cost savings. Electronic cataloging has the potential to improve vendor relationships. Purchasers may create long-term relationships with suppliers and foster trust by pre-approving vendors for specific items or services. This leads to lower rates and better service. However, there might be drawbacks to utilizing an online catalog. It would not, for example, work for transactions involving unusual or specialized items or services. The pre-approval method may decrease competition, resulting in higher pricing. Despite its drawbacks, e-cataloging, a kind of e-procurement, is widely employed. Purchasers will profit from stronger supplier linkages, lower prices, and higher efficiency. Creating and maintaining an online catalog using an e-procurement platform may help firms save time and dollars by simplifying their procurement operations [1].

Many firms utilize electronic invoicing to speed up and automate the billing process. Suppliers immediately provide electronic invoices to purchasers, eliminating the need for paper bills and time-consuming data input. Electronic invoicing has various potential advantages, including labor and cost savings. By removing the need for paper invoices, e-invoicing has the potential to minimize the time and cost involved with invoice processing. Furthermore, invoices submitted through e-invoicing may be reviewed for correctness and completeness in real time, decreasing the chance of mistakes and subsequent disputes. Supplier relationships may benefit from electronic invoicing as well. Automating the invoicing process to guarantee timely and correct payments helps boost suppliers' trust and happiness. This may assist in developing supplier relationships and boost overall efficiency over time.

Nonetheless, there may be some drawbacks to electronic invoicing. For example, it may need large expenditures in new equipment and infrastructure. Furthermore, it may only suit firms with a few suppliers or a complex invoicing procedure. Despite these reservations, electronic invoicing is still widely employed in e-procurement. Purchasers will profit from stronger supplier linkages, lower prices, and higher efficiency. Businesses may save time and effort by outsourcing their accounts payable activities using e-procurement technologies similar to invoicing.

Many firms utilize e-payment as an example of electronic procurement to automate and expedite the payment for products and services. The customer sends payment to the vendor online, eliminating the need for paper checks and time-consuming processing. There are various potential advantages to using electronic payment methods. Businesses save time and money by eliminating the need to process paper checks. Electronic payments, in addition to offering automatic checks for accuracy and completeness, lessen the probability of human mistakes and subsequent disputes. Electronic payments have the potential to boost vendor relationships. Customers' trust and happiness in their service providers may improve due to automated payments. This may assist in developing supplier relationships and boost overall efficiency over time. However, adopting electronic payment methods may have drawbacks. For example, it may need large expenditures in new equipment and infrastructure.

Furthermore, it would only work for organizations with a few suppliers or complicated payment methods. Despite the risks, e-payment, a subset of e-procurement, is widely employed. Purchasers will profit from stronger supplier linkages, lower prices, and higher efficiency. Automating payments using an e-procurement platform is a terrific way for businesses to save time and resources in the accounts payable department. Table 8.1 summarizes the advantages and disadvantages of each e-procurement type.

3 Implementation of E-Procurement

As firms seek more effective ways to handle their procurement operations, e-procurement (or electronic procurement) technologies and techniques are gaining popularity. Organizations may save time and money while strengthening connections with suppliers by adopting and implementing an e-procurement system and process. However, effective implementation requires careful planning and execution. This article discusses best practices for implementing e-procurement systems and procedures in e-business.

The first best practice for developing e-procurement technology and processes is thoroughly examining your current procurement practices. As part of this study, you should analyze your company's procurement rules and procedures and its present procurement systems, processes, and technology. This assessment may identify areas where e-procurement might boost efficiency, save costs, and build vendor relationships. Second, verify that your e-procurement implementation project's

Table 8.1 E-procurement types and their advantages/disadvantages

Type of e-procurement	Advantages	Disadvantages
Online auctions	Competition among suppliers can lead to lower prices Increased efficiency and speed of the procurement process Can reach a larger pool of suppliers	May not always result in the selection of the best supplier May not be appropriate for all types of goods and services Bidders may collude or engage in other unethical behavior
Request for proposals	Can lead to more comprehensive proposals from suppliers Enables the buyer to evaluate proposals systematically May result in the selection of the best supplier	Can be time-consuming and resource-intensive for both buyer and supplier May not be appropriate for all types of goods and services Suppliers may be reluctant to invest resources in preparing a proposal if they believe they have little chance of winning
Reverse auctions	Competition among suppliers can lead to lower prices Increased efficiency and speed of the procurement process Can reach a larger pool of suppliers	May not always result in the selection of the best supplier May not be appropriate for all types of goods and services Bidders may collude or engage in other unethical behavior
E-cataloging	Can simplify the procurement process for both buyer and supplier Enables buyers to search for and purchase items from the catalog easily Can lead to increased efficiency in the procurement process	May not be appropriate for all types of goods and services Catalog may not always be up-to-date or accurate May require significant effort and resources to create and maintain a catalog
E-invoicing	Can reduce errors and inefficiencies associated with the manual processing of invoices Can lead to faster processing and payment of invoices Can lead to cost savings for both buyer and supplier	Requires both buyer and supplier to adopt electronic invoicing systems May require significant effort and resources to implement May face resistance from stakeholders who are used to traditional invoicing processes
E-payment	Can reduce errors and inefficiencies associated with manual payment processing Can lead to faster processing of payments Can lead to cost savings for both buyer and supplier	Requires both buyer and supplier to adopt electronic payment systems May require significant effort and resources to implement May face resistance from stakeholders who are used to traditional payment processes

goals and objectives are adequately stated. These goals should be quantifiable and align with the organization's overall strategy. Having clearly defined goals can assist you in staying on track with your implementation project and achieving the required results. The third point is to include prominent people in the plan's implementation. All parties interested in the subject, including buyers, IT professionals, and suppliers, are represented. To ensure that everyone is on board with the changes, include them from the start of the implementation process [3].

The fourth best practice is to choose the finest e-procurement system for your company. There are several e-procurement solutions, each with its benefits and drawbacks. Choose a solution that fits your organization's requirements, budget, and procurement goals.

The fifth best practice is creating a complete distribution plan with quantifiable goals and timeframes. This plan should contain tasks, responsibilities, and due dates and be reviewed and amended. With careful preparation, you can guarantee that your e-procurement system is implemented on time and within budget. Users of the new e-procurement system should get proper training and assistance. All parties interested in the subject, including buyers, IT professionals, and suppliers, are represented. It is critical to the new system's success that everyone understands how it works and has access to continuing assistance.

Finally, e-procurement technology and practices may help your company in many ways, including better efficiency, lower prices, and stronger vendor connections. For a successful rollout, a thorough examination of current processes, clearly stated goals and objectives, the engagement of critical stakeholders, careful technology selection, a well-defined implementation plan, and enough training and support are essential. By following these rules, your e-business will have the most incredible opportunity to adopt e-procurement effectively and enjoy its numerous advantages. Figure 8.2 summarizes the best practices for implementing e-procurement systems and processes in e-business.

Integrating e-procurement with other systems and processes is critical for enhancing the efficiency and effectiveness of procurement operations in e-business. E-procurement systems can be integrated with other systems and processes, including supply chain management, finance and accounting, and enterprise resource planning (ERP) systems. Integrating e-procurement with these systems and processes can result in numerous benefits, including improved visibility into the procurement process, enhanced control over procurement activities, and increased automation of manual processes. One of the key benefits of integrating e-procurement with other systems and processes is improved visibility into the procurement process. By integrating e-procurement with supply chain management systems, for example, organizations can gain real-time visibility into inventory levels, production schedules, and demand forecasts. This can help procurement teams better plan for and manage procurement activities, ensuring that the correct goods and services are available at the right time and cost. Another benefit of integrating e-procurement with other systems and processes is enhanced control over procurement activities. By integrating e-procurement with finance and accounting systems, organizations can gain better control over spending and budgeting. This can help to reduce the risk of overspending or underspending on procurement activities and can also help to ensure compliance with financial regulations and policies.

Integration of e-procurement with ERP systems can also yield significant benefits. ERP systems are designed to integrate all aspects of business operations, including finance, supply chain management, and human resources. Organizations can

Fig. 8.2 Best practices for implementing e-procurement systems and processes in e-business

optimize procurement processes by integrating e-procurement with an ERP system and improving department communication and collaboration. This can help to reduce errors and delays, improve procurement cycle times, and reduce overall procurement costs. Another key benefit of integrating e-procurement with other systems and processes is increased automation of manual processes. By integrating e-procurement with supply chain management systems, organizations can automate inventory management and order fulfillment processes. This can reduce the time and resources required to manage these processes, enabling procurement teams to focus on more strategic activities [2].

In conclusion, integrating e-procurement with other systems and processes is critical for maximizing the benefits of e-procurement in e-business. By improving visibility into the procurement process, enhancing control over procurement activities, and increasing automation of manual processes, organizations can improve their procurement operations and achieve significant cost savings and operational efficiencies. Therefore, organizations should prioritize the integration of e-procurement with other systems and processes as part of their overall procurement strategy.

4 Benefits and Challenges of E-Procurement

E-procurement, which employs technology to speed up the traditional purchase process, has grown in popularity in recent years. The multiple advantages of electronic procurement, such as lower overhead and increased transparency throughout the procurement process, drive this shift. This article will explain the advantages of e-procurement and the aspects that firms should consider before deploying an e-procurement system.

One key benefit of e-procurement is the potential for cost savings. Organizations may save time and money by deploying an automated procurement management system. Consequently, costs may be lowered, and transactions and processes become more efficient. Using e-procurement to negotiate lower prices with suppliers may assist companies in saving money on purchases.

E-procurement may speed up the procurement process and save time by replacing manual processes with automated ones. Teams may redirect their attention to more strategic undertakings thanks to the time and effort saved by electronic procurement. As a result of reducing the time necessary to complete the procurement cycle, operational efficiency will improve.

E-procurement allows firms to have a better understanding of their buying operations. Acquiring real-time data allows procurement teams to make better judgments by automating the purchase process. Real-time data may help firms become more proactive by showing trends and patterns and improving procurement procedures. E-procurement also allows for better transparency into supplier performance, speeding up problems' discovery and resolution.

Organizations may benefit from e-procurement since it makes complying with procurement standards and procedures more accessible. Electronic procurement systems guarantee that all purchases conform to regulations by enforcing procurement rules and processes. E-procurement helps compliance with financial rules and laws by lowering the probability of fraud and mistakes.

Supplier relationships may be reinforced due to improved buyer-supplier communication and cooperation enabled by e-procurement. Suppliers may better understand what buyers want and accelerate their offers with the help of electronic procurement systems that enable real-time data access. E-procurement may be helpful for companies since it improves communication and collaboration between customers and suppliers, strengthening ties with suppliers.

Organizations that use e-procurement save money, operate more effectively, see more clearly, are more compliant, and have stronger connections with their suppliers. Organizations may boost efficiency, cut costs, and shorten the procurement cycle by automating the operations required. Businesses must examine their goals and objectives before deploying an e-procurement solution, considering the breadth and complexity of their procurement processes, current IT infrastructure, and established procurement rules and procedures. Businesses may tailor an e-procurement system to their requirements, enhancing the system's chances of success.

E-procurement, or electronic procurement, has been an increasingly prevalent method of procuring products and services in recent years. Despite the numerous benefits of electronic procurement, such as time and cost savings and increased oversight, the process is full of obstacles and restrictions. This article will discuss some of the most significant challenges and limitations of e-procurement.

One of the most significant obstacles to e-procurement is the need to invest heavily in technology and infrastructure in advance. Implementing an electronic procurement system necessitates purchasing specialized software, hardware, and other equipment and hiring qualified personnel to supervise the system's administration and maintenance. This may substantially burden resources, particularly for small- and medium-sized enterprises.

The need to safeguard sensitive data is a barrier unique to electronic procurement. Examples of sensitive data that must be transmitted to an e-procurement system include supplier and consumer identities, pricing structures, and purchase order specifics. This information must be protected from cyberattacks and other cyber threats. If organizations take the necessary measures to protect their electronic procurement processes, they can avoid data intrusions, legal liabilities, and reputational damage [1].

Integrating other business processes and systems is an additional challenging aspect of e-procurement. Integrating e-procurement systems with other systems, such as inventory management, finance, and accounting, is necessary for accurate and timely procurement procedures. Data discrepancies, inaccuracies, and delays resulting from inadequately integrated e-procurement systems can significantly impact procurement processes and an organization's overall performance.

E-procurement may only be pertinent to some types of transactions, which is one of its potential disadvantages. Due to the intricacy of the required considerations, traditional procurement procedures may be more appropriate for certain acquisitions, such as those requiring specialized equipment or services. Some purchases, including those with high customization or risk, may not be suitable for e-procurement.

A potential disadvantage of e-procurement is diminished supplier competition. In many e-procurement systems, pre-qualified suppliers are utilized, although this may reduce the number of available vendors and diminish competition. As a result, higher prices and inferior products and services may affect the organization's bottom line. Internal constituents may also be resistant to electronic procurement. Some purchasers can be averse to or intimidated by e-procurement systems. As a result, some individuals may be resistant to e-procurement, which could delay or even stop its prevalent adoption.

In conclusion, e-procurement has many benefits for businesses, but it is also essential to consider the challenges and limitations associated with its implementation and use. Among the obstacles that must be surmounted are substantial upfront costs, data security, and confidentiality concerns, the need for effective system integration, restrictions on the types of purchases eligible for e-procurement, decreased supplier competition, and opposition from internal stakeholders. Businesses will be better prepared to implement and utilize e-procurement systems if they keep these obstacles and limitations in mind. Table 8.2 summarizes the challenges and limitations of e-procurement.

Many firms use electronic procurement solutions to simplify and save money on procurement activities. In this article, we will look at real-world examples of e-procurement in e-business, focusing on the advantages that follow.

One of the most well-known instances of an e-procurement system is Amazon Business. Amazon Business, launched in 2015, is a marketplace where companies can buy and sell anything from office supplies to heavy equipment. Businesses may save time and money by automating their buying procedures and obtaining access to millions of items with Amazon Business. Tax-free transactions and numerous user accounts are among the available choices. A handful of firms have saved substantial money using Amazon Business for their buying requirements.

Table 8.2 Challenges and limitations of e-procurement

Challenge/ limitation	Description
Resistance to change	Employees who are averse to change may need to be trained to utilize the new e-procurement platforms. This may cause delays and cost increases
Security risks	E-procurement systems may be infiltrated, or their data will be stolen. Supplier and pricing list data might be jeopardized
Integration issues	Integrating e-procurement technologies with current business systems, such as ERP and supply chain management, may take time and effort. This may cause delays and cost increases
Technical difficulties	Technical issues with e-procurement systems, such as network outages and software glitches, are always a risk. This may cause a delay in the purchase process and an increase in associated expenditures
Limited supplier participation	The number of prospective suppliers has lowered if certain potential vendors need more infrastructure to participate in e-procurement systems
Dependence on Internet connectivity	Continuous Internet access is required for e-procurement systems to work. If Internet connectivity is unavailable, procurement operations may be delayed or disrupted
Cost	Continuous Internet access is required for e-procurement systems to work. If Internet connectivity is unavailable, procurement operations may be delayed or disrupted

Coupa is an example of e-procurement in e-business; it is a cloud-based spending management software that assists organizations in managing their procurement operations from request to payment. Coupa's features include, among other things, procurement process automation, real-time insight into spending, and supplier management. By automating and optimizing buying and supply chain management operations, Coupa helps organizations save time and money.

SAP Ariba, another well-known e-procurement platform, is used by many businesses globally. SAP Ariba is a procurement and supply chain management software system hosted in the cloud. Its features include a real-time supplier interface, robotic process automation, and spending analysis. SAP Ariba has helped many international firms optimize their buying procedures and save costs.

There are many more e-procurement options available than those listed above. Basware, for example, provides a cloud-based platform for managing procurement processes and spending management. In contrast, Procurify provides procurement software that can be used remotely and used to manage a company's procurement operations.

To summarize, e-procurement is a beneficial tool that assists firms in saving money and increasing buying department efficiency. Amazon Business, Coupa, and SAP Ariba are real-world instances of e-procurement advantages in e-business. Businesses may choose from a wide range of e-procurement systems to choose the one that best suits their needs and greatly enhance their procurement operations.

Summary

E-Procurement Definition

- E-procurement refers to conducting procurement activities electronically using digital technologies and platforms.
- It involves automating and streamlining traditional procurement tasks, such as requisitioning, sourcing, purchasing, and payment, through Internet-based systems.

Types of E-Procurement

- Electronic Catalogs: Involves creating and maintaining online catalogs of products and services offered by suppliers. Buyers can browse, select, and order items directly from the catalog, eliminating the need for manual paperwork.
- E-Auctions: Facilitates the bidding and negotiation process between buyers and suppliers. Suppliers compete in real-time auctions, increasing transparency and potentially lowering buyers' prices.
- E-Sourcing: Focuses on the electronic management of supplier relationships and the sourcing of goods and services. It includes activities such as supplier selection, negotiation, and contract management, all conducted online.
- E-Invoicing: Digitizes invoicing by generating, sending, and receiving invoices electronically. This eliminates manual data entry, reduces errors, and accelerates payment cycles.

Implementation of E-Procurement

- Assessing Readiness: Organizations need to evaluate their current procurement processes and technological capabilities to determine their readiness for e-procurement implementation. This involves identifying key stakeholders, defining objectives, and considering integrating existing systems.
- System Selection and Integration: Organizations must select appropriate e-procurement software or platforms that align with their requirements. Integration with existing enterprise resource planning (ERP) or supply chain management (SCM) systems is crucial for seamless data flow.
- Training and Change Management: Successful implementation requires adequate training and change management efforts to ensure user adoption and smooth transition. Employees should be trained on how to use the e-procurement system effectively, and any resistance to change should be addressed.

Benefits of E-Procurement

- Cost Savings: E-procurement can reduce costs by automating processes, eliminating paperwork, and facilitating better supplier negotiations.
- Increased Efficiency: By digitizing procurement processes, e-procurement streamlines workflows, reduces manual errors, and accelerates cycle times.
- Enhanced Transparency and Compliance: Electronic documentation and audit trails enable improved visibility into procurement activities, ensuring compliance with regulations and policies.
- Supplier Relationship Management: E-procurement systems provide tools for managing supplier relationships, enabling better collaboration and performance tracking.

Challenges of E-Procurement

- Technological Complexity: Implementing and maintaining e-procurement systems can be complex, requiring software integration and security expertise.
- Resistance to Change: Employees may resist adopting new technology and processes, requiring effective change management strategies.
- Supplier Adoption: Encouraging suppliers to participate in e-procurement systems can be challenging, requiring them to adapt their processes and technologies.
- Data Security and Privacy: The digital nature of e-procurement introduces potential risks related to data breaches, unauthorized access, and confidentiality.

Case Study

XYZ Corporation is a global manufacturer that specializes in automotive components. The organization intends to use e-procurement to speed up its purchasing process, reduce costs, and boost overall efficiency. As part of their implementation strategy, they are investigating various types of e-procurement systems and analyzing the benefits and drawbacks of each.

XYZ Corporation employs a traditional procurement method that involves handwritten documentation, phone calls, and in-person interactions with vendors. This procedure is time-consuming, error-prone, and murky. The company's management has implemented an e-procurement system to automate and streamline procurement procedures.

1. How might XYZ Corporation save money and increase efficiency using an e-procurement system?
2. Given the nature of XYZ Corp. and its procurement demands, which e-procurement procedures, such as online auctions, request for proposals (RFPs), and reverse auctions, would be most suitable? Justify your response.
3. What are the potential advantages and disadvantages of integrating the e-procurement system with other XYZ Corporation systems and processes, such as inventory management and accounting systems?
4. When XYZ Corporation adopted e-procurement, supplier selection was one of its challenges. How can the organization manage this issue while using the e-procurement system to ensure the selection of dependable and high-quality suppliers?
5. Legal compliance is another significant issue with e-procurement. What steps should XYZ Corporation take when conducting e-procurement operations to ensure compliance with applicable laws and regulations?

Discussion

1. What is e-procurement and how does it relate to e-business?
2. What are some of the benefits of e-procurement for businesses, and how can it help to streamline procurement processes?
3. What are some of the challenges that businesses face when implementing e-procurement systems, and how can these challenges be addressed?
4. How can businesses ensure that their e-procurement systems are effective and efficient, and what are some best practices for achieving this?
5. What are some of the most common types of e-procurement, and how do they differ from each other?
6. How do online auctions work, and what are some of the advantages and disadvantages of this type of e-procurement?
7. What is a request for proposal (RFP), and how is it used in e-procurement? What are some of the benefits and drawbacks of using RFPs?
8. What is a reverse auction, and how does it work? What are some of the benefits and challenges of using this type of e-procurement?
9. What are some best practices for implementing e-procurement systems and processes within a business?

10. How can businesses ensure that their e-procurement systems are integrated with other systems and processes, such as supply chain management or inventory management?
11. What are some of the key challenges that businesses face when implementing e-procurement, and how can these challenges be addressed?
12. How can businesses measure the success of their e-procurement implementation, and what are some key performance indicators to consider?
13. What are some of the benefits of e-procurement, and how can it help businesses to reduce costs and improve efficiency?
14. How can businesses use e-procurement to improve their relationships with suppliers, and what are some best practices for doing so?
15. What are some of the challenges and limitations of e-procurement, such as supplier selection and legal compliance? How can businesses address these challenges?
16. How can businesses evaluate the effectiveness of their e-procurement systems, and what are some metrics that can be used to measure success?

Multiple-Choice Question

1. What is e-procurement?
   (a) The process of purchasing goods or services through electronic means
   (b) The process of purchasing goods or services through physical stores
   (c) The process of purchasing goods or services through mail-order catalogs
   (d) The process of purchasing goods or services through phone calls
2. Which of the following best defines e-procurement in the context of e-business?
   (a) The use of technology to automate procurement processes
   (b) The use of traditional methods for procurement in online businesses
   (c) The outsourcing of procurement activities to third-party vendors
   (d) The use of social media platforms for purchasing goods or services
3. Why is effective e-procurement important for e-businesses?
   (a) It reduces the need for communication with suppliers.
   (b) It enables better inventory management.
   (c) It improves customer satisfaction.
   (d) It increases profit margins.
4. Which of the following is a common type of e-procurement?
   (a) Online auctions
   (b) In-person negotiations
   (c) Print advertisements
   (d) Cold-calling suppliers

5. What is a request for proposals (RFP)?
   (a) A formal document requesting suppliers to submit proposals for a specific project or requirement
   (b) An online platform for bidding on goods or services
   (c) A reverse auction where suppliers compete to offer the lowest price
   (d) A negotiation process conducted through email exchanges
6. Which of the following is an advantage of online auctions as an e-procurement method?
   (a) Reduced transparency in supplier selection
   (b) Limited access to a wider range of suppliers
   (c) Increased competition leading to lower prices
   (d) Higher transaction costs compared to traditional methods
7. What is one disadvantage of reverse auctions as an e-procurement method?
   (a) Limited supplier participation due to confidentiality concerns
   (b) Higher transparency in pricing and supplier selection
   (c) Increased negotiation power for suppliers
   (d) Longer lead times for procurement activities
8. What are best practices for implementing e-procurement systems and processes?
   (a) Conducting thorough supplier audits before implementation
   (b) Ensuring clear communication and training for all stakeholders involved
   (c) Implementing complex and rigid procurement policies
   (d) Using outdated technology and software systems
9. How can e-procurement be integrated with other systems and processes?
   (a) By keeping e-procurement systems completely separate from other business systems
   (b) By integrating e-procurement systems with enterprise resource planning (ERP) systems
   (c) By relying solely on manual processes for integration
   (d) By outsourcing all e-procurement activities to third-party vendors
10. What is one benefit of e-procurement?
    (a) Increased legal compliance challenges
    (b) Decreased supplier selection options
    (c) Cost savings through improved efficiency
    (d) Reduced transaction transparency
11. What are some challenges and limitations of e-procurement?
    (a) Difficulties in supplier selection and evaluation
    (b) Higher costs compared to traditional procurement methods
    (c) Limited access to global supplier networks
    (d) Reduced efficiency in procurement processes

12. How does e-procurement contribute to cost savings?
    (a) By increasing administrative overhead costs
    (b) By reducing the need for supplier negotiations
    (c) By streamlining procurement processes and reducing manual tasks
    (d) By inflating supplier prices due to increased competition
13. What is a potential limitation of e-procurement related to supplier selection?
    (a) Limited access to global suppliers
    (b) Difficulties in evaluating supplier performance and reliability
    (c) Reduced competition among suppliers
    (d) Higher costs associated with supplier contracts
14. How does e-procurement improve efficiency?
    (a) By increasing manual tasks and paperwork
    (b) By requiring longer lead times for procurement activities
    (c) By automating processes and reducing errors
    (d) By creating additional layers of bureaucracy
15. What is an important factor to consider for e-procurement to ensure legal compliance?
    (a) Regularly updating procurement policies without legal guidance
    (b) Collaborating with suppliers to bypass legal requirements
    (c) Implementing proper data security measures to protect sensitive information
    (d) Ignoring regulations and relying on self-auditing processes

References

1. Croom, S. and A. Brandon-Jones, *Impact of e-procurement: experiences from implementation in the UK public sector.* Journal of Purchasing and Supply Management, 2007. **13**(4): p. 294–303.
2. Croom, S., P. Romano, and M. Giannakis, *Supply chain management: an analytical framework for critical literature review.* European Journal of Purchasing & Supply Management, 2000. **6**(1): p. 67–83.
3. Van Weele, A., *Purchasing and supply chain management*. 2018: UK. Cengage Learning EMEA.

Chapter 9
Digital Marketing

Abstract The current state of e-commerce heavily relies on digital marketing, as it enables companies to connect with their target consumers, enhance their brand image, and drive revenue growth through the utilization of digital platforms. Digital marketing encompasses a broad range of techniques and strategies employed to promote products or services online, including social media, email, PPC, and SEO. Selecting the appropriate channels is crucial for achieving desired outcomes. Developing and executing an effective digital marketing strategy is vital for online businesses. Targeting and personalization are key practices that contribute to success, along with the integration of multiple digital channels for maximum impact. A well-crafted digital marketing strategy facilitates communication with the target audience, resulting in increased revenue and brand loyalty. Furthermore, it helps businesses outpace competitors and adapt to market changes. Metrics, KPIs, and data analytics play a vital role in evaluating and improving digital marketing initiatives. In the upcoming sections, we will provide an overview of digital marketing and its relation to online businesses and explore various channels and their effectiveness. Additionally, we will discuss strategies for developing and implementing digital marketing plans, as well as the significance of metrics, analytics, and data-driven improvements in enhancing digital marketing strategies.

1 Definition and Explanation of Digital Marketing in the Context of E-Business

Companies of all sizes have realized that digital marketing has become an essential tool for promoting and selling their goods or services, and they have begun to realize its potential. When applied to the context of an online company, digital marketing has several distinct benefits that aid firms in expanding their operations and achieving more significant degrees of success in the current day. Because it is capable of doing all of these things, digital marketing enables businesses to tailor their marketing efforts to the specific needs of each customer, reach a large audience,

H. Taherdoost, *E-Business Essentials*, EAI/Springer Innovations in Communication and Computing, https://doi.org/10.1007/978-3-031-39626-7_9

target specific demographics, create a two-way communication channel with customers, monitor performance statistics, and reach a broad audience overall. Everyone, from students to working professionals, will benefit from reading the following overview of digital marketing in e-business, as it will help them better grasp the subject at hand.

One of the most critical advantages online firms may have from using digital marketing is the ability to interact with many people. It has never been easier to reach a worldwide audience than with digital marketing, as many different digital channels, such as search engines, social media, and email, may be used. Digital marketing has also made it possible to reach audiences in various languages. Doing so has always been more complex. Businesses can broaden their client base and raise the brand recognition they have acquired among their target audience thanks to digital marketing, which, in contrast to conventional marketing, was constrained by geographical limitations. Conventional marketing was only effective in certain regions. A firm will have a higher chance of boosting the number of leads, conversions, and sales that it generates if it can interact with a more significant number of individuals.

Thanks to digital marketing, businesses can now direct their attention to a specific demographic of consumers. Firms can build highly focused and successful marketing strategies by harnessing data acquired from many digital platforms. This allows firms to compete effectively in today's marketplace. Because gaining access to the appropriate demographic may considerably influence a company's success, this topic is of the utmost importance for companies that offer specialized goods or services. Businesses can reduce the amount of money they spend on advertising while simultaneously boosting the return on investment (ROI) of those efforts when they focus on a specific demographic to target with their marketing efforts.

Using techniques for digital marketing comes with some significant benefits, one of the most noticeable of which is the possibility of assessing performance indicators. Companies can now monitor various performance measures in real time thanks to the analytics tools that are available to them, such as Google Analytics. This gives them a significant edge over their competitors. This includes website traffic, the percentage of users interacting with the content, and conversion rates. Analyzing this data enables businesses to determine the areas in which they need more development, which enables them to optimize their campaigns to receive better results. When compared to more conventional marketing methods, data-driven marketing provides some key advantages that should be considered. When it came to conventional marketing, keeping track of performance data was often challenging, and in some cases, it was almost impossible [1].

One further key benefit that digital marketing brings to the table is the ability to establish communication channels with target audiences (also known as customers) in which they may communicate back and forth. Businesses can now communicate with consumers and get feedback on the quality of the goods or services they offer by using a variety of digital channels such as social media and email. This kind of communication makes it possible for companies to deepen their connections with their clients, resulting in increased levels of consumer engagement and happiness, leading to higher levels of business. If a business pays attention to the requirements and inclinations of its client base, the company's goods and services will be better, and the customer base will become more loyal to the firm. Another essential benefit of digital marketing is the ability to personalize a business's interactions with its clients. Businesses can develop individualized marketing strategies that are appealing to each unique individual consumer by making use of customer data such as browsing history, purchasing behavior, and demographic information. The usage of data obtained from customers makes this outcome conceivable. Personalized marketing is much more successful than campaigns that use a one-size-fits-all approach because it generates greater consumer engagement and improves conversion rates. Individualized and targeted marketing allows companies to develop their relationships with their client base and build a competitive edge in their market by delivering personalized experiences to each consumer. Suppose individuals want to be successful in the dynamic world of e-business. In that case, they must commit to ongoing education and modify their digital marketing tactics. This will allow them to keep up with the industry's rapid pace of change, which is one of the most crucial things they can do. Because digital marketing is a discipline that is continually undergoing new developments, companies need to remain current on the most recent marketing trends and techniques. Students and professionals can keep one step ahead of the competition and succeed in e-business, provided they demonstrate a dedication to lifelong learning and actively experiment with various approaches.

When it comes to attracting clients and growing sales in today's digital world, e-businesses face significant competition from their conventional counterparts. This competition comes in the form of traditional enterprises that operate in the offline world. Companies that operate traditionally encounter a great deal of difficulty. The proliferation of online shopping has made it simpler for companies of any size to sell their goods and services via the Internet. On the other hand, as a direct consequence of this, there is a growing need for digital marketing strategies that are both successful and time- and resource-efficient. The term "digital marketing" is often used when referring to the process of promoting a product or service via digital methods such as search engines, social media, email, and websites. This is because digital marketing is a commonly used phrase encompassing various strategies, such as pay-per-click (PPC) advertising, search engine optimization (SEO), content marketing, social media marketing, email marketing, and a variety of other techniques. When developing a comprehensive marketing plan for an online business, one of the key objectives of that approach should be digital marketing. Organizations that conduct their operations online have the potential to engage with a particular demographic of customers when they use digital marketing, which is one of the most important advantages that digital marketing offers these organizations. By using marketing strategies such as search engine optimization (SEO) and pay-per-click (PPC) advertising, online companies can target clients actively searching for their goods or services online. This is a significant competitive advantage for online firms. The utilization of search engine optimization (SEO) techniques has made this accomplishment conceivable. This technique targets its viewers with the same precision as this strategy. It has been demonstrated to be significantly more successful than conventional advertising tactics, such as print or television commercials, which do not target their audiences with the same level of specificity. The capacity to adapt one's communications and business dealings to the unique requirements of each consumer is one of the benefits of adopting digital marketing. Organizations that participate in e-commerce can provide clients with messages and offers that are particularly relevant to their unique requirements because these organizations do in-depth analyses of customer data, such as a customer's browsing history and purchase history. This personalized strategy can bring in more money as well as dedicated customers. Another important aspect of digital marketing for companies that conduct online operations is developing a system that facilitates brand identification and recall. E-commerce businesses may potentially boost their customers' familiarity with and recall of their brands by establishing a strong online presence via platforms such as social media and content marketing. Smaller online businesses are less likely to have the same level of brand recognition as bigger, more well-known organizations, and one of the reasons why this is the case is their weaker online presence. Therefore, having a strong online presence is of the utmost significance for smaller Internet businesses. Due to the highly quantitative nature of digital marketing, it is feasible to analyze, in real time, how successful the marketing activities of an online company are. This degree of precision in marketing analytics is another factor that adds to its overall significance. E-commerce companies can fine-tune their marketing strategy to acquire the maximum potential return on investment (ROI) because it is feasible to analyze how successful different digital marketing campaigns are. Another one of the many advantages of doing business online is the likelihood of seeing a reduction in the overhead costs associated with the company's day-to-day operations. This is something that may be done by taking advantage of the many different digital marketing tactics that are available. Because of this, it is now possible for online firms, even those with more limited resources for marketing, to compete successfully with larger organizations with more marketing expenditures. This is made possible by the benefits that come with doing business online. Because of the benefits that digital marketing provides, e-commerce enterprises also can connect with their clients in real time. This ability was made possible by digital marketing. E-commerce firms can discuss with their consumers using social media platforms such as Facebook and Twitter. During these conversations, e-commerce businesses might respond to the issues and concerns made by their customers and request feedback from their customers. This involvement can contribute to developing trust and loyalty among consumers, which may lead to increased sales and the number of customers who return to the firm to do business with it. In conclusion, online businesses need to participate in digital marketing in order for them to continue to be competitive in the digital world that exists today. E-commerce is growing in popularity; as a result, online companies risk losing prospective clients to their rivals if they do not engage in digital marketing. This risk is especially high for those firms that do not have a strong online presence. They do not have the financial resources to take this risk. E-commerce websites can build their customer base, increase their brand awareness, and create leads and sales with the support of digital marketing that is properly implemented [2].

Table 9.1 Benefits of digital marketing for e-businesses

Benefits of digital marketing for e-businesses	Explanation
Targeted audience	Digital marketing allows businesses to precisely target consumers by conducting online product or service investigations
Personalized messaging	Customer data analysis enables e-businesses to transmit highly relevant messages and offers to individual consumers, resulting in increased sales and customer loyalty
Brand recognition and recall	Due to digital marketing, e-businesses can establish a robust online presence through social media and content marketing channels. This has increased both brand awareness and recall
Measurability	Monitoring the efficacy of digital marketing initiatives in real time facilitates data-driven decision-making and the optimization of marketing strategies
Cost-effectiveness	Search engine optimization (SEO), content marketing, and social media marketing have the potential to be significantly less expensive than conventional advertising strategies
Real-time customer engagement	Social media platforms enable online businesses to communicate with customers in real time, fostering customer trust and loyalty
Competitive advantage	Effective digital marketing is essential for online businesses in the digital age to maintain their competitive advantage because it allows them to reach a larger audience and generate leads and sales

Successful digital marketing is essential for the expansion and success of online enterprises in the Internet age. E-businesses can reach their target audience with targeted messages, develop brand recognition and recall, generate leads and sales, build customer trust and loyalty, and maintain a competitive advantage in a market that is becoming increasingly saturated. By utilizing digital marketing strategies such as search engine optimization (SEO), pay-per-click (PPC) advertising, content marketing, social media marketing, and email marketing, e-businesses can construct a robust online presence and advance corporate growth. The use of effective digital marketing strategies is crucial for achieving these goals. Table 9.1 summarizes the benefits of digital marketing for e-business.

2 Digital Marketing Channels

E-commerce companies have access to various digital marketing platforms, which gives them much flexibility when advertising their goods and services to customers. Using tactics optimized for search engines (SEO) is one of the most widespread (Table 9.2).

Table 9.2 Comparison of effectiveness of different digital marketing channels

Digital marketing channel	Reach	Engagement	Sales
SEO	High	Low	High
PPC advertising	Medium	High	High
Social media marketing	High	Medium	Medium
Email marketing	Medium	High	High
Content marketing	Medium	High	High
Mobile marketing	High	High	Medium

2.1 Search Engine Optimization (SEO)

SEO is an essential component of digital marketing for e-businesses. It involves optimizing websites and online content to rank higher on search engine results pages (SERPs), increasing website traffic and revenue.

Reaching a highly targeted audience is one of the most important benefits of SEO for online enterprises. By optimizing content for specific keywords and phrases, e-businesses can attract potential customers actively searching for products and services related to their business. This approach is significantly more effective than conventional advertising methods, which are less precise in targeting. SEO can also help e-businesses establish trust and credibility with prospective consumers. Consumers generally view websites that rank higher on SERPs as more trustworthy and credible.

Additionally, content that ranks higher on SERPs is often viewed as more authoritative and informative. This credibility and trust can result in increased sales and consumer loyalty. However, there are some disadvantages to SEO. It can take time and resources to see results, which is a significant drawback. E-businesses must invest in producing high-quality content and optimizing their website, and it can take several months to see a significant rise in traffic and rankings. Moreover, SEO is a continuous process, and e-businesses must continually optimize and update their content to maintain their rankings and visibility.

To use SEO effectively, e-businesses must conduct keyword research to identify relevant keywords and phrases to optimize their content. This requires analyzing search volume, competition, and business relevance. E-businesses must also ensure that their website is user-friendly, fast-loading, and mobile-responsive to enhance the user experience and improve rankings. Creating informative, high-quality content is also essential for effective SEO. E-businesses should prioritize developing content that provides value to potential customers and solves their problems. This includes blog posts, product descriptions, and other types of online content. E-businesses should also focus on building high-quality backlinks to their website in addition to on-page optimization. Backlinks are links that point to the e-business's website from other sites. They signal to search engines that the content is highly relevant. By engaging in outreach, guest blogging, and other forms of content promotion, e-commerce businesses can build backlinks. Finally, e-businesses must continuously monitor and assess their SEO efforts to ensure they reach their target audience and generate sales. This involves monitoring rankings, traffic, and conversions and adjusting SEO tactics.

2.2 Pay-Per-Click (PPC) Advertising

PPC advertising, or pay-per-click advertising, is a type of Internet marketing that helps online businesses attract customers who are already interested in their products or services. This is achieved by displaying ads on search engine results pages and other online platforms. With PPC ads, companies only pay when customers engage with them, making it a cost-effective and time-efficient marketing strategy. One of the most significant advantages of PPC advertising is its ability to attract qualified visitors more likely to purchase. By targeting specific keywords and demographics, e-commerce enterprises can increase the visibility of their ads and attract interested consumers. Additionally, PPC advertising is highly measurable, allowing businesses to track clicks and conversions with tools like Google Analytics. This data can be used to make informed decisions and adjust campaigns for better results.

Another advantage of PPC advertising is its flexibility. Companies can set their budgets and make changes to the wording and audience of their ads as needed. However, some downsides exist, such as click fraud and increased keyword competition. To overcome these challenges, e-commerce businesses should conduct thorough keyword research and monitor their campaigns closely. To get the most out of PPC advertising, e-commerce businesses should conduct in-depth keyword research to find the most valuable and relevant keywords for their industry. They should also optimize their landing pages and ad content for conversions and target their desired audience. By continuously analyzing the effectiveness of their ads and making adjustments, businesses can generate revenue and attract the customers they want.

2.3 Social Media Marketing

To enhance brand awareness, attract and engage target audiences, and ultimately generate revenue, companies must promote their social media presence as part of their e-business strategy. However, with so many social media options available, it is crucial to establish an efficient social media marketing plan tailored to your business objectives and potential client demographics. Using social media for marketing has two key benefits: expanding audience reach and boosting brand profile. Consistently publishing valuable content, maintaining active communication with followers, and providing frequent updates are all effective ways to develop a robust online presence and connect with your target audience. Additionally, social media offers the opportunity to build personal relationships with existing and potential customers, increasing customer loyalty and retention rates and providing valuable feedback for product and service improvement.

While social media marketing can be more cost-effective than traditional marketing methods, maintaining an active presence on multiple platforms requires significant time and money. Furthermore, negative feedback can harm a company's image if not handled appropriately. To maximize the benefits of social media marketing, it is essential for online companies first to identify their target audience's demographics and choose the social media platforms that best suit their needs. Developing a content strategy that is engaging, valuable, and consistent can also help to build a dedicated audience and increase sales. Regularly monitoring social media analytics is another valuable tool for evaluating the effectiveness of your plan and making necessary adjustments to achieve better outcomes.

2.4 *Email Marketing*

Email marketing is one of the most popular and successful methods for increasing sales for online companies, as well as for increasing engagement with customers and prospects and for building brand recognition. Email marketing is also a marketing tactic with the lowest cost per customer acquisition. However, just as with any other marketing plan, you need to consider both the benefits and drawbacks of email marketing seriously. In this article, we will discuss the benefits and drawbacks of using email marketing for online companies. In addition, we will guide how to make the most of this marketing technique to achieve the most significant possible level of success.

Email marketing is beneficial for several reasons, but one of the most significant benefits is that it lets you personalize your communications to every recipient. This degree of personalization is made possible since the email list may be subdivided depending on criteria such as prior purchase history, interests, and demographics of the recipients of the emails. Emails that have been tailored for the receiver have the potential to develop connections with customers and prospects, as well as boost the possibility that the recipient will convert. This strategy has some benefits, including the relatively low email marketing cost. When compared to other marketing methods, such as print or television commercials, email marketing is a comparatively affordable marketing approach. Other marketing tactics include. While maintaining a broad customer base, e-commerce significantly reduces printing, shipping, and advertising expenses. E-commerce companies can also use email marketing to assess their marketing tactics' effectiveness. E-commerce enterprises may obtain significant insights into the content and message that will most likely connect with their target audience by evaluating open, click-through, and conversion rates. These metrics include open rates, click-through rates, and conversion rates.

On the other hand, this option has certain drawbacks that need to be considered. Regarding email marketing, one of the significant issues is ensuring that the emails you send do not come across as spam to the people who receive them. This may be accomplished by using strategies that call for a double opt-in, providing valuable material, and steering clear language that is excessively promotional. Taking precautions to ensure emails appear appropriately formatted across various devices and email programs is an additional difficult chore. Online companies need to ensure that their emails are optimized for mobile devices and compatible with some email systems to achieve maximum potential engagement and conversion rates.

Suppose online companies want to be successful with email marketing. In that case, they need to focus significantly on building a high-quality email list, segmenting their audience, and producing relevant and engaging content. Only then will they see the kind of results they are looking for. It is vital to consider both the time and frequency of communications sent to recipients to prevent overwhelming them with excessive emails. In conclusion, it is essential for online firms to routinely monitor and evaluate crucial email data in order to hone their company plan and enhance their operational efficacy.

2.5 Content Marketing

Content marketing has proven highly effective for e-commerce companies to engage their target audience, boost brand recognition, and drive sales. However, like any other marketing method, weighing the benefits and drawbacks of using content marketing is essential. This article will discuss the advantages and disadvantages of content marketing and guide how to maximize its potential.

One of the most significant advantages of content marketing is its ability to increase brand awareness among a wider audience. By creating informative and instructional content such as blog articles, videos, and infographics, e-commerce businesses can establish themselves as industry leaders and attract a loyal following on their websites and social media platforms. Furthermore, content marketing can be cost-effective compared to traditional marketing tactics, allowing companies to generate leads and communicate with a larger audience while keeping content generation and delivery costs minimal. Additionally, content marketing can help businesses attract high-quality leads by publishing content that resonates with their ideal consumers and meets the needs of their target audience.

However, content marketing also poses some challenges that businesses should consider. One of the significant challenges is the time and effort required to produce high-quality content consistently. To maintain audience interest and loyalty, e-commerce businesses must consistently post educational and engaging content. Another challenge is that content marketing can take time to yield results. Businesses must be patient and persistent in their content generation and distribution efforts to see a return on investment. To make the most of the potential benefits of content marketing, e-commerce companies must first identify their target audience's demographics and provide tailored content that addresses their concerns and interests. Consistency and regularity are also crucial in content marketing, as they help build a dedicated following. To assess the success of their content marketing plan and make necessary improvements, e-commerce businesses should track website traffic, engagement rates, and lead generation.

2.6 Mobile Marketing

Mobile marketing is essential for Internet companies because it enables them to communicate with their target audience, raise brand awareness, and drive sales. It is necessary to grasp this marketing approach's advantages and downsides to maximize the efficacy achieved via mobile marketing. In this detailed guide, we will cover the benefits and drawbacks of mobile marketing and give guidance on how online companies may use mobile marketing to reach their target audience and improve sales efficiency.

One of the essential advantages of using mobile marketing is that it enables businesses to communicate with many people. E-commerce websites now have a better possibility of interacting with the people they aim to target due to the spread of mobile technologies such as smartphones and tablets. Online companies can give their clients a more customized experience and provide targeted messages to individuals who are most likely to become paying customers if they employ techniques for mobile marketing. The capacity of mobile marketing to bring in new business is only one of its numerous benefits. E-commerce organizations can improve customer trust and loyalty, which may eventually result in more significant sales if they communicate with customers in a customized and relevant way. Mobile marketing could simplify it for online firms to give clients unique offers and discounts, ultimately increasing the possibility that a customer will purchase. However, despite these advantages, mobile marketing does include a few obstacles to overcome. E-commerce companies face various challenges, one of the most important of which is the difficulty in choosing which mobile platform, out of the multiple possibilities available, would be the most successful at reaching their target audience. In addition, mobile marketing necessitates that Internet companies have a significant mobile presence, which may be implemented in a time-consuming and expensive way.

Before attempting to adopt mobile marketing effectively, online companies must first discover their target demographic and the mobile platform(s) their audience uses most to adopt it effectively. In addition, organizations that engage in e-commerce have to optimize their websites for mobile usage and provide versions of their content that are compatible with mobile devices. E-commerce enterprises can also adopt various mobile marketing tactics, such as SMS marketing, mobile advertising, and in-app advertising, to engage with their target customers and enhance sales. These mobile marketing strategies include mobile advertising. Online companies need to keep track of the success rates of their mobile marketing operations. This allows the firms to assess the outcomes of their efforts, discover how they can be improved, and then implement those improvements.

There are many instances of online firms that have utilized digital marketing platforms effectively in order to attract their target customers and boost their sales. One such example is Amazon.com. Take into consideration the following examples:

Airbnb has developed an influential brand using various digital marketing platforms to engage with its target audience. It will be much easier for tourists to find one-of-a-kind hotels if they use their website since it is created with user-friendliness in mind and offers a customized booking experience. They also connect with their audience on social media sites such as Instagram and Twitter, distributing user-generated material and advertising their services. In addition, they have a comprehensive content marketing strategy including blog posts, videos, and interactive material that showcases the unique experiences that customers who use their platform may have.

Warby Parker is a booming e-commerce firm that has caused a stir in the eyeglasses market by offering fashionable eyewear at more affordable costs than traditional retailers on their websites. They have established two-way communication with their clientele via various digital marketing platforms, such as email marketing, and offered individualized product suggestions. They also run aesthetically attractive ads featuring their goods on social media sites to reach the population they are trying to reach via adverts on such platforms. In addition, they work with prominent people to promote their brand and provide customers with unique collections; these activities contribute to a rise in sales and a greater awareness of their brand.

Dollar Shave Club is a well-known online retailer selling men's grooming items and offering services based on a membership model. They advertised their items using digital marketing platforms such as viral videos to raise customer awareness of their brand. In addition, they engage in email marketing, advertising on social media platforms, and partnering with persons with significant influence to enhance sales and reach their target audience. The lighthearted nature of their brand voice and the pertinence of the material they provide are essential factors in developing loyal customers and promoting more purchases.

Each of these examples highlights, in a nutshell, how online firms can make efficient use of a range of digital marketing platforms to strengthen their brands, improve their ability to engage with their target audiences, and boost their overall number of sales.

3 Digital Marketing Strategies

Questionnaires, focus groups, and online data analyses are just a few of the many instruments available to market researchers. The research findings can be used to create consumer personas (idealized representations of the ideal customer). Organizations frequently develop "buyer personas" to better comprehend consumers' desires, needs, and behaviors. Using these profiles, marketers can communicate with target audiences more effectively.

3.1 Define Goals and Objectives

Although "goals" and "objectives" are occasionally used interchangeably, they refer to two distinct concepts. Goals are more general statements of what an organization aims to achieve, whereas objectives are more specific and quantifiable actions it will take to achieve its goals. The company's mission and vision must align with its goals and objectives.

Goals and objectives for digital marketing include increasing website traffic and revenue, generating prospects, and enhancing brand recognition. The targets and objectives must be SMART, specific, measurable, achievable, relevant, and time-bound. The following is a list of suggestions for defining digital marketing goals and objectives (Fig. 9.1):

1. Conduct a SWOT Analysis: Companies benefit from a SWOT analysis because it enables them to identify their strengths, weaknesses, opportunities, and threats. It aids in identifying areas in need of development and provides significant insight into the internal and external factors affecting the company. Businesses can define goals and objectives that align with their strengths and opportunities by conducting a SWOT analysis.

Fig. 9.1 Steps for defining digital marketing goals and objectives

2. Define Your Target Audience: Defining the target audience is crucial in developing practical digital marketing goals and objectives. It facilitates businesses in comprehending their customers' needs, preferences, patterns of behavior, and challenges. Businesses can develop specific, focused goals and objectives when the target audience is identified. These objectives and goals will then resonate with consumers.
3. Utilize Data and Analytics: Data and analytics can better understand customers' behavior, preferences, and trends. Using this information, businesses can establish goals and objectives that are concrete, quantifiable, and aligned with the needs of their customers. For instance, if the data reveals that most consumers originate from social media, the goal may be to increase social media interaction by 50% in 6 months.
4. Company's Purpose and Vision: The company's goals and objectives should align with the organization's overall mission and vision. They should be tailored to the specific industry the organization operates in, the values it maintains, and the function it serves. Businesses can develop goals and objectives that accurately reflect the identity and values of their brand if they keep the company's mission and vision in mind while developing these goals and objectives.
5. Focus on the Return on Investment (ROI): Return on investment (ROI) is an essential metric for measuring the performance of a company's digital marketing initiatives. It evaluates how much profit was generated from digital marketing expenditures. A company's ROI goals and objectives should be explicit, quantifiable, and aligned with the company's broader business objectives.
6. Be Realistic: Your goals and objectives should be attainable and reasonable. Unattainable objectives and aspirations may result in disappointment and a loss of motivation. Businesses must establish goals and objectives that will challenge them but are also attainable within a given time frame.
7. Evaluate and Make Alteration: Because digital marketing is a subject that is constantly evolving, it is essential to assess and modify one's goals and objectives frequently. A company's goals and objectives should be reviewed regularly, and any necessary adjustments should be made in light of data and analytics.

3.2 *Choose the Right Channels*

Today's companies have many distribution channels for developing digital marketing plans. Companies must use the best communication channels to reach their target audience and successfully achieve their marketing goals. Figure 9.2 outlines the steps for choosing the right channel.

1. Know exactly who you are writing for. When developing a digital marketing plan, the target audience is one of the most important considerations. Businesses may make more informed channel decisions if they have a complete picture of their target audience's demographics, interests, and online activity. For instance,

Fig. 9.2 Steps for using the right channel for an online business

if your target demographic skews are young, you could find better successful advertising on platforms like Instagram and TikTok.

2. Find the most promising distribution methods. When choosing channels, businesses should focus on those most likely to assist them in meeting their marketing goals. Before making judgments, companies should investigate which channels have strong engagement rates, traffic levels, and lead generation potential.
3. Give some thought to the content you want to promote. The nature of the company's intended marketable product is a critical consideration in terms of channel choice. Instagram is great for sharing photos and videos, whereas LinkedIn is more suited to sharing articles and other written information. One such platform is Instagram. Businesses must make distribution decisions that give their content an excellent chance of success.
4. Take stock of your financial situation and your options. The amount of money and other resources at your disposal is also a factor in deciding which channels to use. Businesses must weigh the prices and resources when deciding which channels to invest in. When compared to managing a Twitter account, creating and maintaining a YouTube channel may demand more time and effort.
5. Look at the methods your rivals are using: Studying how rivals distribute their products may tell you much about what works and what does not in this space. By analyzing their rivals' channels, businesses may get insight into what works and what does not and better decide which channels to use.

6. The channels need to be reviewed and improved regularly. Businesses must constantly experiment with new channels and evaluate their efficacy. They may use the data to determine which channels are worth keeping and which can be scrapped.

Multichannel digital marketing can expand your reach and pique the interest of your ideal customers. Multichannel marketing aims to convey the same message over as many channels as possible. Organizations may use this tactic to reach different segments of their target audience with a unified message across all channels. It is essential that when businesses use a multichannel marketing approach, the message remains constant across all mediums. The message must be consistent with the goals and ideals of the brand. You need to know your audience, content, budget, and available resources inside and out before choosing the proper channels for your digital marketing plan. By considering the elements above, organizations may build a multichannel marketing strategy that aligns with their marketing goals before determining which channels to utilize. Successful execution of the marketing plan in keeping with the company's overall goal and vision requires constant analysis and optimization of the channels.

3.3 *Create Quality Content*

High-quality content is essential for successful digital marketing campaigns in today's world of advanced computing. By consistently producing engaging material, businesses can increase their market visibility and attract more website visitors. Figure 9.3 illustrates the steps required to create quality content.

1. Know exactly who you are writing for. Having a deep understanding of your target demographic is crucial when crafting engaging content. Understanding the demographics, hobbies, and pain points of one's target audience may help firms develop content that more successfully conveys their messages. Creating a consumer persona may help businesses zero in on their ideal clients and tailor their offerings to their needs.
2. Give the readers a reason to care. The best content answers a question or solves a problem for the reader. Whether the goal of the material is to inform the reader on a specific subject or to provide answers to the difficulties they encounter, it should be informative and actionable. Businesses should provide content that is useful to their intended audience and contains information unavailable elsewhere.
3. Include visuals. Images incorporated into text help readers retain information and find it interesting. Visual components like images, movies, and infographics may make the text more comprehensible and exciting. Visuals not only help the audience better understand the topic, but they also make even the most challenging ideas easier to grasp.

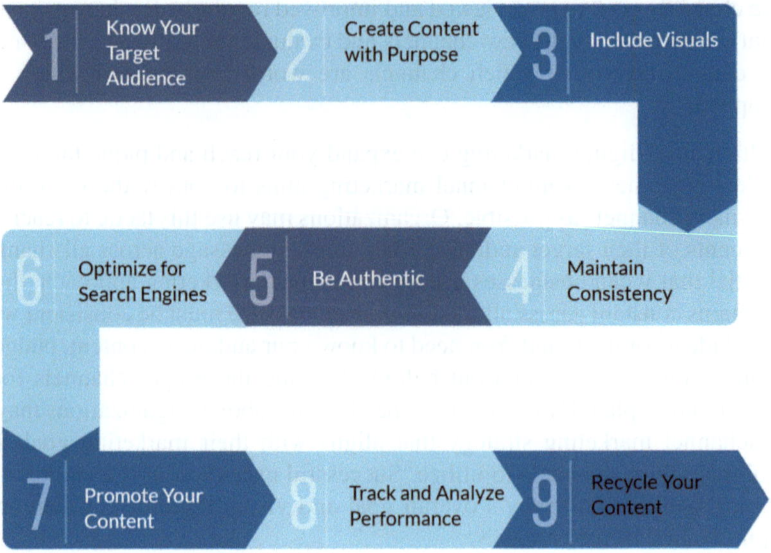

Fig. 9.3 Steps for creating quality content

4. Maintain uniformity: Consistency is one of the essential parts of making excellent content. Businesses should make creating and sticking to a content schedule a top priority. Establishing a consistent schedule, such as once per week or once per month, may assist in building rapport with your audience and keep them coming back for more. This is true whether you use a weekly or monthly calendar.
5. Be honest: To create engaging and informative pieces, authenticity is crucial. Brands must offer content that reflects the company's core beliefs and aims. There should be no overt attempts to make the content seem like a sales pitch. Being yourself is one of the best ways to win over an audience and lay the groundwork for a lasting connection.
6. Optimize your content for search engines. Website traffic may be increased by enhancing content for search engines. For a firm to succeed, it is crucial to communicate with its customers using language they can relate to. They should also use tactics like headings, bullet points, and short paragraphs to make their writing as readable as possible.
7. Promote your material. Promoting one's material is crucial in drawing in one's target demographic. Content promotion is a choice for organizations, and they may choose from channels such as social media, email marketing, and paid advertising. They need more exposure. Thus, they should collaborate with other organizations or influential people.
8. Observe and assess. Tracking and analyzing content's success or failure is crucial for figuring out what works and what does not. The use of analytics solutions to track metrics like site traffic, user engagement, and bounce rate is highly rec-

ommended for businesses. This data might help refine their content strategy and create more exciting pieces for their readers.

9. Repurpose your content. Reusing material is an excellent idea to get the most out of your data. Blog entries may be turned into films, and data can be turned into infographics, which can be used for commercial reasons. By reusing material, organizations might reach a broader audience and strengthen their message across numerous mediums.

Successful digital marketing campaigns always include a focus on content production of the highest quality. Businesses can generate content that engages their audience and drives traffic to their website by first identifying their target audience, then providing value, then incorporating images, then being consistent and genuine, then optimizing for search engine optimization, then marketing the content, then tracking and analyzing its performance, and finally repurposing the content. Consistent analysis and revisions are necessary to align the content strategy with the company's main goals and objectives.

3.4 Monitor and Measure Results

It is vital to monitor and evaluate the outcomes of a digital marketing campaign to identify whether or not the campaign was effective and to make choices on future efforts based on the data gathered. Businesses can find areas for improvement, adjust their strategies, and optimize their campaigns when they monitor and analyze the outcomes of their efforts to see where improvements may be made. As a result, they become closer to accomplishing what they set out to do. In this section, we will review some of the most effective methods for keeping track of and evaluating the outcomes of a digital marketing campaign (Fig. 9.4).

1. Determine your goals and how you will measure them. Before a corporation can begin tracking and analyzing its performance, the organization must determine its goals and metrics. The formulation of SMART goals, that is, specific, measurable, achievable, relevant, and time-bound, is very necessary. The success of the campaign needs to be evaluated using measures that are in line with the aims that were initially articulated. Website traffic, click-through rates (often abbreviated as CTR), conversion rates, cost per acquisition (sometimes abbreviated as CPA), and return on investment (usually abbreviated as ROI) are all examples of metrics.
2. Employ the analytical tools available. Monitoring and analyzing the effectiveness of an organization's digital marketing initiatives may be facilitated by using analytics solutions. Google Analytics is a program used by many people and gives data on website traffic, user behavior, and conversion rates. A target audience's engagement, reach, and demographics may all be analyzed using analytics tools made available by social media platforms. Moz, Ahrefs, and Semrush

Fig. 9.4 Steps for monitoring and measuring results in e-business

are a few examples of other software that fall under search engine optimization (SEO) and keyword ranking analysis tools.

3. Establish a system for tracking conversions. Conversion tracking is a technology that allows companies to monitor certain user activities on their website, such as submitting a contact form or completing an online transaction. The use of conversion tracking is one method that may be used to carry out this monitoring. If a company uses conversion tracking, it will be able to assess the effectiveness of its marketing campaigns and make adjustments to boost the effectiveness of those initiatives. In addition to Google Analytics, some third-party solutions allow users to establish conversion tracking.
4. Monitor the return on investment (ROI) for the campaign. Tracking the return on investment (ROI) is essential to determine whether or not a digital marketing strategy was successful. A campaign's return on investment (ROI) may be calculated by contrasting its revenue with its costs. When companies calculate their

return on investment, also known as ROI, they can assess the profitability of their marketing endeavors and modify their strategy by the findings.
5. Be conscious of your reputation in the Internet world. If you want to maintain an excellent public image for your company, monitoring your Internet reputation is an absolute must. With social listening tools, businesses can monitor brand mentions across social media platforms, review websites, and online forums. Monitoring a company's Internet reputation allows organizations to react to adverse comments and reviews immediately, maintain a good brand image, and respond to customer feedback.
6. Evaluation as well as adaptation of campaigns. The process of monitoring and assessing the success of the campaign should not be a one-time activity but rather one that occurs continuously throughout its whole. Businesses can identify parts of their campaigns that require better, maximize the effectiveness of those efforts, and accomplish their objectives by continually reviewing and making changes to the campaign. For instance, if a campaign needs more leads, firms can increase its success by adjusting the targeting, the messaging, or the offer they are presenting to potential customers.
7. Contrast the campaign's outcomes with the industry's criteria. When the outcomes of the campaign are compared to the norms of the industry, it may give insight into the success of the campaign and show areas that need improvement. Industry standards may be established from various places, including marketing companies, trade organizations, and research organizations. Comparing a firm's outcomes to those of its industry counterparts may help a company determine its strengths and weaknesses and adjust its strategy according to its competitors.
8. Stakeholders should be informed of the results of the campaign. It is vital to convey the outcomes of a digital marketing campaign to stakeholders, including executives, sales teams, and clients. This is done to show the campaign's usefulness and justify the investment. The reporting must be understandable, succinct, and adapted to whatever audience is being addressed. The use of visual aids such as diagrams and charts may increase the clarity of factual information as well as its overall effectiveness.
9. Monitoring and analyzing the outcomes of a digital marketing campaign is necessary to ascertain whether or not the effort was fruitful and improve the campaign's overall efficiency. Businesses can define goals and metrics, use analytics tools, set up conversion tracking, track campaign ROI, monitor online reputation, evaluate and adjust campaigns, compare results to industry benchmarks, and communicate results to stakeholders to achieve their marketing objectives and make data-driven decisions. It is necessary to engage in ongoing analysis and modification of the process to guarantee that the monitoring and measuring procedure is by the main aims and objectives of the firm.

In today's day and age, when it comes to marketing, businesses have access to a wide variety of marketing channels via which they may contact the customers they are trying to get in touch with. Businesses need to include the many different digital marketing channels in their overarching marketing strategy to accomplish their marketing goals and get the most out of their influence. This tactic offers some advantages, including a broader audience as well as an increase in involvement. When a company uses different channels, it can broaden its visibility across different platforms and connect with more people.

It can effectively demonstrate the potential of channel integration by providing real-world examples of successful digital marketing campaigns that include a range of different digital marketing channels. For instance, a firm may leverage social media to raise consumers' knowledge of its brand and interest in its products. After then, the business may employ search engine marketing to produce conversions, after which they could use email marketing to develop leads. These campaigns provide an example of how effectively integrating numerous channels may contribute to success and how those channels can work together to achieve success. An additional advantage of this approach is that it establishes a unified message and a lone voice for the company. This may be accomplished via the integration of numerous digital marketing platforms. This is because a brand's message can be repeated across different media, giving customers a consistent experience across their interactions with the brand. Not only does this contribute to a rise in consumer happiness but it also helps to fortify the brand's identity.

However, to successfully perform the essential responsibilities, it is vital to deploy the marketing budget among the different channels thoughtfully and strategically. Return on investment (ROI) may be maximized for businesses if they first discover which channels are most successful at reaching their target audience and then allocate expenditures by that information. In addition, putting this plan into action helps firms increase the scope of their marketing initiatives while simultaneously using the most productive channels for their campaigns. When optimizing a multichannel digital marketing strategy, data analysis and developing conclusions based on that analysis are crucial steps. As a result, businesses can discover possibilities for improvement and make better informed choices as a direct consequence, eventually resulting in improved outcomes. Using data to assess the efficiency of the different channels makes this objective a realizable goal. This approach helps companies preserve their agility and flexibility, enabling them to adjust their strategy to more effectively satisfy the requirements of the audience they are aiming to attract [3].

In conclusion, companies must integrate different digital marketing platforms to fulfill their marketing objectives and maximize their marketing efforts. Companies can create complete plans that effectively use many channels, provided they present real-world examples, highlight the advantages of integration, provide realistic advice for budget allocation, and explain the need for data analysis. Because of this tactic, companies will be able to communicate with their target audience, which will lead to an increase in engagement and the generation of conversions, ultimately leading to success over the long run.

The "Just Do It" campaign that Nike ran is an excellent example of an outstanding digital marketing campaign that used a variety of platforms. The marketing campaign was launched in 2018, using diverse digital marketing channels to convey a consistent message to the most significant number of people. Some well-known athletes, such as Serena Williams, LeBron James, and Colin Kaepernick, took part in a series of adverts shown on television as part of the campaign. After then, these adverts were published on several social media platforms owned by Nike, such as Facebook, Twitter, and Instagram. In addition, the team designed a simple website and offered visitors the opportunity to craft their "Just Do It" messages, which could subsequently be shared across different social media sites. In addition to these channels, Nike engaged in customer interaction and product promotion via email marketing.

The "Share a Coke" campaign run by Coca-Cola is an exemplary digital marketing push that utilizes multiple channels effectively. In 2011, the promotion urged people to personalize Coca-Cola bottles with their names, aiming to increase both participation and brand awareness. The campaign used various digital marketing channels, including social media, email, and experiential marketing. Coca-Cola developed unique IDs for its #ShareACoke campaign to encourage customers to post online images of their customized bottles and interact with the brand. The company also sent customers customized emails with their names in the subject line, encouraging them to share their images on social media using the hashtag #ShareACoke. In addition, Coca-Cola built vending machines in critical cities worldwide as part of a marketing effort focused on the user's whole experience with the brand. Coca-Cola's marketing strategy included various channels, each with its unique purpose, allowing the company to create a cohesive message and engage customers across multiple touchpoints. The success of the campaign is evidenced by the fact that Coca-Cola's sales increased by 2% in the year following its launch. This example shows how combining multiple digital marketing platforms can generate a consistent message and increase consumer engagement, increasing customer loyalty and income for companies.

4 Metrics and Analytics in Digital Marketing

To determine whether or not a digital marketing campaign was successful, it is necessary to examine several metrics and key performance indicators (KPIs). In this section, we will discuss the typical metrics and KPIs used in digital marketing campaigns, why it is important to align them with business objectives, how to pick the right metrics and KPIs to monitor, and how to utilize the data you gather to make intelligent choices and fine-tune your marketing efforts. In addition, we will go over how to choose the proper metrics and KPIs to track. In digital marketing, key performance indicators and metrics often used include website traffic, conversion rates, return on investment (ROI), cost per acquisition (CPA), and social media activity. Table 9.3 provides a comparison of common KPIs and metrics.

The number of visits a website receives is one of the most critical measures of how successful the website is. Statistics include the total number of users, the number of page visits, the exit rates, and the average amount of time spent on the site. Marketers may use these to understand better how site visitors interact with a website. The total number of pages a single user reads throughout their session is called their "page views," whereas the number of "unique visitors" refers to the number of people visiting the site for the first time. The proportion of site visitors that only look at one page before departing is referred to as the "bounce rate," and time on site is a statistic that is used to measure how engaged an audience is with a website by demonstrating how long they remain there. This statistic is determined by how much time users spend on the website. A digital marketing campaign's effectiveness is determined by looking at the campaign's conversion rate. It does things like counting the number of individuals who visit a website, fill out a form, or complete a purchase, and then it figures out what percentage of those people fit into a particular group. The proportion of site visitors who make the intended purchases or complete the forms of user engagement that the marketer wants them to do may indicate how successful a marketer's website and other digital marketing activities are [4].

Calculating a campaign's return on investment (ROI), which compares its costs with its revenues, is one way to determine whether it will succeed. This key performance indicator (KPI) could assist marketers in better managing their resources and determining the effectiveness of their digital marketing operations. The "cost per acquisition," often known as "CPA," is an essential indicator for determining how much it will cost to attract new customers. To get at this figure, divide the total cost of an advertising campaign by the total number of new customers acquired as a direct result of that campaign. By measuring CPA, marketers can evaluate how successful their attempts to acquire customers have been and make decisions guided by the data to enhance the efficacy of their campaigns. It is possible to get a good idea of how successful digital marketing techniques are by looking at how much interaction they produce on social media. The total number of a user's likes, shares, comments, and followers constitutes that user's total level of social media involvement. By assessing consumers' level of engagement with their social media activities, marketers may get insight into the efficiency of such efforts in terms of engaging with people and eliciting interactions from users with the brand. Marketers need to have a solid understanding of the most popular key performance indicators (KPIs) and metrics used in digital marketing to accurately evaluate their campaigns' effectiveness and make data-driven decisions that will improve performance. Table 9.3 provides a comparison of KPI and metrics utilized in digital marketing.

To ensure the success of digital marketing operations, it is essential to align key performance indicators, also known as KPIs, with the corporate entity's goals and objectives. KPIs, or key performance indicators, are quantitative benchmarks that allow businesses to measure their progress toward attaining their objectives. It is necessary to establish a link between these key performance indicators (KPIs) and the organization's overall strategy to increase their likelihood of generating success and analyze the efficacy of campaigns. Because of this, they will be able to monitor the campaign's efficacy effectively.

The first step in developing relevant and meaningful key performance indicators (KPIs) for measuring the effectiveness of digital marketing campaigns is to define the company's objectives and strategies. To correctly align key performance

Table 9.3 Comparison of common KPIs and metrics used in digital marketing campaigns

KPI/metric	Definition	Importance	Calculation
Website traffic	The number of visitors to a website	The essential metric for measuring digital marketing campaign effectiveness	Unique visitors, page views, bounce rates, and time on site
Conversion rate	The percentage of website visitors who complete a desired action	Essential KPI for measuring the effectiveness of digital marketing campaigns	Number of conversions/number of visitors
Return on investment (ROI)	A critical metric that helps determine the profitability of a campaign by comparing the cost of the campaign to the revenue generated	Essential KPI for determining the success of a digital marketing campaign	(Revenue – cost)/cost
Cost per acquisition (CPA)	The cost of acquiring a customer	An essential metric for measuring the effectiveness of digital marketing campaigns in acquiring new customers	Total cost of the campaign/number of new customers acquired
Social media engagement	The number of likes, shares, comments, and followers on social media platforms	An essential metric for measuring the effectiveness of social media campaigns	Number of likes, shares, comments, and followers on social media platforms

indicators with business objectives and plans, the first step is to establish goals and directives precisely. This stage involves identifying the company's broad aims and objectives and the strategies that will be employed to achieve them. In other terms, this phase involves setting objectives. One example is a company that has decided to increase its income by 20% over the next year and has devised a strategy to achieve this goal by increasing online sales. Following a comprehensive description of the organization's goals and plans, the next step is to identify the key performance indicators (KPIs) that will be used to measure the degree to which the organization is achieving its goals. To accomplish this, it is necessary to select KPIs that are crucial to the organization's goals and strategies, as well as those that can provide actionable insights into the performance of the campaign. Website traffic, conversion rates, and online sales revenue are key performance indicators (KPIs) that may be evaluated by a company seeking, for instance, to increase online sales revenue. Other examples of key performance indicators include customer retention and customer satisfaction.

Moreover, key performance indicators (KPIs) must be SMART, an acronym for specific, measurable, attainable, relevant, and time-bound. This implies that key performance indicators (KPIs) must be meticulously defined, measured, and accompanied by a time-bound objective. The adoption of SMART KPIs ensures that progress can be monitored and that data on the success of campaigns can yield actionable insights. The Smart Key Performance Indicators Collaborative created SMART

KPIs. To ensure the success of digital marketing initiatives, it is essential to align key performance indicators (KPIs) with business objectives and strategic plans. When key performance indicators (KPIs) are aligned with business objectives and plans, they provide a unique road map that can be used to evaluate the efficacy of campaigns and quantify their success. KPIs are also necessary for ensuring that resources are utilized effectively and that digital marketing efforts generate a positive return on investment (ROI). This can be achieved by assuring the tracking of KPIs.

In aligning key performance indicators (KPIs) with business objectives and strategies, it is crucial to establish a balance between short-term and long-term goals. Short-term objectives may be focused on achieving quick victories and generating immediate cash. In contrast, long-term objectives may involve establishing brand recognition and consumer loyalty or developing new products and services. Both of these categories of objectives are essential in the long term. KPIs should be representative of both short- and long-term objectives and be reviewed frequently to ensure that they align with the ever-changing business plans and objectives. When aligning key performance indicators (KPIs) with the organization's goals and strategies, one of the essential factors to consider is ensuring that KPIs are related to the entire consumer journey. This implies that key performance indicators (KPIs) should be selected to depict the various phases of the customer journey, commencing with the customer's initial awareness and interest and continuing through consideration, purchase, and loyalty. In other words, the customer voyage begins with initial customer awareness and interest and concludes with customer loyalty. Businesses can ensure that they are analyzing the efficacy of their digital marketing initiatives throughout the entire customer journey by selecting KPIs that reflect the customer journey and then employing these KPIs to evaluate the efficacy of their digital marketing initiatives. In addition, organizations can establish strategies to improve performance at each stage when they use KPIs that represent the customer's journey. It is essential to continuously monitor and evaluate key performance indicators (KPIs) and select KPIs that align with the company's strategy and objectives. This involves utilizing analytics tools to track the performance of key performance indicators and identify patterns and insights that can be used to improve the success of campaigns. Then, these patterns and insights can be used to enhance the success of future campaigns. The consistent surveillance of a company's critical performance indicator data can significantly aid its ability to determine which aspects of its operations are performing well and which require enhancement. This can also contribute to the ongoing improvement of marketing initiatives over time [2]. Figure 9.5 is a possible flowchart idea for aligning KPIs with business objectives and strategies in digital marketing.

Analyzing and interpreting data is crucial for making decisions in digital marketing. Today, more than ever, it is critical for businesses to use the vast amounts of data at their disposal to get a deeper understanding of their consumers, improve their marketing strategies, and ultimately boost sales. However, the abundance of data available may be overwhelming. Handling and evaluating data meticulously is necessary to produce results that can withstand scrutiny.

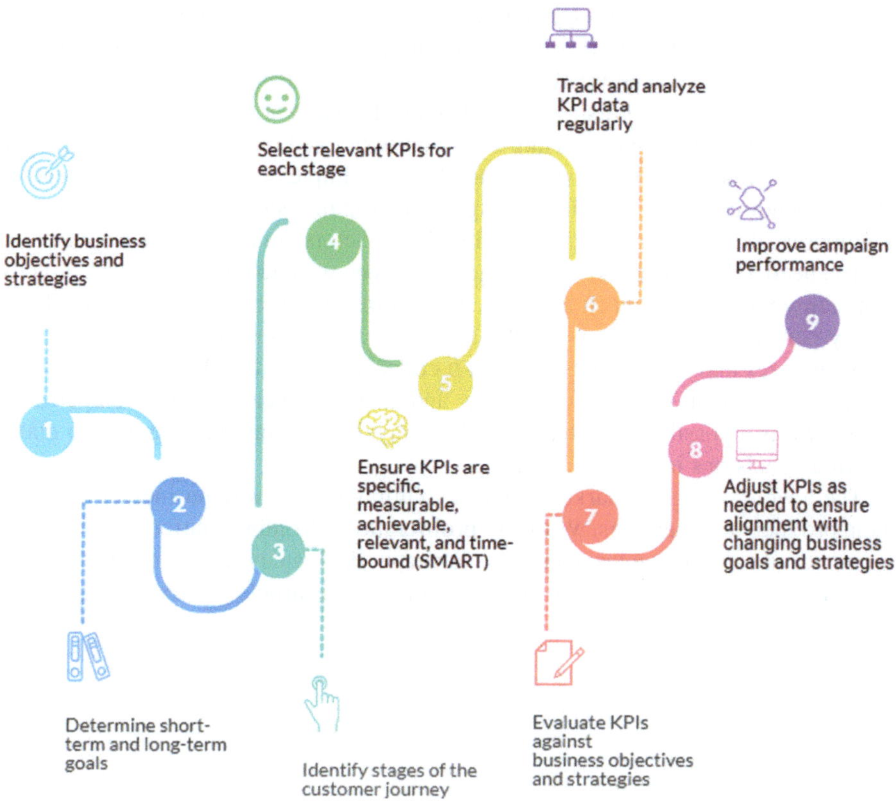

Fig. 9.5 Aligning KPIs with business objectives and strategies in digital marketing

Key performance indicators (KPIs) must be identified as most relevant to achieving the company's goals and strategy before any data analysis can begin. It is common practice for businesses to utilize key performance indicators (KPIs) to measure the success of their marketing campaigns. Since metrics vary depending on the campaign and company goals, each set of KPIs can be used in various ways. Once KPIs have been established, monitoring and evaluating the data regularly is essential to see how things are progressing. It is crucial to any research or interpretation to verify the accuracy and reliability of the data employed. Data can be collected from various places, including online analytics tools, customer relationship management (CRM) programs, and social media platforms. Decisions may only be made with conviction after verifying all available facts' completeness, timeliness, and consistency.

Once the data is collected and verified for accuracy, it may be analyzed for trends and patterns. Regression, correlation, and cluster analyses, among others, may also be helpful. The purpose of data analysis is to provide valuable insights. You may discover which kinds of clients are the most profitable, which marketing strategies work best, and which channels provide the most leads.

The next critical step in obtaining credible results is examining the data. This means using data analysis to inform and direct strategic planning and decision-making processes, such as creating more effective advertising campaigns or shifting resources to more fruitful distribution channels. Looking at the whole picture while evaluating data rather than focusing solely on individual data points is essential. If you see a decline in website traffic 1 month, for example, it is not always indicative of an impending failure of your campaign. The drop in visitors might be seasonal or a temporary drop in search engine results. Other factors are likely to be responsible.

You need to keep an eye on campaign stats if you want to make informed choices that help your business reach its goals. The campaign must be fine-tuned regularly as key performance indicators (KPIs) are tracked and assessed. How well a company analyzes and understands the data it collects daily will influence its ability to make choices that boost performance and keep it ahead of the competition. Figure 9.6 is a possible flowchart idea for analyzing and interpreting data to make informed decisions in digital marketing.

The advent of digital marketing has revolutionized the way businesses promote their products and services. With the proliferation of online platforms and much consumer data, companies can now leverage data analytics to optimize their marketing strategies and achieve better outcomes. This article will explore how data

Fig. 9.6 Steps for analyzing and interpreting data to make informed decisions in digital marketing

analytics and technology can be harnessed to improve digital marketing campaigns. Data analytics involves collecting, processing, and analyzing data to extract insights and guide decision-making. In digital marketing, data analytics is used to assess customer behavior, monitor campaign performance, and identify areas for improvement. Businesses can gain a competitive edge by employing data analytics solutions and staying ahead of the curve in their marketing efforts. One of the most common applications of data analytics in digital marketing is measuring marketing initiatives' effectiveness. By analyzing key performance indicators (KPIs) such as website traffic, conversion rates, and customer engagement, companies can evaluate the impact of their marketing activities and make informed decisions on how to refine their strategies. For instance, if a campaign is underperforming, data analytics tools can help uncover the underlying issues and enable the business to adjust its approach accordingly. Another practical application of data analytics in digital marketing is to gain insights into consumer behavior. By collecting and analyzing consumer data, businesses can better understand their target audience's preferences and behaviors. This can inform the development of more personalized and relevant marketing messages that resonate with customers and enhance engagement. Moreover, by scrutinizing customer data, organizations can identify new market segments or niches they may have overlooked before, allowing them to expand their reach and customer base [4].

Businesses can leverage various data analytics technologies to enhance their digital marketing strategy. Google Analytics is a widely used tool that provides in-depth insights into website traffic, user activity, and conversion rates. Other popular products include Adobe Analytics, which offers extensive analytics capabilities and marketing automation features, and Hootsuite, which enables organizations to monitor social media activity and analyze engagement metrics. When using data analytics tools to devise digital marketing tactics, adhering to data privacy and security regulations is crucial. Businesses must ensure that they acquire and handle consumer data responsibly and ethically and comply with relevant laws and regulations, such as the General Data Protection Regulation (GDPR) of the European Union. Table 9.4 provides data analytics tools for improving digital marketing strategies.

Table 9.4 Data analytics tools for improving digital marketing strategies

Type of data analytics tool	Key features and benefits
Google Analytics	Provides detailed insights into website traffic, user behavior, and conversion rates
Adobe Analytics	Offers advanced analytics capabilities and marketing automation features
Hootsuite	Allows businesses to monitor social media activity and track engagement metrics
CRM tools	Helps businesses manage customer relationships and track interactions
A/B testing tools	Allows businesses to test and compare different versions of marketing messages or campaigns
Marketing automation tools	Helps businesses automate repetitive tasks and streamline marketing workflows
Data visualization tools	Allows businesses to create visual representations of complex data sets for better analysis and decision-making

Summary

Digital Marketing

- Definition: Digital marketing uses digital technologies, such as the Internet, mobile devices, social media, and search engines, to promote products or services and engage with target audiences.

Digital Marketing Channels

- Definition: Digital marketing channels are the various platforms and mediums through which marketers can reach and interact with their target audience. These channels include websites, search engines, social media platforms, email marketing, mobile apps, and display advertising.

Digital Marketing Strategies

- Definition: Digital marketing strategies are the deliberate actions and approaches marketers use to achieve specific objectives in the digital realm. These strategies may include content marketing, search engine optimization (SEO), social media marketing, paid advertising, influencer marketing, email marketing, and more.

Metrics and Analytics in Digital Marketing

- Definition: Metrics and analytics in digital marketing refer to the measurement and analysis of various data points and key performance indicators (KPIs) to evaluate the effectiveness and success of digital marketing campaigns. This includes tracking metrics such as website traffic, conversion, click-through, engagement, and return on investment (ROI) to make data-driven decisions and optimize marketing efforts.

Case Study

XYZ Clothing is a well-known online apparel retailer specializing in fashionable and appealing garments for young people. The company must revise its digital marketing strategies to regain market share and attract new customers, as online sales have declined. As a digital marketing consultant, it is your responsibility to evaluate the current situation and recommend solutions that will assist XYZ Clothing in achieving its e-business goals.

XYZ Clothing has relied heavily on traditional marketing strategies, such as print advertisements and billboards. However, it recognizes the need to embrace digital marketing to remain competitive in the e-commerce industry. They have just launched a brand-new website and wish to use multiple digital marketing channels to increase traffic, conversions, and consumer engagement. However, they are still determining the most effective digital marketing strategies and how to evaluate their effectiveness.

1. XYZ Clothing has identified social media as a viable digital marketing channel. According to you, what are the primary advantages of using social media platforms like Facebook and Instagram for e-commerce?
2. XYZ Clothing's email marketing campaign aims to strengthen client relationships and encourage recurrent purchases. What specific elements should their email marketing strategy include to increase its efficacy?
3. The website of XYZ Clothing has been optimized for search engines (SEO) to increase its visibility in search engine results. How can effective SEO techniques contribute to their e-commerce success and increase organic website traffic?
4. XYZ Clothing is considering pay-per-click (PPC) advertising to generate rapid traffic and conversions. What are the primary benefits of pay-per-click advertising, and what should be considered when establishing and maintaining PPC campaigns?
5. XYZ Clothing desires to use key performance indicators (KPIs) and analytics to evaluate the effectiveness of its digital marketing efforts. What metrics and tools should businesses use to evaluate the efficacy of their digital marketing initiatives, and how can they use this information to refine their strategies?

Discussion

1. How has the rise of e-commerce and mobile technology impacted the growth of digital marketing?
2. What are some examples of successful digital marketing campaigns and how did they contribute to the success of the business?
3. How has the COVID-19 pandemic affected the importance of digital marketing for businesses?
4. What are some emerging trends in digital marketing that businesses should be aware of?
5. How can businesses ensure that their digital marketing efforts align with their overall marketing strategy?
6. Which digital marketing channels do you think are most effective for reaching a specific target audience and why?
7. How can businesses effectively measure the ROI of their digital marketing efforts across different channels?
8. What are some common mistakes businesses make when using social media for digital marketing purposes?
9. How can businesses use email marketing to build relationships with customers and increase sales?
10. What are some best practices for optimizing a website's SEO to improve its visibility in search engine results?
11. What are some common mistakes businesses make when developing digital marketing strategies and how can they be avoided?

12. How can businesses balance the use of automation and personalization in their digital marketing strategies?
13. What are some effective ways to target specific audiences with digital marketing campaigns?
14. How can businesses use content marketing to build brand awareness and attract new customers?
15. What role does storytelling play in successful digital marketing campaigns?
16. How can businesses use data analytics to identify trends and optimize their digital marketing strategies?
17. What are some ethical considerations that businesses should keep in mind when collecting and using customer data for digital marketing purposes?
18. How can businesses use A/B testing to improve the effectiveness of their digital marketing campaigns?
19. What are some key metrics that businesses should track to measure the success of their email marketing campaigns?
20. How can businesses use Google Analytics to gain insights into their website's performance and user behavior?

Multiple-Choice Question

1. What is digital marketing in the context of e-business?

   (a) Traditional marketing techniques applied to online platforms
   (b) Marketing products exclusively through physical stores
   (c) Promoting e-businesses through digital channels
   (d) Selling digital products online

2. Why is effective digital marketing important for e-businesses?

   (a) It helps increase brand awareness and reach a larger audience.
   (b) It reduces operational costs for e-businesses.
   (c) It guarantees higher profits for e-businesses.
   (d) It ensures e-businesses rank higher in search engine results.

3. Which of the following is *not* a common digital marketing channel?

   (a) Social media
   (b) Email marketing
   (c) Print advertising
   (d) Search engine optimization (SEO)

4. Which digital marketing channel is best known for its real-time interaction and engagement with customers?

   (a) Social media
   (b) Email marketing
   (c) Search engine optimization (SEO)
   (d) Pay-per-click (PPC) advertising

5. What is the main goal of search engine optimization (SEO)?
   (a) Increasing website traffic through paid advertising
   (b) Enhancing a website's visibility in search engine results
   (c) Engaging customers through social media platforms
   (d) Creating personalized email campaigns
6. Which digital marketing channel allows businesses to target specific keywords and only pay when their ad is clicked?
   (a) Social media
   (b) Email marketing
   (c) Search engine optimization (SEO)
   (d) Pay-per-click (PPC) advertising
7. What are the best practices for developing and implementing digital marketing strategies?
   (a) Targeting and personalization
   (b) Traditional advertising methods
   (c) Indiscriminate promotion across all channels
   (d) Ignoring data analytics and metrics
8. Which of the following is *not* a way to integrate different digital marketing channels for maximum impact?
   (a) Cross-promoting content across channels
   (b) Coordinating messaging and branding across channels
   (c) Tracking and analyzing data from different channels
   (d) Using different strategies for each channel independently
9. What are key performance indicators (KPIs) used to measure the effectiveness of digital marketing campaigns?
   (a) Metrics that show customer engagement and conversion rates
   (b) The number of employees involved in the campaign
   (c) The total revenue generated by the campaign
   (d) The amount of money spent on the campaign
10. How can data analytics and tools be used to improve digital marketing strategies?
    (a) By identifying customer preferences and behavior patterns
    (b) By increasing the number of digital marketing channels used
    (c) By reducing the budget allocated to digital marketing campaigns
    (d) By eliminating the need for targeted advertising
11. Which of the following statements best defines digital marketing in the context of e-business?
    (a) Utilizing online platforms to promote physical products
    (b) Promoting e-businesses through traditional marketing channels

(c) Implementing marketing strategies specifically designed for digital products
(d) Using digital channels to promote, advertise, and sell products or services

12. Which digital marketing channel allows businesses to interact with their audience through posts, comments, and direct messages?
    (a) Email marketing
    (b) Social media
    (c) Search engine optimization (SEO)
    (d) Pay-per-click (PPC) advertising
13. What is the primary goal of pay-per-click (PPC) advertising?
    (a) Increasing organic search engine rankings
    (b) Driving immediate website traffic and conversions
    (c) Nurturing leads through personalized email campaigns
    (d) Maximizing engagement on social media platforms
14. What is the purpose of targeting in digital marketing strategies?
    (a) Reaching the widest possible audience for a product or service
    (b) Customizing marketing messages for specific customer segments
    (c) Using multiple marketing channels to maximize impact
    (d) Allocating a large budget for digital marketing campaigns
15. How can metrics and analytics improve digital marketing strategies?
    (a) By providing real-time data on customer preferences and behavior
    (b) By reducing the need for targeted advertising campaigns
    (c) By automating the entire digital marketing process
    (d) By eliminating the need for cross-channel integration

References

1. Chaffey, D. and F. Ellis-Chadwick, *Digital marketing*. 2019: Pearson UK.
2. Deepa, N. and S. Deshmukh, *Social media marketing: The next generation of business engagement.* International Journal of Management Research and Reviews, 2013. **3**(2): p. 2461.
3. Kim, A.J. and E. Ko, *Do social media marketing activities enhance customer equity? An empirical study of luxury fashion brand.* Journal of Business research, 2012. **65**(10): p. 1480–1486.
4. Kartajaya, H., P. Kotler, and D.H. Hooi, *Marketing 4.0: moving from traditional to digital.* World Scientific Book Chapters, 2019: p. 99–123.

Chapter 10
Customer Relationship Management

Abstract This chapter explores customer relationship management (CRM) in e-business, focusing on the technology that allows businesses to build and maintain strong customer relationships online. It provides an overview of CRM in e-business, including its components, implementation best practices, and benefits and challenges. The chapter defines and explains CRM in the context of e-business, emphasizing its role in understanding customer needs and preferences. It highlights the significance of CRM for online businesses, enabling them to deliver personalized customer experiences. The typical components of CRM in e-business, such as customer data management, marketing automation, and customer care, are discussed along with their individual and combined advantages and disadvantages. Best practices for implementing CRM systems and integrating them with other processes are examined, followed by suggestions for future research. The benefits of CRM in e-business, such as increased customer satisfaction and loyalty, are explored, as well as potential challenges related to data privacy and security. By the end of this chapter, readers will have a comprehensive understanding of CRM in online enterprises and its significance in modern business operations.

1 Introduction to CRM in E-Business

The term "customer relationship management" (or "CRM") refers to a subset of business intelligence that focuses on customer interactions and data over a company's whole customer lifetime. In order to keep consumers coming back, this tactic emphasizes going above and beyond what they anticipate. customer relationship management (CRM) has emerged as a valuable resource for e-commerce enterprises as the necessity of offering a satisfying client experience has grown in the current online market. This portion of the article will define customer relationship management and explain how it might be used in electronic commerce. Using customer relationship management (CRM) software, online stores may streamline consumer interactions. Technology is used to collect, analyze, and utilize information

H. Taherdoost, *E-Business Essentials*, EAI/Springer Innovations in Communication and Computing, https://doi.org/10.1007/978-3-031-39626-7_10

about customers to boost business and enhance client satisfaction. In e-commerce, customer relationship management (CRM) exists primarily to boost customer happiness and loyalty. To do this, businesses must cater to customers by meeting their unique requirements and satisfying their tastes. Methods and tools, including customer service management, marketing automation, and data analytics, are used to accomplish this goal [1].

A 360-degree picture of a company's customers is one of the most valuable features of customer relationship management (CRM) software for online enterprises. This shows that organizations may learn more about their customers by collecting and analyzing data from many channels, including social media, email, and website interactions. Businesses may use this information to tailor their marketing strategies and provide customers with a more bespoke service. Businesses may boost customer satisfaction and retention by catering to individual customers' needs and offering them relevant, personalized experiences. Implementing CRM in digital businesses also improves customer service management. Using customer relationship management (CRM) software, organizations may coordinate their communications with customers across several mediums, including phone, email, instant message, and social media. This paves the way for organizations to deliver constant, high-quality support across all client interaction channels. Several customer care tasks may be automated using CRM software. These tasks include answering commonly asked questions and allowing customers to solve their problems.

E-business lead generation and sales administration benefit greatly from customer relationship management (CRM). Using customer relationship management (CRM) software, firms may track and study customer actions to spot new leads and chances for upselling and cross-selling. Using this data, you may create marketing and sales tactics more likely to result in closed deals and, hence, more income. By providing the data needed to complete deals and create connections with customers, CRM solutions may also assist sales teams in managing client contacts more successfully. Because of this, sales teams can close more deals and invest more time in developing connections with customers. Figure 10.1 illustrates some significant benefits of effective CRM for e-business [2].

2 Components of CRM in E-Business

Customer relationship management, often known as CRM, is essential to any successful online business approach. Customer relationship management (CRM) is the process of managing client interactions in such a manner as to establish long-term relationships and encourage company growth. At its foundation, CRM corresponds to the word "customer relationship management." Customer relationship management in the digital sphere involves using a large variety of programs and processes meant to aid companies in understanding their clientele's requirements and preferences and provide personalized experiences. This is necessary for effective customer relationship management. E-commerce contact management, sales

Fig. 10.1 Benefits of effective CRM for e-business

management, marketing automation, analytics, customer care, support, collaboration, and communication are the main components of customer relationship management, often known as CRM. In the next part, we will analyze each of these components in further depth and illustrate how they might aid firms in developing strong customer connections and achieving success in the world of e-business, which is constantly evolving.

The success of online businesses needs to have an efficient customer relationship management (CRM) system. The administration of contacts is necessary for customer relationship management, often known as CRM. This requires maintaining a record of all data and contacts to understand customers' requirements and preferences better. Consumers today have access to a more excellent range of goods and services than ever before as a direct consequence of the expansion of digital platforms. They can research their choices, compare costs and features effortlessly, and purchase from the same streamlined website. Organizations must thoroughly comprehend their customers' requirements and preferences to develop long-term connections with their audience successfully. Providing insights that can be utilized to modify offerings and create personalized experiences is one of how contact management can support companies in accomplishing this purpose. This is one of the many ways contact management can help organizations.

The capacity to monitor client contacts across numerous channels, such as email, phone, and social media, is one of the most critical parts of contact management, which also involves several other vital qualities, such as the ability to monitor client contacts across multiple channels. In addition, it is necessary to have accurate and up-to-date records of customers, including the customers' contact information and purchase history. Contact management software may assist organizations in segmenting their current client base and developing targeted marketing campaigns depending on the preferences and behaviors of clients. This can be accomplished by using the data collected by the program. As a result of how it conducts its business, Amazon is a corporation that serves as a model for efficient contact management. Amazon maintains accurate records of its contacts with its customers and information about those consumers thanks to an advanced contact management system. Amazon can present customers with individualized suggestions based on their prior purchases and browsing behaviors thanks to an algorithm that the company has developed. Amazon has emerged as one of the most successful online retailers in the world because of its ability to cultivate client loyalty by providing individualized shopping experiences for individual customers [3].

If a company efficiently manages its contacts, it can also increase customers' happiness and loyalty to the brand. Businesses are better positioned to immediately handle customer concerns and give timely help when accessing accurate and up-to-date consumer data. This will make the development of long-term relationships with customers more accessible and cultivate trust between the many parties involved. The capability of contact management to allow the streamlining of company procedures and the development of productivity is another significant benefit of using this technology. Businesses may free up significant time and resources to concentrate on strategic goals if they automate routine, repetitive processes like data input and customer follow-up. The following is a list of examples of these types of jobs. Figure 10.2 illustrates the critical steps needed for implementing CRM in an e-business.

Effective sales administration is an absolute necessity for success in e-business. The objective of sales management is to improve the efficacy of the sales process in order to increase revenue. Within the framework of customer relationship management (CRM), sales management is one of the critical components that aids businesses in establishing and maintaining healthy client relationships.

One of the primary goals of sales management in an e-business environment is to increase sales efficiency by automating various tasks and procedures. This includes the generation of leads, the monitoring of contacts, and order processing. By

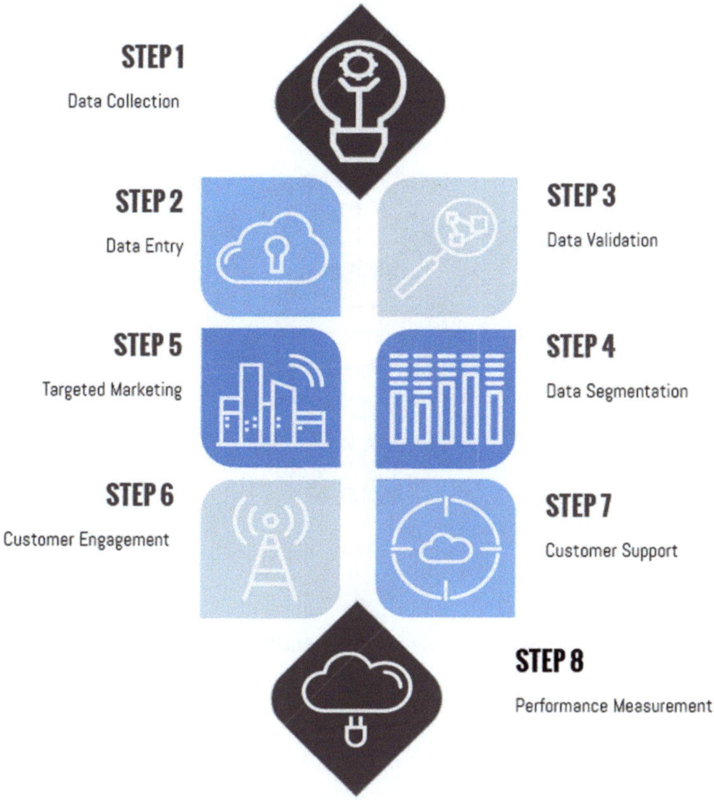

Fig. 10.2 Implementing an effective contact management strategy in e-business

automating these processes, businesses can increase their productivity, simplify their sales process, and reduce errors. Another essential component of effective sales administration is the practice of anticipating sales. When companies analyze sales patterns and data, they can make more informed decisions regarding inventory management, pricing, and marketing strategies. This may assist organizations in maintaining a competitive edge and fostering sustainable growth in their operations. Effective sales management requires establishing and maintaining stable relationships with customers. In addition to the use of automation and forecasting, this is implemented. This requires an awareness of consumer needs, the delivery of personalized experiences, and the prompt resolution of any complaints or issues that may arise. By doing so, businesses can cultivate loyal customers likelier to make additional purchases and recommend the business to others.

Amazon's sales method is a real-world example of how effective sales management in online enterprises can be. Amazon's sophisticated sales management system provides consumers personalized recommendations based on their perusing history and previous purchases. In addition, Amazon utilizes sales forecasting to optimize its inventory management and pricing strategies. Amazon is one of the

Fig. 10.3 Various components of sales management in e-business

most successful companies in the annals of e-commerce due to its emphasis on sales management. Figure 10.3 demonstrates various components of sales management in e-business.

Marketing automation is integral to customer relationship management (CRM) in an e-business environment. A component of this strategy is automating repetitive marketing tasks with technology and transmitting personalized communications to many customers simultaneously. This helps businesses save time and money and enables them to communicate highly targeted and relevantly with consumers, which may increase customer engagement and ultimately contribute to higher sales.

A vital aspect of marketing automation is the nurturing of leads. This involves sending specific communications to prospective clients to cultivate a relationship with them and encourage them to purchase. When a client engages in specific actions, such as opening an email or visiting a particular website page, marketing automation software can automate transmitting these messages; examples of such conduct include the following. The segmentation process is another essential element of marketing automation. When companies segment their client base based on demographics, behaviors, or other characteristics, they can send consumers highly customized communications. These messages are more likely to resonate with consumers and inspire action. A company may deliver a special message to customers who have already purchased and those who still need to purchase. In order to encourage recurrent commerce from existing consumers, this may be done. Email marketing campaigns are one type of marketing that can be made more efficient and successful through marketing automation. Businesses can increase their open rates, click-through rates, and conversion rates by automating the distribution of messages and personalizing them based on their customers' actions and preferences. This can generate additional revenue and increase marketing efforts' return on investment (ROI).

The Salesforce case demonstrates the effectiveness of marketing automation by providing a real-world illustration of its application. Salesforce, the leading provider of CRM software, uses marketing automation to deliver highly targeted and pertinent communications to individuals who have expressed an interest in purchasing their products. By segmenting leads based on industry and firm size variables, Salesforce can provide customized communications to prospective customers. These personalized messages are more likely to resonate with prospective consumers and elicit action. This has allowed Salesforce to generate high-quality prospects, contributing to revenue growth consistently.

Karmaloop, an online retailer, serves as an additional illustration of how effective marketing automation can be when implemented. Karmaloop uses marketing automation to segment its email list and send personalized communications to different groups based on characteristics such as previous purchases and email engagement. These considerations were taken into account when constructing the email list. As a direct result of adopting these actions, Karmaloop achieved a 50% increase in open email rates and a 300% increase in click-through rates, which led to a substantial increase in revenue.

In addition, to lead nurturing, segmentation, and email marketing, additional marketing tasks such as social media management, content marketing, and advertising can be automated with marketing automation. Organizations can save time and money by automating these processes while delivering result-generating marketing messages. Artificial intelligence enables this to be feasible. Figure 10.4 indicates various steps in marketing automation.

Analytics is a crucial element of customer relationship management (CRM) in e-business. It enables organizations to acquire and analyze customer data to improve their marketing strategies and customer experiences. Analytics aims to gain insights into consumer behaviors and preferences by accumulating, processing, and analyzing vast data. E-businesses can gain a competitive advantage by making decisions based on data, which enables them to optimize their marketing strategies, discover new opportunities, and increase client retention. This is possible through the use of analytics tools.

Regarding customer relationship management (CRM), one of the primary benefits of analytics is that it enables businesses to understand better the customers they serve. Suppose businesses monitor customer activity across various channels and touchpoints. In that case, they can identify patterns and trends in customer behavior that can assist them in tailoring their marketing efforts to the needs and preferences

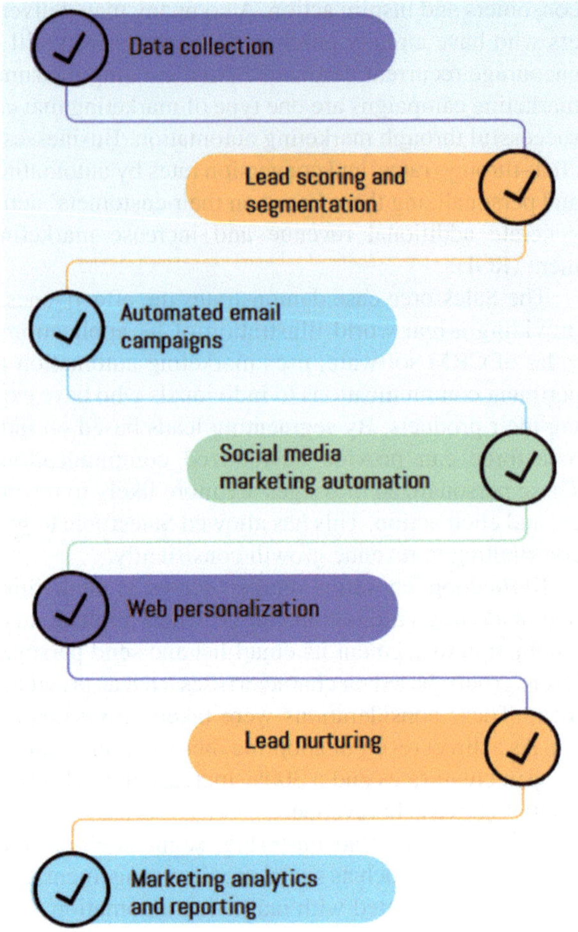

Fig. 10.4 Various steps involved in marketing automation

of their consumers. This includes tracking consumer interactions with a company's website, social media platforms, and customer support channels. Thanks to analytics, businesses can also evaluate the efficacy of their marketing initiatives and consumer interactions. By measuring key performance indicators (KPIs) such as conversion, click-through, and customer retention rates, businesses can identify areas where they can improve their marketing efforts and the experiences they provide to their customers. For instance, an e-commerce company may use analytics to track the conversion rates of its website's various product pages. This would allow the company to enhance the efficacy of its product pages, thereby increasing sales. Another essential advantage of using analytics in CRM is that it facilitates identifying new expansion and product development opportunities. Businesses must analyze consumer data to discover new market segments, products, and services. A company may, for instance, use analytics to identify a new consumer segment that is highly interested in its products but has yet to make a purchase. This would allow the company to better target this group with its marketing efforts by tailoring those efforts to meet the requirements of the new customer segment [4].

Amazon is one of the largest online purchasing companies in the world, and its success has been founded on data-driven decisions. Amazon employs various analytics technologies, including website analytics, customer evaluations, and purchase data, to monitor and improve its customers' behaviors and enhance its marketing strategies. Thanks to the data collected and analyzed, Amazon can improve the user experience, alter its product recommendations, and customize its marketing efforts. For example, Amazon's recommendation engine combines previous purchases and perusing history to determine which products will likely interest each customer. As a result, there has been a discernible increase in consumer participation and purchases. In addition, Amazon utilizes user evaluations to improve product descriptions, determine whether a product may have potential issues, and assess overall consumer satisfaction. Figure 10.5 illustrates an analytical process for CRM in e-business.

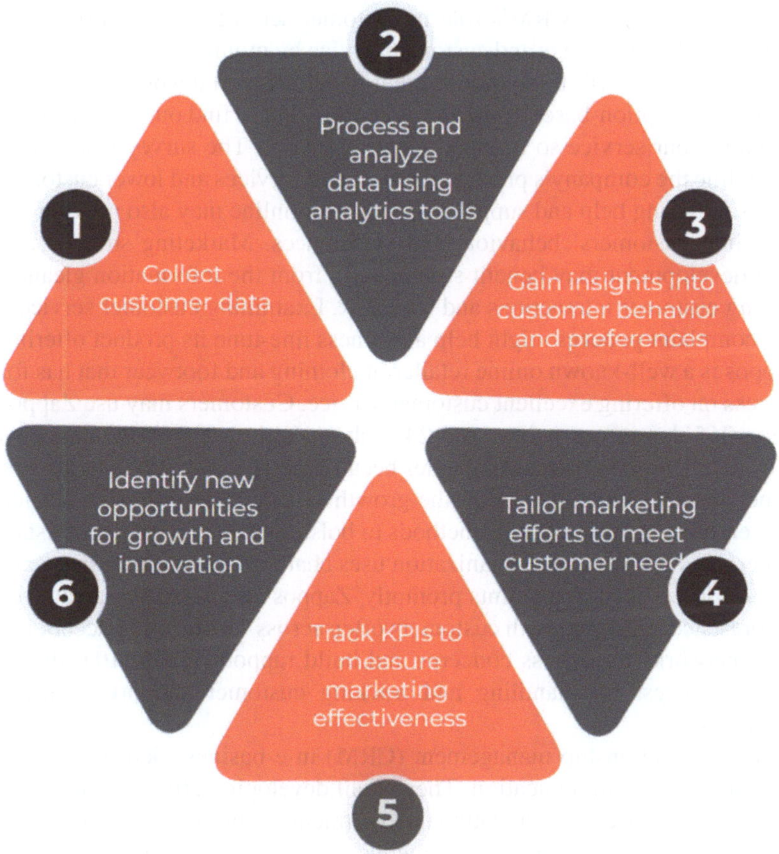

Fig. 10.5 Analytics process for CRM in e-business

CRM features like customer care and assistance are essential for modern e-commerce businesses. As e-commerce and online shopping have grown in popularity, companies have come to appreciate the need to offer first-rate customer service to retain their clientele. This part will explain why CRM benefits organizations and how they may improve their customer service strategy by offering outstanding care and assistance to their clientele. Customer service and support include everything from first contact to follow-up questions and resolved issues with a product or service purchased from a business. Chatbots, email support, phone help, and social media support are all examples of what people may mean when they talk about "customer service" in the context of an online company. Businesses may win over customers and earn loyalty by responding to their needs quickly and effectively.

The ability to foster meaningful connections with customers is a crucial advantage of CRM systems' customer care and support features. Businesses may increase customer loyalty and advocacy by responding quickly and effectively to customer concerns and offering personalized support. A live chat option on a website is one way an online retailer may speedily address customer concerns and win their confidence. The ability to aid businesses in discovering and addressing customers' pain points is also crucial to CRM's role in customer service and support. Businesses may improve their offerings and customer service by monitoring customer activities and responses. This will likely boost client happiness and devotion to the brand. If you run a subscription-based company, you may want to find out why customers are abandoning your service so that you may improve it. The survey results might be used to refine the company's present products and services and lower customer attrition rates. Offering help and support to customers online may also provide valuable insights into customers' behaviors and preferences. Marketing strategy and new product development may benefit significantly from the information gleaned from evaluating customer interactions and feedback. Examining customer service tickets to spot complaint patterns might help a business fine-tune its product offering.

Zappos is a well-known online retailer of clothing and footwear that has founded its success on offering excellent customer service. Customers may use Zappos' free delivery, 365-day return policy, and 24/7 phone and chat support. Zappos stands apart from other e-commerce companies because of its focus on delivering excellent customer service, which has led to the growth of a devoted following. In addition, Zappos employs cutting-edge IT methods to bolster its already stellar assistance for customers. For instance, the organization uses chatbots to answer customers' questions and address their complaints promptly. Zappos also maintains an active social media presence, engaging with customers often across Twitter and Facebook as well as other networks to address concerns and build rapport. Figure 10.6 illustrates a simplified process for handling and tracking customer support inquiries in a CRM system.

Customer relationship management (CRM) in e-business depends on effective collaboration and communication. They entail developing efficient communication channels across various organizational departments to promote departmental cooperation and satisfy client requirements. By utilizing solutions facilitating

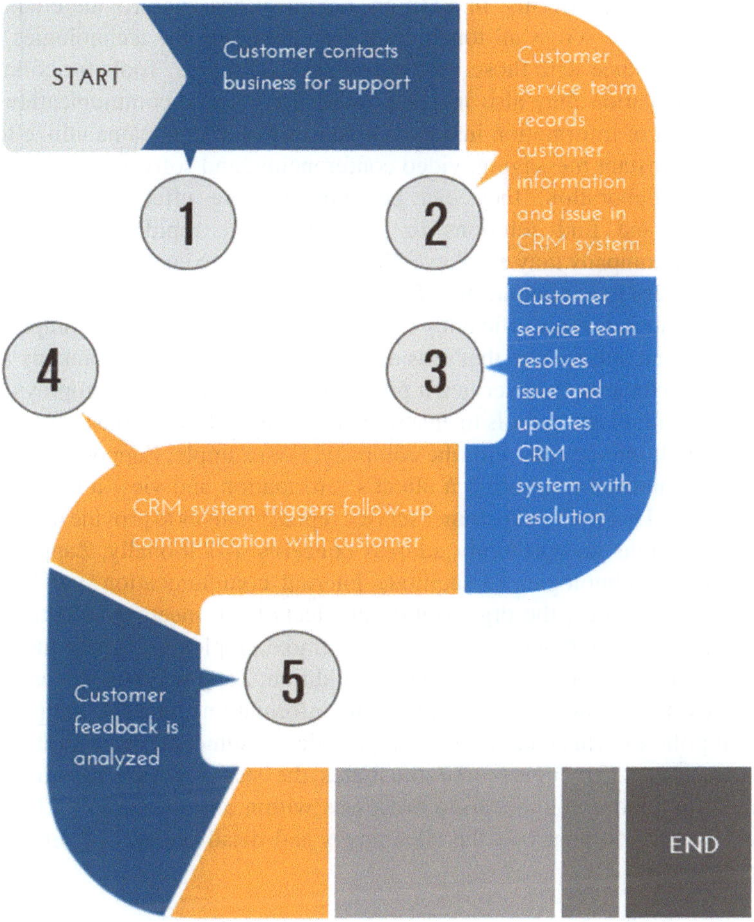

Fig. 10.6 Simplified process for handling customer support inquiries and tracking them in a CRM system

cooperation and communication, e-businesses can operate more efficiently, share knowledge, and provide a superior consumer experience.

One of the most significant benefits of collaboration and communication in CRM is that they enable companies to break down previously existing barriers between departments, one of these features' most essential benefits. This may help businesses improve their decision-making processes, operate more efficiently, and provide consumers a more satisfying experience. If businesses encourage the exchange of customer data and insights across their departments, they can better understand their customer's needs and preferences. Providing prompt and effective customer service necessitates the provider's ability to communicate effectively. Businesses can provide a personalized customer experience tailored to their needs by providing many communication channels to their customers. These include email, live chat, and social media platforms. In addition, organizations can provide clients with a unified experience across all touchpoints by integrating the technologies used for customer assistance with those used by other departments. Tools for collaboration and communication may also increase internal corporate communication and the dissemination of information inside an organization. When teams utilize technologies such as instant messaging, video conferencing, and software designed specifically for collaboration, they can collaborate more effectively, communicate information, and find solutions to problems more rapidly. For instance, an e-commerce company may employ collaboration tools to enable its sales and fulfillment employees to communicate information on its inventory. This would allow the teams to collaborate more efficiently and reduce the time required to complete orders.

Zappos is an online store that sells apparel and shoes, and the company attributes its success to the excellent service it provides customers. Zappos employs cooperation and communication tools to ensure that customers have a consistent experience throughout all interactions with the company. For example, Zappos's customer service representatives can access a client's information and view a customer's purchase history. This enables customer service representatives to provide individualized recommendations and promptly address concerns. Additionally, Zappos employs collaborative technologies to facilitate internal communication and information sharing. For instance, the organization conducts team meetings and training sessions via videoconferencing, allowing its remote employees to remain in contact and efficiently collaborate on projects. In addition, Zappos customer service agents have access to a knowledge base containing information about the company's products and policies. This enables them to provide consumers with accurate and expedient assistance [3]. Figure 10.7 illustrates the steps in implementing effective collaboration and communication processes within an e-business's CRM system. Table 10.1 also summarizes the advantages and disadvantages of different CRM components.

3 Implementation of CRM in E-Business

A customer relationship management (CRM) system setup in a web-based company may be a laborious process calling for careful preparation and execution. To guarantee the smooth rollout of their CRM system and get the most out of it, businesses should follow several best practices (Table 10.2).

Defining your objectives is the first step in developing a successful customer relationship management strategy. Determining the business operations and consumer interactions that will be supported by the system and specifying their intended results falls under this category. For instance, if the objective is to boost customer satisfaction, the CRM system should focus on expediting responses, customizing support, and fixing problems efficiently.

Another crucial best practice that must be followed is to include all relevant parties in the planning and execution stages. Staff from other departments, such as

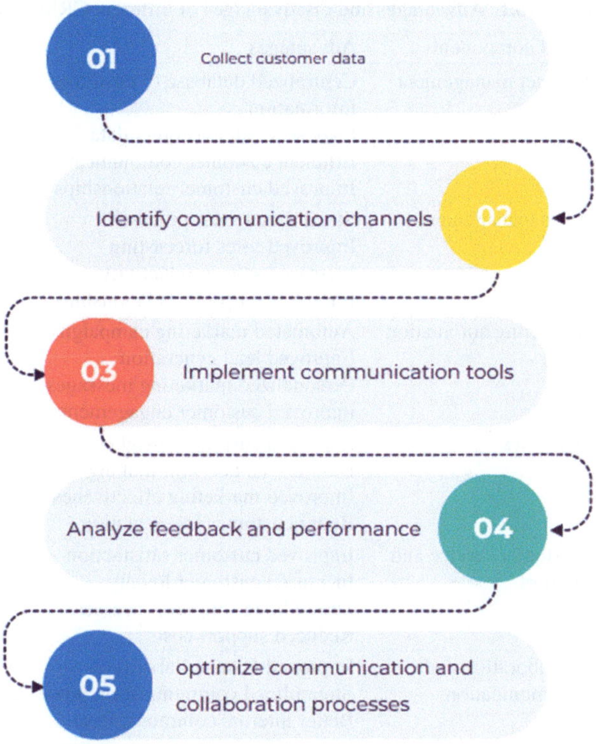

Fig. 10.7 Steps involved in implementing effective collaboration and communication processes within an e-business's CRM system

management, marketing, and customer service, are also included here. By including all relevant parties, firms can ensure their CRM system is tailor-made for their requirements and aligned with their overarching business goals.

In addition, organizations must ensure they are investing in the proper customer relationship management (CRM) system. To achieve this goal, one must evaluate all viable alternatives thoroughly before settling on a flexible method that can be scaled up or down as needed. Additional essential features include user-friendliness, system compatibility, and support for existing customers.

After choosing a customer relationship management system, the following step is to create a detailed rollout strategy. Step one is figuring out when things must be done, how long they will take, and what resources you need. It is also essential to ensure everyone involved knows the strategy and has the same idea of the timeline and expectations.

Companies should prioritize testing and training throughout the implementation process. This involves ensuring that all workers have had enough training on the CRM system and doing thorough reviews to detect any possible flaws or inaccuracies. Data management and security standards must be established to guarantee the privacy of all customer information at all times. This is a significant stage.

The success of a company's customer relationship management (CRM) system requires constant attention and, if required, alterations. This entails monitoring

Table 10.1 Advantages and disadvantages of different CRM components

CRM component	Advantages	Disadvantages
Contact management	Centralized database of customer information Easy access to customer data Efficient customer communication Improved customer relationships	Limited functionality beyond contact management No automation features
Sales management	Streamlined sales process Improved sales forecasting Increased sales productivity Improved customer relationships	Limited functionality beyond sales management No marketing automation features
Marketing automation	Automated marketing campaigns Improved lead generation Personalized marketing messages—improved customer engagement	Can be expensive Requires significant upfront investment Requires technical expertise
Analytics	Improved customer insights Data-driven decision-making Improved marketing effectiveness Better customer segmentation	Requires significant technical expertise Can be time-consuming and resource-intensive
Customer service and support	Improved customer satisfaction Increased customer loyalty Streamlined support processes Reduced support costs	Can be resource-intensive Requires significant customer service training Requires technical expertise
Collaboration and communication	Improved team collaboration Streamlined communication processes Better internal communication Improved customer relationships	Can be difficult to integrate with other systems Can be time-consuming and resource-intensive

metrics like customer retention, new sale conversions, and general happiness with the system and using that information to guide choices about improving the system and the organization.

Online shoe and clothing shop Zappos is a great case study in customer relationship management (CRM) success. Zappos's customer relationship management (CRM) system is mainly responsible for the company's illustrious reputation for excellent service. The organization uses customer relationship management (CRM) technologies to monitor customer interactions and preferences. Additionally, the organization provides tailored guidance and support for each client. Zappos's success may be attributed to the company's dedication to its customers and the efficient use of its customer relationship management (CRM) system, contributing to its rapid growth and increased profitability.

Systems for customer relationship management (CRM) have become an indispensable tool for e-commerce businesses. With these systems, businesses can manage their interactions and connections with consumers more efficiently, increasing customer satisfaction and sales. However, the system must be integrated with the enterprise's many other crucial processes and systems to maximize CRM's benefits.

An e-commerce platform is a robust software that should be integrated with CRM. Integrating CRM with an e-commerce platform may provide businesses with

Table 10.2 Best practices for implementing CRM systems in e-businesses

Best practices for implementing CRM systems in e-businesses	Advantages	Disadvantages
Clearly define objectives and goals	Ensures the system meets specific business needs and desired outcomes	Inadequate planning may lead to system failure
Involve all relevant stakeholders	Ensures CRM system meets the needs of all stakeholders and is aligned with overall business strategy	Too many stakeholders may lead to conflicting priorities
Carefully select the right CRM solution	Scalable, flexible, customizable, and easy to use	The costly and time-consuming evaluation process
Develop a clear plan for implementation	Ensures project milestones, timelines, and resources are identified and communicated to all stakeholders	Inadequate communication may lead to confusion and delays
Focus on testing and training	Ensures CRM system is free of issues and all employees are adequately trained on how to use it	Testing and training may be time-consuming and resource-intensive
Continuously monitor and evaluate the performance	Provides data to make informed decisions about the system and overall business strategy	Inadequate monitoring and evaluation may lead to missed opportunities or system failure

significant consumer behavior and preference data. E-commerce platforms are the backbone of every online business, and CRM can provide these insights to companies. For instance, if a company integrates CRM with an e-commerce platform, it will be able to track the purchases and preferences of its customers and use the resulting data to develop more targeted marketing campaigns and provide customized offers. Integration of the marketing automation platform with CRM is yet another essential software component. Marketing automation platforms enable businesses to automate various marketing processes, including email campaigns and social media monitoring. By integrating customer relationship management software with marketing automation platforms, businesses can develop marketing campaigns that are more customized and targeted based on the actions and preferences of customers. For instance, a company could use the CRM system to identify consumers who have abandoned their purchasing baskets and then send them a personalized email reminder to complete their purchase. This can be accomplished by identifying consumers who have abandoned purchasing trolleys.

In addition to interacting with many other systems, customer relationship management (CRM) should be integrated with essential business processes. The sales process is crucial to integrate with CRM due to its significance. Incorporating CRM into the sales process enables businesses to monitor each transaction's progression and identify any obstacles or problems that must be resolved. For example, suppose a customer has been stalled in the sales process for an extended period. In that case, the CRM system may alert a sales professional to follow up with the customer and provide additional assistance. The customer would receive additional assistance in this scenario. Customer service is an additional essential process that should be integrated with CRM. Interactions with customer service provide organizations with valuable information regarding the preferences and behaviors of customers; integrating CRM with the process of providing customer service may enable businesses to monitor these interactions more efficiently. For instance, if a customer calls customer service with a complaint, the CRM system may document the call and provide a record of the problem and how it was resolved. This data can then be used to improve the overall customer experience and to address recurring issues [1].

In addition to ensuring that the customer relationship management (CRM) system is integrated with all other processes and systems, it is essential to ensure that it is connected to the overall business strategy. This involves aligning the CRM system's objectives and goals with the organization's overall objectives and goals. For example, the company aims to increase the proportion of consumers loyal to the brand over time. In that case, the CRM system should focus on identifying the customer preferences and actions most likely to result in continued patronage. Data administration is one of the most critical factors when integrating CRM with other systems and processes. In order to avoid issues such as data loss and security vulnerabilities, it is necessary to ensure that data is handled and stored appropriately. This requires establishing direct data access and security standards and implementing appropriate data backup and recovery processes. Table 10.3 provides a comparison of CRM solutions on critical features.

Amazon is a prime example of a corporation that has effectively fused CRM with other systems and procedures. One of the reasons Amazon has been so successful is because of its use of customer relationship management (CRM). Amazon's e-commerce platform, marketing automation platform, and customer support routine are just a few systems and processes tightly coupled with the company's customer relationship management (CRM) platform. Thanks to integrating numerous platforms, Amazon is better positioned to track customer activity, identify trends, and target specific demographics with its advertising. For example, Amazon's customer relationship management (CRM) system is integrated with its web storefront. The business may thus keep tabs on consumer spending habits and inclinations. Individualized product suggestions may be made for each customer based on

Table 10.3 Comparison of CRM solutions A, B, and C on key features

System/process	Integration level with CRM	Benefits of integration
E-commerce platform	High	Customer behavior and preference insights for targeted marketing campaigns and personalized offers
Marketing automation platform	High	Automated personalized marketing campaigns based on customer behaviors and preferences
Sales process	Medium	Improved tracking of each sale and identification of bottlenecks or issues
Customer service process	Medium	Improved tracking of customer interactions and resolution of recurring issues

purchases and browsing behaviors. Customers are more satisfied and loyal to Amazon and more inclined to make subsequent purchases. Amazon's e-commerce and marketing automation systems use the same customer relationship management (CRM) database. This allows the company to streamline its marketing efforts by automating chores like email and social media management and developing more specific campaigns for individual consumers based on their preferences and behaviors. Amazon's website and mobile app are only two examples of how the company's customer relationship management system is used. The corporation then uses this information to send customers timed, personalized emails based on their browsing and purchasing habits.

Consequently, customers are more invested in the brand, and the firm sees more sales. Amazon's CRM is integrated with many other essential systems, including its customer support operation. Interactions with customer care departments provide valuable information about consumer tastes and habits. Amazon's ability to keep tabs on these contacts and deliver top-notch support has improved because of the company's use of a customer relationship management system. If a client has an issue and calls Amazon's customer care department, the CRM system will keep track of the conversation and detail the issue and the resolution. The data is then used to improve the service and fix any problems that have shown up repeatedly.

Amazon's proficiency with CRM and smooth integration with other crucial procedures and systems likely contribute to its success in e-commerce. Because of its improved capacity to monitor consumer behaviors and preferences, Amazon can produce more personalized and focused marketing efforts, boost user engagement, and deliver better customer care.

4 Benefits and Challenges of CRM in E-Business

Customer relationship management (CRM) systems have emerged as an integral component of online business operations. These systems enable businesses to monitor and analyze consumer interactions, information, and preferences. This allows businesses to design and implement targeted marketing strategies, improving customer service and increasing sales. In this section, we will discuss the benefits of CRM in e-business and provide examples of companies that have implemented CRM to improve their operations. Additionally, we will provide examples of businesses that have implemented CRM (Table 10.4).

Customer satisfaction is crucial to any online business's success, directly impacting customer loyalty, client retention, and revenue. Utilizing a customer relationship management (CRM) system in an online business can significantly increase customer satisfaction by allowing businesses to comprehend better and accommodate customers' needs and preferences. The ability of a customer relationship management (CRM) system to collect and analyze client interactions, behaviors, and preferences provides organizations with valuable information that can aid them in identifying areas for improvement and providing superior customer service. By

Table 10.4 The benefits of CRM in e-business

Benefits	Description	Real-life example
Improved customer satisfaction	Analyzing customer interactions and preferences to tailor products and services to meet customer needs	Amazon makes personalized recommendations for products and services based on customer behavior and preferences
Improved customer retention	Identifying customers at risk of leaving and taking proactive steps to retain them	Netflix monitors customer viewing habits and preferences to make personalized recommendations for content
Streamlined operations and increased efficiency	Automating specific tasks and processes, such as data entry and reporting, to save time and resources	Salesforce automates sales processes and provides real-time data analytics to sales teams
Identification of new opportunities for growth	Analyzing customer data to identify trends and patterns in customer behaviors and preferences	Apple tracks customer feedback and preferences to develop new products and services tailored to customer needs
Insights into the effectiveness of marketing campaigns	Tracking customer responses to marketing efforts to identify the most effective campaigns and adjust strategies accordingly	PepsiCo analyzes customer responses to marketing campaigns to adjust real-time marketing strategies
Improved customer service operations	Identifying areas where customer service needs improvement, such as response times or issue resolution	American Express monitors customer interactions to identify areas where it can improve its customer service operations

utilizing customer relationship management (CRM) software, businesses can provide a more personalized experience for their clients and customers. This may result in increased client satisfaction. For example, a customer relationship management (CRM) system can segment customers based on their preferences, demographics, and purchase histories. This allows businesses to better tailor their marketing campaigns and promotions to the specific needs of their consumers. Customers are more likely to respond positively to marketing communications pertinent to their interests; therefore, this approach may result in greater customer engagement and satisfaction with the product or service.

In order to increase customer satisfaction, companies must provide more efficient and effective customer service; a CRM system may help organizations provide this service more efficiently and effectively. Businesses can respond to consumer inquiries and concerns with incredible speed and effectiveness when centralizing customer information and interactions. This may include automated responses, self-service portals, and real-time messaging, which can improve the overall customer experience and lead to higher satisfaction levels. In addition to enhancing customer service and customization, a CRM system may provide organizations with valuable feedback on their products and services. This input can result in increased levels of consumer satisfaction, which can then contribute to an increase in client retention. For instance, a customer relationship management (CRM) system may be used to collect consumer feedback via surveys and other types of input, which can assist businesses in identifying areas for development and modifying their products and services accordingly.

Airbnb is an excellent example of how customer relationship management (CRM) can enhance customer satisfaction in a real-world setting. Airbnb provided more efficient and effective customer service due to the implementation of a customer relationship management (CRM) system, which centralized client information and interactions. The company also uses its CRM system to collect customer feedback, allowing them to identify improvement areas and modify its platform and services following these findings. Due to this emphasis on customer satisfaction, Airbnb has seen increased customer loyalty and positive reviews, contributing to the company's rising sales and income.

CRM systems have proven their worth by substantially contributing to an increase in the percentage of online businesses' retained customers. Using customer relationship management (CRM) data to determine which customers are most likely to defect is one strategy companies employ to retain customers. This data may include consumers who last purchased a while ago or have expressed dissatisfaction with the products or services. When businesses can identify these customers at an early stage, they are better able to take preventative measures to address their concerns and retain their loyalty. Netflix is a fantastic example of an online business that has effectively increased customer retention rates through customer relationship management (CRM) data. Netflix makes recommendations for the content likely to interest its customers based on the viewing patterns and preferences garnered from its CRM system, which tracks consumer viewing patterns and preferences. This strategy has contributed significantly to Netflix's success in retaining customers and expanding its subscriber base. In actuality, Netflix acquired a record 37 million subscribers in 2020, bringing the total number of Netflix users to more than 200 million. CRM data allows businesses to tailor customer communication, which may be essential to retaining loyalty. By analyzing customer interactions, businesses can understand how their consumers prefer to be reached and which messages are the most effective. For instance, if a customer indicates that they would prefer to be contacted via email rather than by phone, a business may tailor their approach to communication accordingly. Customer relationship management (CRM) data may also aid businesses in tailoring their products and services to the preferences and needs of customers. By observing consumers' interactions and preferences, businesses can better understand which products and services are in high demand. This allows the companies to modify their offers to market demand. This not only increases the percentage of returning consumers, but it also has the potential to increase sales and revenue [2].

CRM systems can increase an organization's overall productivity and streamline its functions. Automating routine operations and procedures, such as data entry and report generation, can save time and resources, allowing personnel to focus on more essential company tasks. Salesforce automates its sales operations using its CRM system, offering sales teams real-time data analytics. This enables sales teams to make informed assessments of leads and prospects, allowing them to concentrate their efforts and quickly close deals. CRM systems can also optimize supply chain management, reducing the time and money spent on procurement and delivery operations. CRM systems can provide real-time updates on stock levels and delivery schedules because they integrate with various other systems and procedures, such as inventory management and logistics. For example, Walmart uses its CRM system to monitor inventory levels and analyze real-time sales data. This enables the company to make informed decisions regarding stock replenishment and delivery schedule.

Moreover, customer relationship management (CRM) solutions may enhance internal communication and collaboration, enabling teams to operate more effectively and productively. A common platform is provided for all customer-related data and interactions, allowing teams to access and share information effortlessly. This reduces the possibility of misunderstandings and the quantity of duplicated labor. Microsoft Dynamics, for instance, utilizes its CRM system to improve communication between its sales, marketing, and customer support departments. This enables the teams to collaborate efficiently and effectively.

CRM (customer relationship management) solutions are indispensable for businesses seeking new expansion opportunities. Businesses can gain insights into consumer preferences and behaviors by collecting and analyzing customer data—these insights aid organizations to develop new products and services to meet customer demands. Apple, for instance, obtains consumer feedback via its CRM system, which aids the company in identifying areas for development and creating new products and services. Customer relationship management (CRM) systems can assist businesses in identifying expansion opportunities by analyzing customer data to identify behavioral patterns and trends. By observing customers' interactions with the company and analyzing the purchase histories of those customers, businesses can determine which products and services are in high demand and develop new offerings to meet those demands. For instance, Amazon uses its CRM system to track the preferences and activities of its customers. This method has enabled the company to recognize expansion opportunities in markets such as home automation and personal assistants. CRM systems can also help businesses discover new growth opportunities by providing insights into consumer desires and preferences that may not be apparent from other sources. This can benefit enterprises that would otherwise struggle to collect this information. For instance, a company may use its customer relationship management (CRM) system to evaluate customer feedback and identify areas where its products or services are lacking. This data can then be utilized to develop new products and services that are more adapted to the needs and preferences of customers. In addition to conducting customer data analyses, customer relationship management (CRM) systems can also aid businesses in identifying new expansion opportunities by providing insights into the dynamics of market trends and competition. By observing the actions of their competitors and following the trends in their respective industries, businesses can identify areas in which they can differentiate themselves and gain a competitive advantage. For instance, a company may use a CRM system to monitor consumer comments on social media platforms. This may provide insights into market trends and aid the firm in developing new products and services to meet consumers' ever-changing needs [3].

In the highly competitive field of online business, it is essential to run successful marketing campaigns to acquire new customers and retain existing ones. Utilizing a customer relationship management (CRM) system allows businesses to gain significant insights into the efficacy of their marketing activities. This is because CRM systems enable businesses to track how consumers respond to various marketing activities, such as email campaigns, social media promotions, and other digital marketing initiatives. For instance, PepsiCo uses its customer relationship management (CRM) system to track how customers respond to its numerous marketing initiatives. The company monitors consumer behavior, such as the number of people who click on a particular link or purchase after receiving a promotional email. It uses this data to enhance its products and services. After collecting and analyzing this data, the next stage is determining which advertisements are the most effective at engaging consumers and driving sales. PepsiCo can now make real-time adjustments to its marketing strategies, allowing it to increase the efficacy of its numerous initiatives. For instance, the company may decide to increase its investment in a successful marketing channel, or it may decide to alter the message of a campaign that is failing to resonate with consumers. Using a customer relationship management (CRM) system to track the success of its numerous marketing initiatives, PepsiCo can enhance the performance of its marketing initiatives and increase revenue. This helps the company acquire new customers and retain existing ones and improves the company's overall business performance by increasing revenue and sales. In conclusion, using a customer relationship management (CRM) system to monitor consumer responses to marketing campaigns is an indispensable tool for e-business companies seeking to maintain their position as market leaders.

Implementing a CRM system in e-business may provide several benefits, one of the most significant being enhancing customer support operations. Suppose companies monitor and analyze the interactions and preferences of their customers. In that case, they can gain valuable insights into the efficacy of their customer service operations and identify areas where they can make improvements. Customer relationship management (CRM) systems can improve customer service through various methods. One of these methods is by providing a centralized database of client data. This facilitates customer service agents' access to the information they need to proficiently and promptly address customer concerns. American Express, for instance, uses its CRM system to monitor how its customers interact with the company and identify vulnerable points in its customer care operations to strengthen them. American Express can offer its high-value clients a more personalized level of service because its customer service representatives have access to a complete view of a customer's interactions with the company. This view contains the client's account history and previous interactions with the customer service department. This level of customization may facilitate the development of deeper client relationships and the promotion of increased brand loyalty among those clients. CRM systems provide a centralized customer information database and can improve customer service by automating tasks like email responses and customer follow-ups. In addition,

CRM systems can provide a central repository for consumer data. This may help ensure that customer inquiries are responded to quickly and effectively, enhancing the customer's overall experience with the company.

Moreover, customer relationship management (CRM) systems may aid businesses in identifying patterns and trends in customer complaints, allowing organizations to address common complaints before they become prevalent issues proactively. Businesses can identify opportunities to improve their products and services by analyzing consumer interactions and feedback. This reduces the number of consumer complaints and increases overall customer satisfaction.

In the e-commerce sector, customer relationship management (CRM) systems have developed into a crucial asset. While customer relationship management (CRM) systems have numerous advantages, they also have several downsides and restrictions that firms should consider (Table 10.5).

The expense of customer relationship management systems is often identified as a significant obstacle for firms considering an implementation. It may be difficult for smaller and medium-sized firms to justify the high expense of implementing and maintaining a customer relationship management (CRM) system. Businesses may be concerned about the initial setup expenses and maintenance and staff training expenditures. Implementing a customer relationship management system presents several obstacles, not the least of which is ensuring the accuracy of the data input. The system's insights may only be trustworthy if the data put into it is accurate and updated often enough. Employees who provide the data may not bother checking its veracity, which might lead to incorrect conclusions.

Furthermore, organizations with a broad customer base may need help establishing a CRM system. For instance, building a complete picture of a client may be challenging if the customer has many roles inside the organization or works in different departments. When this happens, businesses may need to invest more time and effort into customizing their CRM platforms to meet their consumers' demands. The possibility of an overload of data is another drawback of customer relationship

Table 10.5 Challenges and limitations of CRM in e-business

Challenge/ limitation	Description	Potential solutions
Data quality and accuracy	Only accurate or complete data can lead to accurate insights and decision-making	Regular data cleaning and validation, training for data entry personnel
Integration with other systems	Integration with other systems (e.g., ERP, marketing automation) can be complex and time-consuming	Adopting standardized protocols for data exchange and working with vendors to improve integration
Cost and complexity	Implementing and maintaining a CRM system can be expensive and complex	Conducting a cost-benefit analysis before implementation, investing in training and support for staff
Privacy and security	Collecting and storing customer data can raise privacy and security concerns	Establishing clear policies and procedures for data handling and investing in secure storage and encryption

management systems. Businesses may need help making sense of the data and drawing valid conclusions. Organizations may need to hire data analysts or invest in AI-powered technologies to help them sift through massive amounts of data.

Another obstacle businesses must overcome when introducing CRM systems is customers' right to privacy. There are potential ethical and legal issues because of the inclusion of personally identifiable information in the data acquired by these systems. Businesses must have robust data protection and security procedures to safeguard their customers' private information.

Online businesses have unique challenges that prevent the widespread adoption of customer relationship management (CRM) solutions. For instance, the usefulness of customer relationship management (CRM) systems is directly proportional to the quality of the data that is put into them. The system's ability to draw reliable conclusions depends on the data's accuracy and timeliness. In addition, CRM systems are not a silver bullet that will fix everything wrong with a company. They can be utilized to boost efficiency, but they need help fixing all of a company's issues. Some firms have overcome these challenges and constraints using customer relationship management (CRM) solutions. In 2015, for instance, IBM's sales team used a CRM system to understand consumers' wants and preferences better. Thanks to the system's real-time consumer behavior data, sales representatives could tailor their pitches to each buyer. Consequently, IBM saw an increase in happy customers and financial gains.

Finally, while customer relationship management (CRM) systems have many advantages, they also have several disadvantages and restrictions that online organizations should be aware of. Companies need to consider the abovementioned things to make a well-informed choice about using a CRM system. By taking this step, companies can ensure a smooth CRM rollout and eliminate any problems that could arise.

Summary

Definition of CRM in E-Business

- Customer relationship management (CRM) in e-business refers to the strategic approach and technology-driven system that enables businesses to manage and nurture their relationships with customers online effectively.

Components of CRM in E-Business

- Customer data management: Customer data management is a crucial component of CRM in e-business that involves collecting, storing, and analyzing customer information to gain insights into their preferences, behaviors, and needs.
- Sales and marketing automation: Sales and marketing automation tools within CRM in e-business streamline and automate tasks, such as lead generation, email marketing campaigns, and customer segmentation, to enhance efficiency and effectiveness in sales and marketing activities.

- Customer service and support: CRM in e-business includes customer service and support features that enable businesses to provide prompt and personalized assistance through various channels, such as live chat, email, and social media, to enhance customer satisfaction and loyalty.
- Analytics and reporting: The analytics and reporting component of CRM in e-business allows businesses to leverage data-driven insights to measure performance, track customer interactions, and identify trends, enabling informed decision-making and continuous improvement.

Implementation of CRM in E-Business

- Pre-implementation planning: Before implementing CRM in e-business, organizations should define their goals, assess existing systems and processes, identify key stakeholders, and establish a clear implementation strategy to ensure a successful deployment.
- System integration: Integrating CRM systems with other business applications and databases is crucial for seamless data flow and information exchange, ensuring a unified view of customer interactions across different touchpoints.
- User training and adoption: Proper user training and change management initiatives are essential to facilitate user adoption of the CRM system, ensuring that employees understand its benefits and features and how to utilize it effectively to improve customer relationships.

Benefits and Challenges of CRM in E-Business

- Benefits of CRM in e-business: CRM in e-business offers numerous benefits, including enhanced customer satisfaction, personalized marketing campaigns, improved customer retention, streamlined sales processes, data-driven decision-making, and increased overall business profitability.
- Challenges of CRM in e-business: Implementing CRM in e-business may pose challenges such as data quality and integration issues, resistance to change from employees, privacy and security concerns, and ongoing maintenance and updates to ensure system effectiveness.

Case Study

ABC Electronics is an online electronics retailer that sells a range of electronic devices, including mobile phones, laptop computers, and household appliances. The company desires to improve its customer relationship management (CRM) abilities to increase customer satisfaction, loyalty, and overall business success. ABC Electronics plans to implement a CRM system to enhance customer data administration, marketing automation, and customer service. However, the organization faces several obstacles and must thoroughly analyze the installation of CRM components.

1. What potential hazards and issues could ABC Electronics face when deploying a CRM system concerning data privacy and security? How will ABC Electronics ensure that customer data is adequately protected and applicable data privacy regulations are adhered to?
2. How can ABC Electronics efficiently collect, manage, and store customer data to support its CRM initiatives? What potential benefits could ABC Electronics derive from a robust customer data management system?
3. How can ABC Electronics use marketing automation within its CRM system to personalize customer contacts and boost the success of marketing campaigns? What challenges may ABC Electronics encounter when adopting marketing automation, and how can these be overcome?
4. How can CRM assist ABC Electronics in enhancing its customer service operations? How can ABC Electronics provide a seamless customer experience by integrating its CRM system with additional customer support channels and tools, such as live chat and ticketing systems?
5. In terms of increased consumer contentment and loyalty, how can ABC Electronics quantify the effectiveness of its CRM implementation? Which key performance indicators (KPIs) should ABC Electronics monitor to determine the impact of CRM on its e-business operations?

Discussion

1. What is the definition of CRM in the context of e-business?
2. How does CRM differ in e-business compared to traditional business models?
3. Why is effective CRM important for e-businesses?
4. What are the common components of CRM in e-business?
5. How does customer data management contribute to effective CRM?
6. What are the advantages and disadvantages of marketing automation in CRM?
7. How can customer service be integrated into a CRM system?
8. What are the best practices for implementing CRM systems and processes in e-businesses?
9. How can CRM be effectively integrated with other systems and processes?
10. What are some common challenges that arise during the implementation of CRM in e-business?
11. What are the benefits of implementing CRM in e-business, such as improved customer satisfaction and loyalty?
12. How can data privacy and security be addressed when implementing CRM in e-business?
13. What are some potential challenges or limitations of CRM in e-business?

Multiple-Choice Question

1. What does CRM stand for in the context of e-business?
   (a) Customer relationship management
   (b) Customer resource management
   (c) Customer revenue management
   (d) Customer retention marketing
2. Why is effective CRM important for e-businesses?
   (a) To increase data privacy and security
   (b) To automate marketing processes
   (c) To improve customer satisfaction and loyalty
   (d) To reduce operational costs
3. Which of the following is a component of CRM in e-business?
   (a) Inventory management
   (b) Supply chain optimization
   (c) Customer data management
   (d) Financial forecasting
4. What is the advantage of marketing automation in CRM?
   (a) Improved customer service
   (b) Enhanced data privacy and security
   (c) Increased operational costs
   (d) Streamlined lead generation and nurturing
5. What is a disadvantage of customer data management in CRM?
   (a) Improved customer segmentation
   (b) Limited access to customer information
   (c) Efficient tracking of customer interactions
   (d) Enhanced personalized marketing campaigns
6. What are the best practices for implementing CRM systems in e-businesses?
   (a) Centralizing customer data
   (b) Ignoring data privacy regulations
   (c) Involving all employees in the process
   (d) Using a generic CRM software
7. How can CRM be integrated with other systems and processes?
   (a) By using separate software for each process
   (b) By manually transferring data between systems
   (c) By integrating CRM with ERP and e-commerce platforms
   (d) By ignoring the integration process altogether

8. Which of the following is a benefit of CRM in e-business?
   (a) Increased operational costs
   (b) Reduced customer satisfaction
   (c) Improved customer loyalty
   (d) Decreased customer retention
9. What is a challenge of CRM in e-business regarding data privacy and security?
   (a) Increased trust and transparency
   (b) Improved data accuracy and consistency
   (c) Compliance with data protection regulations
   (d) Decreased need for data encryption
10. How can CRM contribute to improved customer satisfaction?
   (a) By ignoring customer feedback
   (b) By providing personalized experiences
   (c) By increasing response time to customer inquiries
   (d) By disregarding customer preferences
11. Which component of CRM helps in tracking and resolving customer issues?
   (a) Marketing automation
   (b) Customer data management
   (c) Customer service
   (d) Supply chain optimization
12. What is a limitation of CRM in terms of scalability?
   (a) Difficulty in storing customer data
   (b) Limited number of users
   (c) Inability to handle large volumes of data
   (d) Lack of integration with other systems
13. How can CRM assist in customer retention?
   (a) By offering generic products and services
   (b) By providing excellent customer service
   (c) By ignoring customer feedback
   (d) By increasing prices regularly
14. Which of the following is a potential benefit of CRM for e-businesses?
   (a) Decreased customer loyalty
   (b) Improved lead generation
   (c) Reduced customer satisfaction
   (d) Higher operational costs

15. What is a challenge of CRM in e-business regarding user adoption?
    (a) Limited availability of CRM software
    (b) Resistance from employees to adopt new processes
    (c) Lack of customer data for analysis
    (d) Inability to track customer interactions

References

1. Rai, A.K., *Customer relationship management: Concepts and cases*. 2012: PHI Learning Pvt. Ltd.
2. Chen, I.J. and K. Popovich, *Understanding customer relationship management (CRM): People, process and technology.* Business process management journal, 2003. **9**(5): p. 672-688.
3. Chai, L. and P. Pavlou, *Customer relationship management. com: a cross-cultural empirical investigation of electronic commerce.* 2002.
4. Payne, A. and P. Frow, *Customer relationship management: from strategy to implementation.* Journal of marketing management, 2006. **22**(1–2): p. 135-168

Chapter 11
Business Intelligence

Abstract Business intelligence (BI) plays a crucial role in modern e-businesses, enabling informed decision-making through data analysis. In the digital era, businesses face overwhelming amounts of data from diverse sources like customer data, social media, and website analytics. BI encompasses tools and techniques that capture, analyze, and transform data into valuable information for business decisions. This chapter introduces the fundamentals of BI and its significance in e-commerce. It defines and describes BI, explaining the instruments and techniques used for data acquisition and analysis. Emphasis is placed on the importance of BI for online businesses, highlighting its role in enhancing decision-making, operational efficiency, and customer satisfaction. The chapter further explores the core components of BI, including data warehousing, data mining, and predictive analytics. Each component is outlined, discussing their advantages and disadvantages and their integration to form BI solutions. Best practices for implementing BI systems and procedures in online businesses are also covered, including the integration of BI with other systems. The chapter concludes by evaluating both the positive and negative aspects of BI in e-business. The benefits of BI, such as improved decision-making and operational efficiency, are highlighted with real-world examples. Additionally, businesses are advised to consider factors like data integrity, privacy concerns, and other challenges associated with BI implementation. Upon completing this chapter, readers will gain a comprehensive understanding of essential BI concepts, components, and challenges in the context of e-businesses.

1 Introduction to Business Intelligence

By "business intelligence" (BI), we mean the tools, processes, and strategies that help companies sift through mountains of raw data in search of actionable insights. BI involves gathering and analyzing information from various sources to make better decisions, spot patterns, and draw conclusions to enhance business outcomes. "Business intelligence" (BI) refers to gathering and analyzing information from

H. Taherdoost, *E-Business Essentials*, EAI/Springer Innovations in Communication and Computing, https://doi.org/10.1007/978-3-031-39626-7_11

several sources. Sales, financial, and operational data are examples of what may be found in internal systems. In contrast, market research, consumer data, and social media data are all examples of what can be found in external systems. When this information is collected, it is analyzed with the help of business intelligence (BI) technologies and tools like data visualization programs, dashboards, and reporting programs. BI is based on analyzing data, which helps firms learn and make better choices. Business intelligence (BI) helps companies find new areas for expansion, enhance existing processes' efficiency, and provide customers with a better product or service. In addition, businesses may use business intelligence to assess their progress toward their objectives and measure their performance against key performance indicators (KPIs).

Data collection, integration, analysis, and visualization are all facets of the business intelligence (BI) process. During data collection, several sources of information are gathered and standardized into a central database. The data are validated, transformed, and made ready for analysis during the data integration stage. In the process known as "data analysis," statistical and mathematical models are used to examine the data for discernible patterns, trends, and linkages. Data visualization is the last step in the process, and it involves presenting the results of the analysis in an understandable style, such as charts, graphs, and dashboards.

Business intelligence (BI) is becoming more critical to the operations of e-businesses since it gives vital insights and analysis of corporate data. This is because BI allows for more intelligent decision-making. This section will examine how business intelligence (BI) may aid e-commerce organizations by illuminating new revenue streams, streamlining processes, enriching the customer experience, and quantifying KPIs. In addition, we will share case studies of successful e-businesses that have used BI strategies for operational growth and success.

In business intelligence (BI), unstructured data is transformed into useful information that may fuel a company's growth using different technologies, processes, and strategies. Online businesses may benefit from business intelligence (BI), including discovering new market potential and trends. For instance, an online retailer may examine its sales data to see which things sell well in various regions. The corporation may use this data to decide better what goods to keep in stock and where to focus its marketing efforts.

The everyday operations of Internet firms may also benefit from BI. The management of a company's supply chain may be managed, and the processes for fulfilling orders can be enhanced by analyzing data on inventory levels, manufacturing schedules, and customer demand. For example, an online retailer may utilize BI to track stock and adjust its ordering and restocking practices to reduce out-of-stocks and surpluses.

Using BI in an e-commerce setting also helps improve the quality of service provided to customers. Businesses may learn more about their target market's tastes and habits by examining data such as their customers' clickstreams, purchases, and demographics. This data might help companies better understand their customers and develop products and advertising campaigns that appeal to them. For instance, an online retailer might use business intelligence (BI) to provide product suggestions to shoppers based on their past actions.

With BI tools, online firms may assess and track their progress. By keeping tabs on KPIs, or key performance indicators, firms can see how far down the path to success they are. Revenue growth, client retention rate, and website traffic are all examples of KPIs. This data might also help companies zero in on improvement areas and fine-tune their approaches for maximum efficacy [1].

When it comes to real-world examples of businesses that have successfully used BI strategies, look no further than Amazon. Amazon uses various business intelligence (BI) software and hardware to understand its customers better, track sales, and manage its supply chain. For instance, Amazon uses data analytics to foretell customer demand, which in turn aids in optimizing stock levels and customer service. Amazon also utilizes BI to tailor its marketing strategies and provide a better overall experience for its customers. Amazon is in a position to provide product suggestions and offer personalized discounts thanks to the analysis of client data. These recommendations and advertisements are tailored to each individual's tastes and behaviors. Amazon has seen a rise in sales and consumer loyalty due to this approach.

2 Components of Business Intelligence

Business intelligence (BI) is a system that enables online businesses to make more informed decisions by accumulating and analyzing data from multiple sources. The following are the foundations of business intelligence in online transactions (Table 11.1).

Warehousing data: Information is essential to the success of any online business. It is crucial to have a reliable method for managing and analyzing the ever-increasing data generated by online companies. Here, data warehousing as an integral component of BI enters into action. "Data warehousing" refers to consolidating data from multiple databases into a single, readily accessible repository. This is data collected through various channels, including sales, consumer interactions, and website analytics. Online businesses need access to the centralized storage of all pertinent data to make informed business decisions.

Integrating data from multiple sources is a significant advantage of data warehousing in e-commerce. An online retailer's website, social media, and customer service interactions are all potential information sources. By incorporating this information into a data warehouse, the company can comprehensively view client activity across all channels. In addition, data warehousing enables online businesses to conduct sophisticated analyses and queries. The storage of all pertinent information in one location facilitates data mining and analysis. This is essential for understanding consumer behaviors, identifying market shifts, and improving internal operations.

Table 11.1 Components of business intelligence in e-business

Component	Definition	Benefits	Tools
Data warehousing	Collecting and storing data from various sources in a centralized repository	Integration of data from different systems into a single source of truth	Amazon Redshift, Google BigQuery, Microsoft Azure Synapse Analytics
Data mining	Analyzing large datasets to discover hidden patterns and relationships	Identifying customer behavior patterns, market trends, and product preferences	IBM SPSS Modeler, SAS Enterprise Miner, RapidMiner
Online analytical processing (OLAP)	Technology that enables users to analyze data from multiple perspectives	Analyzing sales data by different dimensions, such as geography, product, and time	Microsoft SQL Server Analysis Services, Oracle OLAP, SAP BusinessObjects Analysis
Reporting and visualization	Presenting data in an easy-to-understand format, such as charts, graphs, and dashboards	Monitoring business performance and identifying trends and issues	Tableau, Power BI, QlikView
Predictive analytics	Using statistical models to predict future outcomes based on historical data	Forecasting sales, identifying potential customers, and optimizing marketing campaigns	IBM SPSS Modeler, SAS Predictive Analytics, Oracle Predictive Analytics
Data quality management	Ensuring that data is accurate, complete, and consistent	Improving the accuracy of customer data, product data, and sales data	Informatica Data Quality, Talend Data Quality, IBM InfoSphere Information Server
Data governance	Policies and procedures that ensure the security and privacy of data	Ensuring compliance with regulations, such as GDPR and CCPA	Collibra, Informatica Axon, Talend Data Fabric

Amazon is a prosperous online retailer with numerous benefits from implementing a data repository. Amazon is an outstanding example of a company that employs data warehousing strategies to remain competitive. Amazon maintains track of consumer purchases, website traffic, inventory levels, and logistics, among other things. Thanks to the data stored in its repository, Amazon can learn much about its customers' behaviors and market tendencies.

Additionally, data warehousing in e-business facilitates real-time decision-making. In the fast-paced realm of online commerce, decisions must be made with accuracy and swiftness. With data warehousing, organizations can always have access to the most recent data, enabling them to make better decisions. This is essential for e-commerce businesses due to the rate at which the market and consumer preferences can change. Regarding data governance and security, the significance of data warehousing cannot be exaggerated. The General Data Protection Regulation (GDPR) and the California Consumer Privacy Act (CCPA) both emphasize the importance of maintaining the confidentiality and security of personal data. This is essential for online businesses that store confidential client information.

Business intelligence (BI) in e-commerce is mainly dependent on data mining. It requires sophisticated analytic techniques to extract useful information from vast databases. Without data mining within the context of an online business, it is impossible to comprehend consumer behavior, identify market trends, and increase operational efficiency. Recognizing patterns and tendencies in customer behavior is one of the most practical applications of data mining for online businesses. By analyzing extensive databases, e-commerce sites can learn more about their customers' likes, preferences, purchasing behaviors, and the factors that keep them returning. This data can benefit marketing efforts, product development, and customer service experience. In addition, data mining assists online businesses in identifying emerging industry trends and potential competitive threats. Multiple data channels, including social media, customer evaluations, and industry research, can help online retailers identify new opportunities and risks. This information can influence strategic decisions and maintain a competitive advantage [2].

Netflix is a successful online business that utilizes data analytics. Netflix bases its production and acquisition decisions on data extraction of user behaviors and preferences. By analyzing the viewing patterns of its millions of subscribers, Netflix can develop popular content with its audience.

Through data mining, e-business operations can also be made more efficient. By analyzing data collected from diverse sources, such as supply chain management systems and customer support contacts, e-commerce operations can be enhanced. This information can streamline processes, reduce expenses, and increase productivity. Another advantage of data mining in online enterprises is the ability to personalize services for each client. By analyzing data collected from their customers, e-commerce sites can provide more personalized recommendations and offers to their customers. This, in turn, increases customer retention and satisfaction. Data mining is also essential for protecting an online business from fraudulent activity. Online companies can detect and stop fraud by scrutinizing transaction data and other pertinent information. This protects the company and its customers from financial and other potential damage.

Business intelligence (BI) in e-business environments relies heavily on online analytical processing (OLAP). It facilitates the quick and effective analysis of massive datasets, assisting Internet businesses in learning about client habits, market tendencies, and operational efficacy. Online companies may examine their data from different perspectives and at varying levels of detail thanks to the multidimensional view provided by OLAP.

One of the main benefits of OLAP for online businesses is the ability to do real-time data analysis. Online firms may use online analytical processing (OLAP) to keep track of key performance metrics in real time, allowing for more timely data-driven decision-making. In the fast-paced environments of e-business, when time is of the highest importance, and decisions must be made as quickly as possible, this is of the utmost value. In addition, OLAP allows e-commerce to analyze data from several sources. Combining data from many sources, such as sales data, website analytics, and social media data, may help e-commerce businesses understand their customers and the trends in their target markets. This data may refine the company's advertising strategies, enhance its product offerings, and boost its bottom line.

When discussing the value of online analytical processing (OLAP) for e-commerce, Amazon makes a great case study. By analyzing sales data from its e-commerce platform, Amazon uses online analytical processing (OLAP) to get insights into customer behavior and industry trends. We can use this data to create more targeted marketing strategies, fine-tune our product offerings, and provide better customer service.

Furthermore, OLAP enables Internet businesses to do complex modeling and analysis. OLAP provides a multidimensional data perspective, allowing online firms to analyze it from various angles. As a result, they can see trends and patterns that wouldn't be so apparent with more traditional analysis. Strategic decisions and the development of prediction models based on this data may be utilized to foretell future consumer and market behaviors and trends. The capacity to "drill down" into data for a deeper understanding of performance indicators is another perk of using OLAP in e-business. To better understand where they may make changes or seize opportunities, online organizations can use OLAP to evaluate data at different levels of detail. This data has the potential to be used to boost earnings, streamline internal processes, and provide a better experience for customers. OLAP is also crucial in the management and security of information in the context of online businesses. Since OLAP offers a consolidated view of the data, it allows e-businesses to keep tabs on data consumption and guarantee adherence to data governance regulations. Furthermore, OLAP provides various security features, such as role-based access control and data encryption, to protect sensitive business data [1].

Information display and reporting "business intelligence" (BI) is an umbrella word in electronic commerce, including various subsystems that aid businesses in making educated decisions via data analysis and presentation. Business intelligence relies heavily on reporting and data visualization. The following section will examine how reporting and visualization tools help businesses conduct data analysis and make sound decisions. Reporting is gathering data into summaries that provide insights into an organization's operations. On the other hand, data visualization is the practice of visual representations of data. E-business reporting and visualization technologies let companies quickly and accurately assess datasets, yielding actionable insights for strategic decision-making. These tools simplify previously incomprehensible data by creating aesthetically attractive reports, dashboards, and scorecards.

There are a variety of ways in which businesses might benefit from using reporting and visualization tools. One of the most significant benefits is the ability to keep tabs on business operations in real time. Data analytics allows businesses to get real-time insights into their processes, allowing them to make decisions at the moment that might improve their efficiency and productivity. In addition, these technologies provide businesses with a straightforward means of evaluating their data, which aids in discovering patterns and trends that may inform future decisions.

Netflix is a great case study of a company that uses reporting and visualization tools well. The streaming giant uses data analytics to monitor its users' behaviors and preferences so it may provide relevant content to its paying subscribers. Netflix's user interaction data analysis provides insights into how its users engage with the service. These discoveries can direct new content creation and improve the user experience. Netflix's use of data analytics has been a critical factor in the company's rise to prominence as one of the world's most successful media firms.

The online retail giant Amazon is just one more company that exemplifies the successful use of reporting and visualization technologies. The online retailer uses analytics data to learn about consumers' habits and interests. The findings from this research are utilized as a foundation for future product development and advertising campaigns. Amazon also uses data analytics to improve the efficiency of its supply chain, which helps the company save money while producing more goods. As the largest online retailer in the world, Amazon has relied on several reporting and data visualization tools to keep it at the top.

Companies must have reporting and visualization tools to meet regulatory requirements. Since there are rules in place that necessitate the reporting and analysis of financial data, financial institutions are expected to comply with these rules. Financial institutions can use reporting and visualization tools to analyze massive amounts of data and provide insights into their financial performance. Financial institutions may meet the necessary standards for regulation in this way.

Analytics potentially future-forcing: Predictive analytics, a core component of business intelligence, helps Internet businesses make educated guesses about the future by analyzing the patterns present in historical data. With intricate algorithms, predictive analytics can make educated guesses about how a system will behave in the future. It might also identify possible dangers and provide advice on growth opportunities. In the following paragraphs, we'll discuss the value of predictive analytics in e-business and how it may help businesses stay ahead of the curve.

One of the most essential benefits of predictive analytics in e-business is the ability to foresee consumers' activities. Customer loyalty and churn may be predicted with the help of predictive analytics by looking at past interactions and purchases to see who is most likely to remain a customer and who is most likely to defect. Then, you can put this information to use by creating targeted advertising campaigns, offering personalized product recommendations, and coming up with other strategies to increase customer loyalty and repeat business. Predictive analytics has several vital applications in e-business, including inventory management. By analyzing past sales trends and seasonality patterns, online retailers may anticipate consumer demand and adjust stock accordingly. Overstocking and understocking may be avoided, resulting in cost savings and increased customer satisfaction. Predictive analytics may be used for more than only studying customer behavior and managing inventory; they can also be used to fine-tune pricing strategies. By studying market trends, their rivals' pricing tactics, and their clients' buying habits, e-commerce businesses can establish the optimal selling prices for their products and services. This helps companies maximize profits and ensures they remain competitive in the market [3].

Online retailer Amazon and streaming service Netflix are two companies that have successfully used predictive analytics. Using data from customers' browsing and buying habits, Amazon uses predictive analytics to offer products to its users. Not only has Amazon experienced an increase in revenue due to this, but consumer satisfaction and brand loyalty have also increased. Like Amazon suggests books based on your reading habits, Netflix presents shows and movies depending on what you've already seen. This has helped Netflix retain customers and given it an advantage in the rapidly growing but more competitive streaming sector.

Quality control over data collection: For businesses operating in the digital sphere, data quality management (DQM) is a crucial component of business intelligence (BI). Data quality management (DQM) is the process by which a company verifies that its data is accurate, consistent, and complete before it puts it to use. DQM is essential to business intelligence since it affects the accuracy of the inferences and decisions businesses make based on their data.

DQM is beneficial for e-businesses because it improves the reliability and accuracy of data. This is crucial because BI systems rely heavily on accurate data to analyze and make decisions. If the data are consistent and dependable, organizations can rely on the conclusions and decisions they draw from them. DQM is particularly useful for e-businesses because it can help identify and resolve data quality issues before they escalate. Data quality issues can significantly impact the dependability of BI insights and decisions, so this is crucial. By identifying and resolving these issues as soon as possible, organizations can save money and increase the reliability of their BI insights.

Amazon is an excellent example of an e-commerce website that utilizes DQM effectively. Amazon is one of the world's largest online retailers; data is essential to their operations. To ensure the integrity of its data, Amazon has implemented several DQM procedures, such as automatic data validation and cleansing and human data review techniques. By adhering to these procedures, you can ensure that the data your Amazon BI tools utilize are accurate and exhaustive.

DQM in e-businesses can help improve operational efficiency, data accuracy, and dependability. Organizations can save time and money on data gathering, processing, and analysis by standardizing their data collection and management procedures. Consequently, businesses can reinvest the savings in other areas of their operations or expand.

PayPal is an online business that has implemented DQM to increase productivity. An online payment system, PayPal must rely heavily on data to succeed. By employing DQM practices, which streamlined its data collection and administration procedures, PayPal was able to save time and money. As a consequence, PayPal was able to reinvest these savings in other company expansion initiatives.

DQM in online businesses has the potential to enhance the overall consumer experience. Suppose the data used to assess consumer behavior and preferences is reliable and exhaustive. In that case, organizations can better comprehend their customers' needs and desires and design products and services to meet those demands. This has the potential to improve customer service, thereby increasing repeat business.

Netflix is an example of a company that has utilized DQM to enhance the user experience. Netflix is a subscription-based streaming service that makes personalized recommendations based on user information. Netflix's ability to provide consumers with trustworthy content recommendations is mainly attributable to the company's use of DQM procedures. Consequently, consumers received special requests, increasing their satisfaction and loyalty.

Data governance is an essential business intelligence (BI) component for e-businesses. It is a set of rules and regulations designed to maintain the security and dependability of information throughout an organization. Data governance may benefit organizations by providing a transparent framework for administering data and ensuring it conforms to all applicable standards. This section will explain why data governance is essential for online businesses and demonstrate how some organizations have employed it to enhance their business intelligence (BI) tools.

Due to data governance in e-businesses, organizations may benefit from improved data quality. When businesses implement a data governance strategy, they can identify and address data quality issues before escalating. Data governance is establishing norms and standards to ensure high-quality data (i.e., data that is accurate, complete, and consistent). By keeping accurate product descriptions, prices, and stock levels, a retail e-commerce website can provide its customers with a positive online purchasing experience. Compliance with regulations is another essential concern for online businesses, which can be aided by effective data governance. As more and more information is created and stored, companies must ensure compliance with privacy regulations. The General Data Protection Regulation (GDPR) is one such law that requires organizations to know precisely what data they possess, where it resides, and how it is managed. Data governance regulations and procedures enable online businesses to comply with legal requirements, avoid penalties, and safeguard their reputations. Data governance in e-businesses also facilitates the making of informed decisions. Providing access to transparent, consistent data and data governance facilitates improved decision-making. Data governance enables the incorporation of data from multiple sources for an online business, providing a complete image of client routines and preferences. This may help the online store create more targeted marketing campaigns, which can increase sales and customer loyalty [1].

Starbucks provides a concrete example of how data governance can enhance BI capabilities. Starbucks was able to create a unified picture of consumer activity across all channels by establishing a single source of truth for customer data through the use of data governance. Using this method, Starbucks was able to understand its consumers better and tailor its marketing to their preferences. Both customer retention and acquisition skyrocketed, which increased the company's net income.

Data governance improves interdepartmental cooperation and decision-making efficacy. By establishing a unified data management strategy, data governance enhances internal communication and collaboration. Data governance enables an e-business, for example, to ensure that all divisions have access to the same data, thereby fostering collaborative strategy development. Moreover, data governance assists e-businesses with risk management by providing data integrity. Using policies and processes for data governance enables organizations to better prepare for and respond to data breaches, data loss, and cyberattacks. With data governance, online businesses can protect consumer information and avoid reputational harm. Table 11.2 summarizes the advantages and disadvantages of different business intelligence components.

Table 11.2 Advantages and disadvantages of different business intelligence Components

Business intelligence component	Advantages	Disadvantages
Data warehousing	Centralized repository for data from various sources Improved data quality Enhanced data integration and analysis	Requires significant upfront investment It can be complex and time-consuming to implement It may require ongoing maintenance and updates
Data mining	Discover hidden patterns and relationships Identify market trends and product preferences Improve decision-making based on insights from data	It may require expertise in statistical analysis and programming Requires access to large datasets May yield unexpected or irrelevant results
Online analytical processing (OLAP)	Multidimensional analysis of data Quickly answer complex business queries Efficient use of data storage	Requires significant upfront investment It can be challenging to implement and maintain It may not be suitable for all types of data analysis
Reporting and visualization	Easy-to-understand presentation of data Real-time access to data insights Improved decision-making based on data	It may require significant customization to meet specific business needs It may not provide all necessary data insights It may not be suitable for all types of data analysis
Predictive analytics	Accurate forecasting of future outcomes Identification of potential customers and opportunities Improved decision-making based on data-driven insights	Requires access to large datasets and historical data It may require expertise in statistical analysis and programming May yield unexpected or unreliable results
Data quality management	Improved accuracy and completeness of data Better decision-making based on reliable data Improved operational efficiency	It may be time-consuming and resource-intensive to implement It may require significant changes to existing data processes It may require ongoing maintenance and updates
Data governance	Ensures data security and privacy Compliance with regulations and standards Improved data quality and reliability	It may be complex and time-consuming to implement It may require significant changes to existing data processes May limit accessibility to data

3 Implementation of Business Intelligence

By employing and implementing BI tools and processes, e-businesses may gain beneficial insights and momentum. However, the implementation procedure may need to be simplified and problematic. Here, we will discuss some guidelines for how e-businesses can effectively utilize BI systems and processes (Fig. 11.1).

3.1 Define the Business Problem and Objectives

Successful business intelligence (BI) system implementation begins with explicitly defining the organization's challenges and objectives. With a clear understanding of the issue to be addressed and the desired outcomes, the implementation process may become focused and successful.

First, you must pinpoint the specific business issue that requires resolution with the aid of the BI system. Falling revenues, inefficient operations, and a need for more understanding of the company's target market are examples of opportunities

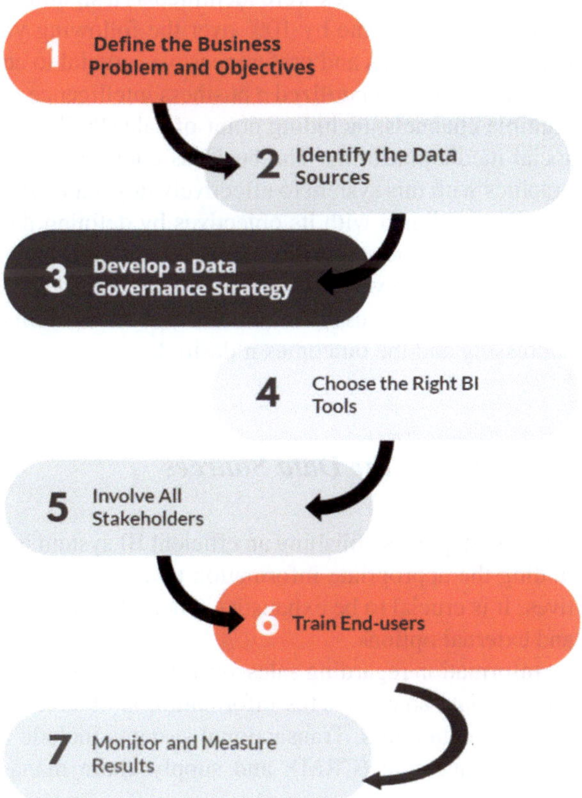

Fig. 11.1 Steps for implementation of business intelligence in e-business

and hazards that the business may face. It is essential to have a detailed account of the issue and ensure all parties understand its nature and scope. Once the problem has been identified, the objectives the BI system will help achieve can be specified. You might resolve to increase your income by a certain percentage or decrease your expenses by a certain amount. It may also involve determining where the company has knowledge deficits regarding its customers or competitors that prevent it from making optimal strategic decisions. Involving key stakeholders across the organization is essential for accurately defining the business challenge and objectives. In addition to those from finance, marketing, and operations, the upper management and the information technology team members could be present. By involving a diverse group of stakeholders in the process, it is possible to gain a deeper understanding of the issue and ensure that the objectives align with the organization's overall strategy. After establishing the problem and the desired outcomes, it is time to place pen to paper. This information should be accessible to all relevant stakeholders, as it will serve as a reference point during implementation. To maintain the BI system aligned with the organization's objectives, it should be revised as necessary as the project progresses.

When implementing a business intelligence (BI) system at a large retail organization, it is essential to define the business problem and the organization's objectives. As declining sales were identified as a significant issue, the company resolved to increase its income by 10% over the following year. Understanding customers' purchasing patterns and preferences was crucial to achieving this objective.

The organization utilized a business intelligence (BI) system to collect data from multiple channels, including point-of-sale (POS) terminals, consumer surveys, and social media platforms. The business could monitor client preferences and demographics with this system to effectively target advertising. The company ensured the BI system aligned with its objectives by defining the business challenge and goals at the project's outset. Within the first year of implementation, the system increased the company's revenue by 12%, which exceeded expectations. In large part, the organization's success can be attributed to its ability to define the problem it was addressing and the outcomes it desired.

3.2 *Identify the Data Sources*

The first step in establishing an efficient BI system is identifying the data source and finding the appropriate information to address the organization's issues and objectives. It is crucial to be exhaustive when identifying data sources, including internal and external options.

Information regarding sales, customers, finances, and operations are examples of internal data sources. This information is obtained from databases, data repositories, and data marts. Transactional systems include accounting, customer relationship management (CRM), and supply chain management (SCM), whereas data warehouses and data marts are used to hold historical data for reporting and analysis. On the other hand, external data sources include market, social media, and demographic data. This information is available from various external sources, including private research organizations, public organizations, and online community forums. Ensure external data is accurate, complete, and up to date before incorporating it into the BI system.

Alongside the data's sources, its proprietors and custodians, who should be held accountable for the data's quality and integrity, should be identified. This helps ensure the data is accurate, consistent, and compliant with the company's established norms and guidelines. Data owners and custodians are responsible for its completeness, accuracy, and timeliness, as well as its security and upkeep. After identifying the data sources, the next step is extracting, transforming, and loading (ETL) the data into the BI system. Data is removed from source systems, converted into the desired format and structure, and inserted into the business intelligence system. The ETL procedure is crucial to the success of the BI system because it ensures that the data is clear, accurate, and available for analysis.

3.3 Develop a Data Governance Strategy

A data governance plan must first be developed to implement an effective BI system in an online business—development of data quality, security, and regulatory compliance policies and processes. A data governance policy must be developed to ensure accurate, exhaustive, and consistent data. In this plan, data quality, data security, and data privacy should all be addressed.

Improving data integrity is one of data governance's primary objectives. The effective operation of a business intelligence (BI) system requires accurate, complete, and consistent information. Creating data validation criteria, data cleansing methods, and data training for staff are all methods for enhancing data quality. By implementing a data quality framework, you can ensure that your data are always accurate and trustworthy. A strategy for data governance should also prioritize data security. The data collected by online businesses must be protected from theft and misuse. Restrictions on data access, encryption, and backup and recovery protocols are essential to any effective data security plan. With the help of a well-implemented framework, data security and protection against cyberattacks can be enhanced. Preserving sensitive data is another essential component of any data governance strategy. Online businesses must take precautions to prevent the abduction or misuse of consumer information. As part of a comprehensive data privacy policy, data access rules and data handling procedures must be established. A data privacy strategy can prevent sensitive data from straying into the incorrect hands.

3.4 Choose the Right BI Tools

When deciding which BI products to use, several factors must be considered. Important considerations include the instruments' capacity to manage the quantity and type of data being evaluated. There is a vast selection of BI tools, each optimized for a specific type of data or compilation of data. Choose a system that can accommodate the particular data requirements of your business. The dependability and timeliness of the data provided by your BI tools are also essential considerations. To make decisions based on the most recent information, the relevant instruments must be able to provide accurate and timely data. Additionally, tools should provide data quickly, allowing for quicker decision-making.

Choosing BI tools based on their ease of use is also essential. The instruments should be straightforward to ensure users can obtain the required information without extensive training or technical expertise. This is especially important in e-business, where managers and decision-makers need immediate data access and analysis. Business intelligence (BI) technologies include reporting, data visualization, online analytical processing (OLAP), and predictive analytics applications. Each instrument has advantages and disadvantages; selecting one will depend on the organization's circumstances.

For instance, reporting tools are designed to generate operational status reports swiftly and effortlessly. These resources are helpful because they give users instantaneous access to essential data such as sales figures and customer satisfaction rankings. Conversely, data visualization technologies seek to provide more intricate, dynamic data visuals. Finding trends and patterns in enormous datasets is a frequent application of these methods. OLAP software lets users view data from multiple perspectives and conduct extensive research. Sales data segmented by region, product, and period are a few examples of complex datasets that may benefit from these techniques. Lastly, predictive analytics tools are designed to predict using statistical models and historical data. Utilizing these instruments can be advantageous for sales forecasting, prospect profiling, and campaign optimization.

3.5 Involve All Stakeholders

For the project to be successful, all stakeholders must be consulted before, during, and after implementing a BI system. This group comprises, in addition to executives and the IT department, the end users who will use the system daily. Participation of stakeholders must begin during the planning phase and continue uninterrupted throughout the implementation phase. During the planning phase, stakeholders should be identified, and their responsibilities in the project's execution should be mapped out in great detail. Executives should be involved in delineating the issue, determining a course of action, and assessing potential BI solutions. IT personnel should be involved in both the formulation of decisions and the implementation of the resultant technical framework. During the business requirements development and system testing phases, end user input is crucial. At each stage of the implementation process, stakeholders should receive regular progress updates and opportunities to provide feedback. Scheduled status updates and other communication channels, such as discussion groups and email, may be helpful here. Additionally, regular meetings should be conducted to address concerns and ensure the project remains on track.

Involving all relevant parties guarantees that the system will meet the business's requirements and be well received by its end consumers. This is crucial because consumers will only utilize the system if they perceive a purpose. Involving end users early on is essential to integrate user feedback into the final product. As a consequence of these modifications, the system will be easier to use, and its adoption rate will increase.

American Eagle Outfitters (AEO), an apparel and accessories retailer, exemplifies how all required stakeholders can be involved in implementing a BI system. AEO implemented a BI system to enhance inventory management and sales forecasting. During the implementation, several distinct categories of teams worked together. These groups comprised IT professionals, business analysts, and end consumers. Teams collaborated to identify essential business metrics, evaluate prospective business intelligence solutions, and design intuitive dashboards. AEO improved inventory management and increased sales by involving all stakeholders and utilizing the BI system's insights.

3.6 Train End Users

End user training is necessary to successfully implement business intelligence (BI) technology in an online enterprise. End users must be well versed in the system's functionality, navigation, and data interpretation. The system's benefits may be improved if end users are adequately trained in utilizing a business intelligence (BI) system.

The first stage in educating end users is to develop a training strategy. Plans for training should include not only the program's objectives but also its content and methodology. Additionally, the training plan should specify who will be trained and when. End users may be instructed in several ways. One option for training involves an instructor lecturing end users in a classroom setting. Another option is to provide customers access to online training courses they can complete at their own pace. Both methods may be effective; therefore, the situation and resources of the organization will determine which one is used.

The training should be engaging and enjoyable, including hands-on exercises that give participants expertise with the system. This method increases students' likelihood of retaining what they have learned in class. End users should receive training not only on how to use the BI system but also on how to interpret the data it generates. It is essential to be able to interpret measurements and see how they relate to the larger organizational picture. Users of the BI system must be instructed on how to make sensible decisions based on these insights. Providing users with continued access to materials is an excellent method for reinforcing what they have learned in training. One example is a support center or help station where consumers can seek answers to their questions. Another option is to provide digital materials such as how-to guides and video tutorials.

Amazon has effectively educated its customers about its business intelligence (BI) system. Amazon employees receive comprehensive training on the company's business intelligence system, "Spend Management." There are both instructor-led classroom sessions and independent online study. Employees can apply their newly acquired knowledge through entertaining and interactive hands-on activities. Amazon employees have round-the-clock access to a support desk and abundant online resources. As a result of this training, Amazon has realized substantial cost reductions due to employee proficiency with the Spend Management system.

3.7 Monitor and Measure Results

A BI system's efficacy significantly depends on its performance monitoring and measurement capacity. It is crucial to periodically monitor the organization's established key performance indicators specific to the issue and objectives. Indicators include sales volume, customer satisfaction, and operational effectiveness. In addition to monitoring key performance indicators, user input is crucial for ensuring the system serves its intended purpose. Techniques such as online polls, in-person interviews, and focus groups are helpful. User input may be used to fine-tune the BI system and address any potential issues. Providing regular reports is a vital element of monitoring and evaluating accomplishments. Regular reports should be generated, with the format and content adapted to satisfy the needs of the various interested parties. Data-driven decisions can be made using the information provided by these reports, which can be used to identify patterns, highlight areas of concern, and obtain insights. It is essential to have a plan in place for addressing problems as they arise. This may include a method for managing data quality issues and a helpdesk or support personnel handling technological matters. Organizations can ensure that their BI installation is providing value and generating success by monitoring and assessing the results of the BI system and having a plan for resolving any issues that arise.

Without business intelligence (BI), e-commerce would not be feasible. However, it must be integrated with the existing company infrastructure for maximum efficiency. Integration with CRM systems is a crucial component of the whole. Business intelligence can be applied to CRM databases to extract relevant information. Business intelligence can be used to analyze aspects such as customers' purchasing behaviors and preferences. Analyzing this data could result in more precise marketing strategies, improved customer service, and increased customer retention rates. The ERP system is another area where BI can be integrated with other e-business systems and processes. A variety of data types, including sales, inventory, and purchasing data, can be maintained in an ERP system. When business intelligence (BI) and enterprise resource planning (ERP) are combined, businesses can enhance their operations and make better decisions. For instance, BI can be used to identify cost-cutting opportunities, improve supply chain management, and increase manufacturing output. Supply chain management (SCM) and business intelligence (BI) can be combined to increase efficiency. Systems for supply chain management (SCM) manage the distribution of goods and services. By integrating BI and SCM, businesses can see their supply chain processes more plainly and identify areas for improvement. Supply chain management, inventory control, and time-sensitive delivery are a few of the applications of BI. Human resource management (HRM) systems are another area where BI can be linked. Human resource management systems compile employee qualifications, output, and present data [4].

Businesses can better understand their employees and make more informed decisions when combining BI and HRM. Business intelligence (BI) can be used to enhance personnel planning, identify high-performing employees, and identify areas where additional training is required. Financial management systems and BI can also be interconnected. A financial management system is in place to monitor the company's income, expenses, and earnings. When business intelligence (BI) is combined with financial management, a company's financial performance can be better understood, and decisions can be made more confidently. For instance, BI can help identify successful product lines, enhance pricing strategies, and identify areas where costs can be reduced. Table 11.3 provides an overview of integrating business intelligence with other systems and processes.

4 Benefits and Challenges of Business Intelligence in E-Business

Business intelligence (BI) is now an essential element of e-business. In a market that changes swiftly and is highly competitive, the ability to derive insights from data and make informed decisions is crucial. Implementing BI in e-business has numerous advantages, including enhanced decision-making, increased productivity, an enhanced consumer experience, and cost reductions (Table 11.4).

Enhanced decision-making is one of the most essential advantages of BI in e-business. Business intelligence provides a platform for analyzing and interpreting data, which enables business executives to make decisions based on data. By identifying trends, patterns, and opportunities, business intelligence allows e-businesses to respond rapidly to market changes and customer demands. Amazon, one of the largest e-commerce companies in the world, employs BI to analyze consumer behavior and make personalized product recommendations, leading to increased sales and customer satisfaction.

Table 11.3 Integration of business intelligence with other systems and processes in e-business

System/process	Description	Integration with BI	Benefits
E-commerce platform	The software platform used to sell products or services online	Integrating BI with e-commerce platforms can provide insights into customer behavior, sales patterns, and product performance	Improved marketing strategies, better customer experiences, and increased revenue
Customer relationship management (CRM)	The system is used to manage interactions with customers, including sales, service, and marketing activities	Integrating BI with CRM systems can provide insights into customer behaviors, preferences, and needs	More personalized customer interactions, improved customer retention, and increased revenue
Supply chain management (SCM)	The process of managing the flow of goods and services from suppliers to customers	Integrating BI with SCM systems can provide insights into inventory levels, supplier performance, and logistics	Improved inventory management, reduced costs, and improved supplier relationships
Financial management	The process of managing financial transactions and operations within an organization	Integrating BI with financial management systems can provide insights into financial performance, cash flow, and budgeting	Improved financial planning, more informed decision-making, and reduced financial risk
Human resources (HR)	The process of managing employee-related activities, such as recruitment, training, and performance evaluation	Integrating BI with HR systems can provide insights into workforce demographics, performance, and engagement	Improved hiring and training strategies, better employee retention, and increased productivity

An additional advantage of BI in e-business is enhanced productivity. BI enables e-businesses to automate processes, reduce manual duties, and eliminate redundancies. E-businesses can save time and resources by streamlining their operations, allowing them to focus on strategic initiatives. Google uses BI, for instance, to optimize its data centers and reduce energy consumption, resulting in substantial cost savings and increased efficiency.

The importance of BI in enhancing the consumer experience cannot be overstated. By analyzing customer data, e-businesses can comprehend consumer preferences, behaviors, and requirements, enabling them to provide customized products and services. Popular streaming service Netflix, for instance, utilizes BI to analyze user viewing behavior and provide personalized recommendations, resulting in increased customer retention and satisfaction.

BI can assist e-businesses in achieving cost reductions. E-businesses can optimize operations, reduce expenses, and increase profits by identifying inefficiencies, redundancies, and wastes. Consumer goods company Procter & Gamble employs BI to analyze supply chain data and identify cost-saving opportunities, resulting in significant savings. In conclusion, the benefits of business intelligence in e-commerce

Table 11.4 Benefits of BI in e-business

The advantage of BI in e-business	Description	Example
Improved decision-making	BI can provide real-time insights into business operations, allowing for more informed and timely decision-making	A retail company uses BI to track sales data and adjust real-time inventory levels based on demand
Increased efficiency	BI can automate processes and streamline workflows, reducing manual effort and increasing productivity	A manufacturing company uses BI to monitor machine performance and schedule maintenance, reducing downtime and increasing output
Enhanced customer experience	BI can provide insights into customer behaviors and preferences, allowing businesses to tailor their offerings and improve customer satisfaction	A hospitality company uses BI to track guest preferences and personalize their experience, resulting in higher customer satisfaction ratings
Competitive advantage	BI can give businesses a competitive edge by enabling them to identify trends and opportunities before competitors	A financial services company uses BI to identify market trends and adjust investment strategies, resulting in higher client returns

are numerous and substantial. By implementing BI, e-businesses can achieve multiple benefits, including enhanced decision-making, increased productivity, an improved consumer experience, and cost reductions. With the assistance of BI, e-businesses can obtain a competitive advantage in an ever-changing market and provide value to their consumers. Figure 11.2 illustrates some steps for evaluating the benefits of business intelligence.

E-commerce has incorporated business intelligence (BI) tools to comprehend consumer behaviors better, improve operational efficiency, and base decisions on factual data. While implementing BI in an online business environment has numerous benefits, it is also subject to challenges and restrictions (Table 11.5).

Implementing BI in an e-business environment is challenging due to data availability and quality concerns. Client interactions, online purchases, and social media generate copious amounts of data for e-commerce operations. These data could be more reliable and accurate, which may lead to incorrect conclusions and actions. Ineffective marketing strategies and poor sales performance may result, for instance, from inaccurate customer preferences and behavior data. BI software's high cost and technical complexity are additional obstacles to its widespread adoption in e-business. The time, money, and effort required to implement a BI system are substantial. In addition, the technology may be challenging to install and require a high level of expertise. This may be particularly challenging for online SMB businesses that need more financial resources to invest in BI systems.

The implementation of BI in e-business is complicated by data privacy and security concerns. With the increasing importance of consumer data in BI, businesses must take additional measures to protect sensitive information and ensure full compliance with all applicable privacy laws and regulations. Customers' loss of trust,

Fig. 11.2 Evaluating the benefits of business intelligence in e-business

legal penalties, and a tarnished reputation are all possible outcomes of noncompliance. In 2018, Facebook was heavily criticized when it was discovered that the personal information of millions of users had been collected without their consent or knowledge.

In addition, more qualified personnel are needed to ensure the adoption of BI in e-business. Online businesses may need help finding employees with expertise in data analytics, computer programming, and statistical analysis. The system may need to be more utilized, and as a result, discernment may suffer.

Table 11.5 Challenges of BI in e-business

Challenge	Description	Example
Data quality issues	BI systems are only as good as the data they are built on. Data quality issues, such as incomplete or inaccurate data, can lead to incorrect insights and decisions	An e-commerce company relies on customer reviews to make decisions about product improvements, but some thoughts are fake and misrepresentative
Integration issues	Integrating data from multiple sources can be challenging, especially when stored in different formats or systems	A retail company uses one system to track sales and another to track inventory, making it difficult to get a complete view of its operations
Scalability issues	As the amount of data and users increase, BI systems may need help to keep up with the demand, leading to slow response times and decreased efficiency	A healthcare company's BI system cannot handle the volume of patient data, resulting in slow loading times and delayed decision-making
User adoption issues	Even the most advanced BI system will only succeed if users adopt it. Training and support may be necessary to encourage user adoption	A financial services company implements a new BI system but needs to provide adequate training for end users, leading to low adoption rates and little usage
Security concerns	BI systems can be a target for cyberattacks, and sensitive business information can be at risk if not adequately secured	A manufacturing company's BI system is hacked, and sensitive customer data is stolen, resulting in a loss of trust and damage to the company's reputation

With the proper strategy, e-businesses can succeed despite these constraints. For instance, purifying and validating data before analysis may aid in drawing accurate conclusions from the analyzed data. Cloud-based BI solutions also facilitate and reduce the overall cost of IT for online businesses. Encryption and access restrictions are necessary to protect sensitive consumer data from intruders.

E-businesses may gain much from BI, but it has challenges and restrictions. For online businesses to reap the maximum benefits of BI, they must surmount these obstacles. This will enable them to understand their clientele and internal operations better, resulting in process enhancements and data-driven decisions that increase productivity and revenue.

Summary

Definition of Business Intelligence (BI)

- Business intelligence refers to collecting, analyzing, and presenting data to support informed decision-making in an organization.
- It involves using technologies, tools, and methodologies to extract valuable insights from data and transform them into actionable information.

Components of Business Intelligence

- Data sources: BI relies on various data sources, such as databases, data warehouses, and external sources, to gather relevant information.
- Data integration: BI systems integrate data from multiple sources to create a unified view, ensuring consistency and accuracy.
- Data warehousing: BI employs data warehousing techniques to store and manage large volumes of structured and unstructured data.
- Analytics and reporting: BI tools enable the analysis and visualization of data, allowing users to generate reports, dashboards, and visual representations.
- Data mining: BI leverages data mining techniques to discover patterns, correlations, and trends within datasets, revealing hidden insights.

Implementation of Business Intelligence

- Planning and strategy: Successful BI implementation starts with defining clear goals, identifying key metrics, and aligning them with business objectives.
- Data collection and integration: Relevant data is collected from various sources, transformed, and integrated into a centralized data repository.
- Data modeling and analysis: Data is modeled to facilitate analysis and exploration, enabling users to derive meaningful insights.
- Reporting and visualization: BI tools create reports, visualizations, and interactive dashboards to present data comprehensibly.
- Deployment and maintenance: The BI system is deployed, and ongoing maintenance ensures data accuracy, system performance, and security.

Benefits of Business Intelligence in E-Business

- Enhanced decision-making: BI provides timely, accurate, and relevant information, empowering e-businesses to make data-driven decisions.
- Competitive advantage: BI enables e-businesses to gain a deeper understanding of their market, customers, and competitors, facilitating better positioning and differentiation.
- Improved operational efficiency: BI helps optimize business processes, identify bottlenecks, and streamline operations, increasing efficiency and cost savings.
- Customer insights: BI enables e-businesses to analyze customer behaviors, preferences, and buying patterns, enabling personalized marketing and improved customer satisfaction.
- Forecasting and planning: BI supports forecasting and predictive analysis, aiding e-businesses in anticipating market trends and planning future strategies.

Challenges of Business Intelligence in E-Business

- Data quality and integration: Ensuring data accuracy, consistency, and integration from multiple sources can be complex and challenging.
- Data security and privacy: Protecting sensitive data and complying with data privacy regulations pose challenges in BI implementations.
- Scalability and performance: As data volumes grow, ensuring the scalability and performance of BI systems can be a significant challenge.

- User adoption and training: Encouraging user adoption of BI tools and providing adequate training to utilize the system effectively can be a hurdle.
- Changing business needs: Adapting BI systems to evolving business requirements and incorporating new data sources and technologies can be challenging.

Case Study

The well-known e-commerce company XYZ goods specializes in the online sale of consumer goods. The organization operates in numerous countries and interacts with vast amounts of data from various sources, including client transactions, Internet traffic, inventory management, and marketing activities. As XYZ Electronics grows, the management team recognizes the need for a robust business intelligence (BI) system to obtain actionable insights, enhance decision-making, and boost operational efficiency.

1. How can XYZ Electronics define and describe business intelligence, and how would BI integration benefit their e-commerce operations?
2. How can XYZ Electronics use the frequently employed business intelligence components, such as data warehousing, data mining, and predictive analytics, to improve its e-commerce processes?
3. Discuss the advantages and disadvantages of data warehousing, data mining, and predictive analytics for XYZ Electronics. How can they minimize the disadvantages and maximize the advantages of each component?
4. How can e-commerce companies like XYZ Electronics best develop business intelligence tools and processes? How can the organization ensure BI's successful implementation and adoption across all departments?
5. How can XYZ Electronics integrate its business intelligence system with other existing systems and processes, such as inventory management and marketing campaigns? What obstacles might they face during the integration process, and how might they surmount them?

Discussion

1. How does business intelligence differ from traditional data analysis?
2. What are some examples of how business intelligence has been used to drive business success?
3. How has the growth of big data impacted the importance of business intelligence for e-businesses?
4. What are some emerging trends in business intelligence that e-businesses should be aware of?

5. How can e-businesses ensure that their business intelligence efforts align with their overall business strategy?
6. What is data warehousing, and how does it contribute to business intelligence?
7. How does data mining help e-businesses uncover valuable insights from their data?
8. What is predictive analytics, and how can it be used by e-businesses to make more informed decisions?
9. What are some advantages and disadvantages of using different components of business intelligence?
10. What are some best practices for implementing a business intelligence system in an e-business?
11. How can e-businesses ensure that their business intelligence system is integrated with other systems and processes?
12. What are some common challenges that e-businesses face when implementing a business intelligence system, and how can they be overcome?
13. How can business intelligence help e-businesses make better decisions and improve operational efficiency?
14. What are some potential drawbacks of using business intelligence, such as concerns around data quality and privacy?
15. How can e-businesses address these challenges and ensure that they are using business intelligence in an ethical and responsible way?

Multiple-Choice Question

1. What is the definition of business intelligence?
   (a) The use of computers and technology in business operations
   (b) The process of analyzing business data to gain insights and make informed decisions
   (c) The management of business resources and assets
   (d) The development of strategies for business growth and expansion
2. Why is business intelligence important for e-businesses?
   (a) It helps in designing attractive websites for customers.
   (b) It enables e-businesses to analyze customer behaviors and preferences.
   (c) It provides financial support for e-businesses.
   (d) It automates business processes and reduces costs.
3. Which of the following is a component of business intelligence?
   (a) Customer relationship management (CRM)
   (b) Supply chain management (SCM)
   (c) Enterprise resource planning (ERP)
   (d) All of the above

4. Data warehousing is a component of business intelligence that involves:
   (a) Collecting and storing large amounts of structured and unstructured data
   (b) Analyzing data to identify patterns and trends
   (c) Predicting future outcomes based on historical data
   (d) Extracting data from various sources for analysis
5. What is one advantage of data mining in business intelligence?
   (a) It helps in identifying customer preferences and behavior patterns.
   (b) It enables real-time monitoring of business operations.
   (c) It improves data storage and management.
   (d) It automates routine business tasks.
6. Predictive analytics is a component of business intelligence that:
   (a) Analyzes historical data to make future predictions and forecasts
   (b) Extracts data from multiple sources for analysis
   (c) Identifies patterns and trends in data
   (d) Provides real-time insights into business operations
7. What are the advantages of implementing business intelligence systems in e-businesses?
   (a) Improved decision-making and operational efficiency
   (b) Increased marketing costs and reduced customer satisfaction
   (c) Enhanced employee training programs
   (d) Greater focus on administrative tasks
8. Best practices for implementing business intelligence systems include:
   (a) Defining clear business objectives and goals
   (b) Collecting and storing data without a specific purpose
   (c) Relying solely on intuition for decision-making
   (d) Keeping data analysis restricted to a few employees
9. How can business intelligence be integrated with other systems and processes?
   (a) By sharing data and insights across departments and teams
   (b) By isolating business intelligence systems from other systems
   (c) By avoiding collaboration with other departments
   (d) By using separate databases for Business Intelligence purposes
10. What is one benefit of business intelligence in e-businesses?
   (a) Improved customer service and satisfaction
   (b) Increased production costs
   (c) Reduced employee productivity
   (d) Limited access to business data

11. What is one challenge or limitation of business intelligence?
    (a) Data quality and accuracy concerns
    (b) Increased operational inefficiencies
    (c) Inability to analyze large volumes of data
    (d) Lack of security measures for data protection
12. How does business intelligence contribute to improved decision-making?
    (a) By providing real-time insights and analytics
    (b) By minimizing the need for human involvement in decision-making
    (c) By relying on gut feelings and intuition
    (d) By reducing the amount of data available for analysis
13. Which component of business intelligence helps in identifying trends and patterns in data?
    (a) Data mining
    (b) Data warehousing
    (c) Predictive analytics
    (d) Customer relationship management
14. What is one disadvantage of data mining in business intelligence?
    (a) Limited availability of data for analysis
    (b) Increased complexity in data storage and management
    (c) Inability to handle large volumes of data
    (d) Overreliance on automated decision-making
15. What is one challenge related to data quality in business intelligence?
    (a) Lack of data privacy and security measures
    (b) Inconsistent and inaccurate data across multiple sources
    (c) Limited availability of data for analysis
    (d) Difficulties in integrating data from various systems

References

1. Chaudhuri, S. and U. Dayal, *An overview of data warehousing and OLAP technology*. ACM Sigmod record, 1997. **26**(1): p. 65-74.
2. Han, J., J. Pei, and H. Tong, *Data mining: concepts and techniques*. 2022: Morgan kaufmann.
3. Larose, D.T. and C.D. Larose, *Discovering knowledge in data: an introduction to data mining*. Vol. 4. 2014: John Wiley & Sons.
4. Power, D.J., *Decision support systems: concepts and resources for managers*. 2002: Greenwood Publishing Group.

Chapter 12
E-Business Adoption

Abstract The digital age has revolutionized organizational operations, making e-business practices indispensable for maintaining a competitive edge in today's marketplace. Incorporating technology into daily operations enhances efficiency and productivity. This chapter provides a brief introduction to e-business adoption, highlighting its significance for modern businesses and the factors that influence its spread. Various factors shape the implementation of e-business, including technical, organizational, and environmental aspects. This chapter explores their impact on adopting e-business and offers insights on overcoming associated challenges. The adoption process itself plays a crucial role, starting with awareness and curiosity, followed by evaluation, trial, and adoption. The subsequent chapter discusses effective adoption procedures for each phase. Although e-business brings numerous benefits, businesses encounter obstacles and challenges during its implementation. This chapter identifies and analyzes these hurdles, providing strategies for overcoming them. Additionally, it concludes by examining the pros and cons of embracing an e-business model and offers a glimpse into the future of e-business adoption. By the end of this chapter, readers will gain a comprehensive understanding of e-business adoption and its potential to benefit contemporary businesses.

1 Introduction to E-Business Adoption

E-business adoption, or the adoption of electronic business practices, incorporates these practices into a company's daily operations. In this context, "e-commerce" refers to using digital technologies, such as the Internet, intranets, and other online platforms, to conduct business activities, such as marketing, sales, and communication. In recent years, e-business adoption has become more crucial to keep up with the competition in a global market that is becoming increasingly digital.

Businesses are highly motivated to implement e-business strategies due to the ability to increase output while reducing administrative costs. By selling their products online, e-commerce businesses can reduce the costs of maintaining a physical

H. Taherdoost, *E-Business Essentials*, EAI/Springer Innovations in Communication and Computing, https://doi.org/10.1007/978-3-031-39626-7_12

storefront. Online payment processing and automated inventory management are two examples of how digital technology assists the e-commerce industry in reducing the need for human labor and accelerating business processes. E-commerce may also assist businesses in expanding their market presence and attracting more customers. Businesses can use digital technology to advertise their products to a larger audience and attract consumers from various locations. For instance, a clothing retailer may use social media advertisements to reach people in specific cities or countries, thereby increasing brand awareness and sales. In addition to increasing efficiency and expanding market reach, e-commerce may give businesses a competitive advantage. When businesses utilize digital technology, they can quickly adapt to the changing requirements of their customers and the market landscape. Suppose a company specializes in e-commerce and uses data analytics to monitor its customers' purchases and perusing patterns. In that case, it can rapidly alter its inventory and advertising strategies to better meet its customers' requirements [1].

Despite the numerous advantages of conducting business online, many organizations need help to employ digital technologies due to the perceived risks and implementation difficulties. Many individuals are concerned that conducting business online is not secure regarding confidential or financial information. To combat these obstacles, businesses must implement safeguards to secure the privacy of their customer's personal information and online transactions. Adopting e-business also necessitates overcoming the difficulty of locating individuals with the required skills and experience. When establishing a new e-commerce platform, having expertise in web design and programming and online marketing and distribution techniques could be advantageous. To ensure a seamless transition to e-business best practices, companies may need to invest in training current employees or engaging external consultants. Despite these obstacles, companies that implement e-business stand to gain substantial benefits. For example, a business that successfully integrates e-commerce into its operations may increase its overall efficacy, decrease its expenses, and increase its profits. Adopting e-business practices can assist businesses in adapting to the quickly evolving digital marketplace and meeting the rising demand for convenient and customizable purchasing options among customers (Fig. 12.1).

2 Factors Influencing E-Business Adoption

The adoption of digital business technologies is known as "e-business adoption." However, several internal and external factors influence the adoption of e-business processes. In this section, we will examine some of the considerations that should go into the e-business strategy of an organization.

Cost: Businesses that utilize e-commerce platforms typically experience increased productivity, consumer base, and market share. However, the cost of implementing e-business strategies is significant for many organizations. Businesses must carefully weigh the costs and benefits of investing in the implementation of

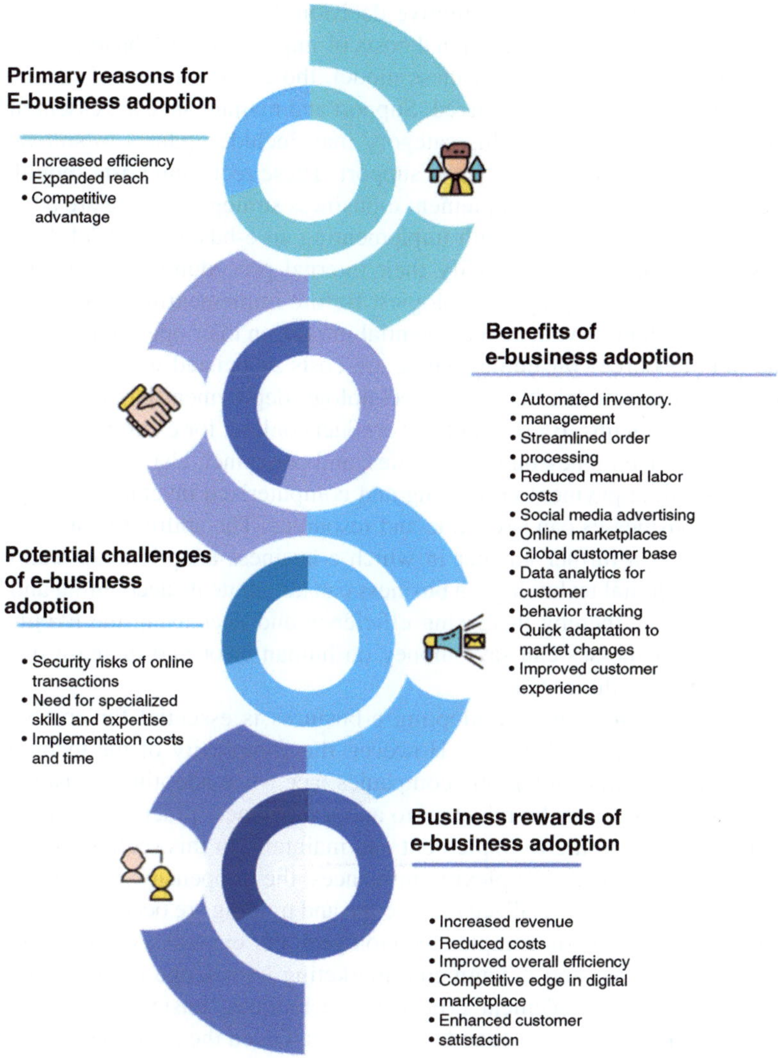

Fig. 12.1 Importance of e-business adoption for modern businesses

digital technology. The cost of computers and related software is one of the numerous concealed costs associated with going digital. The organization may need to acquire new servers, software licenses, and other hardware and software components to implement a new e-commerce platform. Before purchasing, businesses should compare their budget to these components' anticipated return on investment. The time and money required to establish an online business is another expense. Businesses may need to invest in employee training or hire outside consultants to guide them through the complexities and potential pitfalls of implementing new technology. Before reaching a definitive decision, businesses should evaluate the potential benefits against the additional costs of implementing e-business.

When transitioning to an e-business model, the continual cost of maintenance and support should also be considered. Support and maintenance are essential when introducing new technology. This category may include updates to software and hardware, as well as regular technical support. These recurring costs should factor into a company's decision to implement e-business strategies. There are additional, less apparent costs associated with implementing an e-business model. For example, businesses may need to modify their internal procedures and organizational structure to exploit emerging technologies. Before implementing e-business methods, businesses should evaluate the potential impact on their operations, which may be time- and resource-consuming. There are costs associated with transitioning to an online business model, but digital technology deployment has the potential to generate financial gains. By selling their products online, for example, e-commerce businesses can save money on rent, utilities, and personnel. Using digital technologies such as online payment processing and computerized inventory management, e-commerce businesses can save time and resources. The ability to automate business processes is a further manner in which e-business adoption can lead to cost savings. Using digital technology, a business can automate its accounting and bookkeeping processes, thereby increasing efficiency and decreasing the likelihood of human error. Businesses can save money on human error and increase output by automating specific duties [2].

In today's global economy, adopting e-business is essential for companies to maintain a competitive advantage. However, the complexity of the technology is one of the most significant barriers companies face when adopting e-business technologies. Businesses may be reluctant to adopt e-business practices if the required technology is challenging to implement and maintain. In this section, we will discuss how technological complexity influences the propensity of businesses to embrace e-commerce. Specialized knowledge and training are occasionally required to implement and maintain complex technologies. For example, you may need web design, programming, and e-commerce marketing knowledge to create an online store. To effectively implement e-business techniques, businesses may need to invest in staff training or hire outside consultants. Given the overwhelming number of alternatives, technological solutions may take time to comprehend. There are numerous e-commerce platforms, payment processors, and marketing automation tools, each with distinct advantages. Because of this, it may be difficult for businesses to select the appropriate technologies to achieve their objectives.

Due to the complexity of the technology, it may be more expensive to deploy and maintain. Creating an e-commerce platform may require investments in technology, software, and technological expertise. Additionally, firms may incur additional expenses for routine maintenance and software upgrades. Due to the complexity of technology and its application, e-business advantages are occasionally delayed. Implementing and integrating multiple technologies to function effectively and efficiently can be time-consuming for businesses.

Additionally, technological complexity may hinder an organization's adaptability. Expanding companies might need to adapt their e-business systems to accommodate new standards. The complexity of modern technology may impede the ability to quickly adapt to altering market conditions or customer demands. Despite the obstacles presented by complex technology, businesses can still successfully implement e-business strategies. One strategy is initially implementing only essential technological features and adding more as the business grows. Firms may commence with a pilot program to reduce uncertainty and establish familiarity with the technology. One alternative is to employ third-party vendors or service providers to manage specific technical aspects of online business operations. A third-party service provider can free up internal resources and facilitate the development and administration of an e-commerce platform.

Access to information technology resources: Companies' adoption of best practices for e-business depends on the availability of IT workers with the necessary capabilities. Web development, programming, database administration, online marketing, and cybersecurity expertise are all required for e-business technology. The availability of competent IT employees inside a firm may substantially influence the success of its e-business activities. For example, organizations with in-house IT expertise may need help integrating e-business technology due to possible problems. To effectively negotiate the challenges of e-business adoption, they may need to engage in staff training or seek advice from outside specialists. As a result, there is the prospect of increased costs and implementation delays.

On the other hand, businesses that have access to competent IT personnel may successfully incorporate e-business processes. For example, a firm with a specialized IT department may create e-commerce systems suited to its unique requirements. This guarantees that the company's customers have a faultless online buying experience. The availability of IT experts with the necessary skill sets is critical to administering and maintaining e-business technology. Companies must keep up with current advancements and trends to remain competitive in today's dynamic digital industry. This necessitates continual expenditures on workers' information technology abilities and continuing training for these personnel. For example, a firm that depends on e-commerce as its principal sales channel must regularly improve its website's search engine performance, user experience, and mobile responsiveness. This requires a thorough awareness of the processes involved in site design and online marketing. This can only be attained via ongoing training and chances for employees to enhance their talents [3].

Furthermore, IT skills availability is critical for ensuring data confidentiality and preventing cybersecurity breaches in e-business activities. When doing business online, it is usual practice to communicate private customer information such as credit card numbers, addresses, and other contact and identity data. Consequently, companies must secure the security of their online transactions while protecting the privacy of their consumers' information. For example, an e-commerce firm with insufficient IT skills risks being targeted by hackers, which might result in customer distrust and the theft of critical data. A corporation with educated IT specialists on staff may lessen the danger of cyber assaults by using extensive security measures such as data encryption, firewalls, and intrusion detection systems.

Security concerns are a significant factor preventing companies from implementing e-business best practices. E-commerce may have provided businesses with new opportunities to expand their customer base and revenue streams, but it has also spawned security issues that must be addressed. The possibility of a cyberattack is one of the primary concerns regarding the implementation of e-business. Businesses that indulge in e-business operations expose themselves to the possibility of an online attack, which could result in the seizure of sensitive data, financial losses, and a damaged reputation. There have been numerous high-profile cyberattacks on corporations in recent years, each resulting in significant financial losses and brand damage. These assaults have resulted in significant monetary losses and reputational harm.

Regarding data protection, the security of client's personal information is another concern. Due to the increasing amount of confidential information transferred online, businesses are responsible for protecting their customers' information. This involves instituting safeguards such as encryption, access restrictions, and regular data backups to prevent data loss or theft.

In addition, using digital technology in e-business procedures may create new opportunities for deceptive and unethical behavior. Phishing schemes, in which assailants attempt to trick consumers into revealing their confidential information, are one example of growing concern. Companies that wish to protect their consumers from these scams and educate them on recognizing and avoiding them must take the necessary precautions. Companies raise concerns about data security and privacy in the legal and regulatory climate surrounding online commerce. To remain compliant with laws and regulations such as the General Data Protection Regulation (GDPR) of the European Union and the California Consumer Privacy Act (CCPA), businesses must adhere to stringent data protection regulations. If one does not comply with these requirements, he or she may incur significant financial penalties and reputational damage.

In order to address these security concerns, businesses must employ a more proactive e-business strategy. This entails installing firewalls, antivirus software, and intrusion detection systems to protect their networks and data. In addition, businesses owe it to their employees to educate them on the best practices for maintaining data security and to ensure their employees know the risks associated with conducting business online. In addition, it is the responsibility of businesses to keep abreast of the most recent security hazards and to take precautions against them. This entails regularly conducting vulnerability testing and security assessments to identify and address any potential system vulnerabilities. E-business adoption is slowed by security concerns, which can be mitigated through meticulous planning and the implementation of appropriate security measures. Companies that take a proactive approach to e-commerce security may realize the benefits of conducting business online. In addition, these businesses can reduce the risks associated with conducting business online.

Legal and regulatory considerations: Before implementing e-business best practices, businesses are required to consider several crucial factors carefully. Among these factors are legal and regulatory considerations. The use of digital technologies and the Internet generates new legal and regulatory obstacles that businesses must surmount to avoid incurring monetary and other penalties. Legally speaking, data privacy is one of the most significant obstacles companies face when instituting practices related to electronic commerce. In order to protect the personal information of their customers, businesses are required to adhere to stringent data privacy regulations as they continue to collect and store increasing amounts of consumer data. For instance, the General Data Protection Regulation (GDPR) of the European Union mandates that businesses obtain consumers' express consent prior to collecting and processing their data, as well as implement adequate security measures to safeguard this data. This regulation also requires businesses to implement adequate data security measures. The legal framework governing the adoption of e-business must also incorporate the protection of intellectual property rights. When businesses use digital technology to produce and distribute their products and services, they are obligated to respect the intellectual property rights of others. This includes intellectual property rights such as copyrights, patents, and trademarks, which may be enforced through legal action in the event of infringement [1].

In addition to those pertinent to preserving personal information and intellectual property rights, the adoption of e-business also presents several additional legal and regulatory challenges. For example, businesses may be required to conform to laws governing electronic signatures, online payment processing, and online advertising. If a company does not comply with these requirements, they risk incurring legal and financial penalties and damaging their reputation. To successfully navigate these legal and regulatory challenges, businesses must invest in the education of legal professionals and ensure that they thoroughly understand the laws pertinent to their operations. In addition, it mandates that businesses implement the necessary policies and procedures to ensure compliance with the rules. For a company to legally conduct business in a particular jurisdiction, for instance, it may be required to establish data protection policies, implement secure payment processing systems, and obtain the appropriate certifications or licenses. Although legal and regulatory obstacles are difficult to overcome, businesses that can navigate them successfully stand to gain various advantages. For instance, adhering to data privacy standards can aid in building consumer trust and loyalty, whereas adhering to intellectual property rights can aid in protecting a company's intellectual property and preventing legal issues. These two factors contribute to the prosperity of a business.

Organizational culture and resistance to change: Both organizational culture and resistance to change are important factors that could influence an organization's adoption of e-business practices. An organization's culture consists of shared values, beliefs, and practices. This culture can influence how employees perceive and respond to new working methods and technological developments. On the other hand, resistance to change is the natural inclination of both individuals and organizations to oppose new methods of carrying out responsibilities that have been previously established.

One aspect of a company's culture that may influence the adoption of e-business is the degree to which the company values technological innovation and advancement. Businesses with a robust innovation culture may be more receptive to incorporating new technology, including e-business best practices. In contrast, businesses with a more traditional or risk-averse culture may be more resistant to change, especially the incorporation of practices associated with conducting business online. The extent to which a company values collaboration and open communication is yet another aspect of organizational culture that may influence the adoption of e-business. Best practices in e-business frequently require cross-functional cooperation and communication, which may be difficult for organizations to achieve if departments are compartmentalized, or an emphasis is placed on the individual. Businesses prioritizing collaboration and open communication have a greater chance of successfully implementing e-business procedures.

Change resistance is an additional factor that could influence the adoption of e-business. Employees may resist change for various reasons, including fear of job loss, ignorance of the benefits of e-business, and job security concerns. All of these worries and concerns are valid reasons. Effective communication and training programs emphasizing the benefits of e-business may help individuals overcome resistance to change in certain instances. On the other hand, there are situations in which resistance to change may be deeply ingrained in the organization's culture and may require substantial effort to surmount. A company with a long history of manual procedures and a culture that places a high value on tradition may be more resistant to adopting e-business than one with a more recent history and a more adaptable and flexible culture. Businesses may need to invest in change management programs to surmount resistance to change among their employees. These programs help employees recognize the benefits of implementing e-business practices and provide the necessary training and support for employees to adapt to new processes and technologies [1].

The history of Blockbuster serves as a real-world illustration of how corporate culture and resistance to change may impact the adoption of e-business. Throughout the 1990s and into the early 2000s, the Blockbuster network of stores dominated the video rental industry. On the other hand, Blockbuster was slow to adopt e-business strategies and stuck to its traditional retail business model as the industry shifted to online video streaming. Consequently, Blockbuster could not leverage this market shift in 2010; Blockbuster filed for insolvency as a direct result of its inability to adapt to the industry's altering demands.

Competition and market pressure: Competition and market pressure are crucial factors that drive the adoption of e-business. The advent of the digital era has made it easier for businesses to expand their customer base and recruit customers from various geographies, increasing competition. In order to maintain their competitive advantage and meet consumers' high expectations, businesses must employ digital technology and techniques for conducting business online.

One way market competition and pressure influence the adoption of e-business is by compelling businesses to improve the quality of the customer experience. For example, if a company's competitors offer online ordering and delivery services, it may need to implement similar services to meet consumer demands and remain competitive. E-business adoption is also driven by market competition and pressure in another way, namely, by enterprises' need to differentiate themselves from their competitors. By providing them with one-of-a-kind digital experiences, businesses can attract and retain consumers searching for the novel and unusual. A clothing store, for instance, can differentiate itself from rivals by providing consumers with the option of virtually trying on different garments via a "virtual dressing room" experience.

E-business adoption may also be influenced by market competition and pressure, which can instill a sense of urgency in businesses regarding adopting new technology. Suppose one of a competitor's innovative e-business practices is proving successful. In that case, other businesses may feel compelled to implement similar methods to maintain their competitive position. For example, the engagement and conversion rates of a competitor's mobile app are high. In that case, other businesses in the same industry may feel compelled to create a similar app to maintain market leadership. The development of new digital channels is one of the ways that market competition and pressure may encourage the adoption of e-business. New channels, such as social media, smartphone applications, and voice assistants, are emerging as potential methods for businesses to communicate with consumers. This consumer familiarity with digital technology is propelling the expansion of these new channels. To maintain their competitive advantage, businesses must investigate and implement these new channels. However, the burden imposed on businesses to incorporate new e-business practices and technologies can also create difficulties for those businesses. It may be challenging to keep up with the constantly evolving technology and trends, and it may take time for businesses to acquire the resources and knowledge required to implement new practices effectively. Additionally, companies may need more support to change from employees or constituents willing to adopt new technologies or procedures. This may present a difficulty for businesses [3].

The adoption of e-business is significantly influenced by potential returns on investment and potential returns on investment (ROI). Businesses invest in e-business technologies due to their potential benefits and anticipated return on investment (ROI). These benefits include a reduction in overall costs, an increase in overall efficiency, a boost in overall customer service, and an increase in overall competitiveness. However, the success of e-business adoption depends on several factors that can affect both the technology's adoption and the realization of its benefits.

E-business adoption may result in expense savings, which is one of its potential benefits. Traditional corporate activities, such as paper-based transactions and physical storage, may incur significantly higher costs than those associated with e-business technology, which can significantly reduce these expenses. For instance, businesses may utilize e-commerce platforms to enhance the efficacy of their supply chain procedures and reduce transaction costs. Using e-business solutions can increase a company's productivity while decreasing labor costs. The use of e-business may also result in improved consumer service. Supporting technologies for e-business enables businesses to meet customers' requirements around the clock. The fact that consumers can, for instance, place orders, verify the status of their shipment, and make payments online contributes to an improvement in the overall quality of their customer experience. Adopting technologies for conducting business online allows businesses to reduce the time required to respond to customer questions and complaints. In turn, this can assist the business in increasing consumer loyalty and satisfaction.

Moreover, the implementation of an e-business model may lead to an increase in productivity. E-business technologies enable businesses to automate their business operations, reducing the required manual labor. For instance, businesses can use computerized inventory management systems to monitor their stock levels and substitute items when they run out of supplies. This eliminates manual inventory management and ensures that items are always available for purchase, increasing overall productivity. Moreover, using technology that supports e-business may help businesses become more competitive in their respective markets. E-business technologies enable organizations to communicate with more individuals and expand their customer bases. Using the numerous social media channels available today, businesses may advertise their products and services to potential customers. Companies that have adopted e-business technology can respond quickly to market changes, thereby maintaining their competitive advantage. However, the success of e-business adoption depends on several factors that can affect both the technology's adoption and the realization of its benefits. For example, businesses may need help with technological infrastructure, information technology (IT) knowledge, or security concerns. Moreover, incorporating e-business may necessitate substantial time, money, and effort investments, affecting the anticipated return on investment (ROI).

Customers' requirements and expectations have shifted significantly since the advent of e-business, which has led to the proliferation of e-business. Customers have come to expect to be able to communicate with businesses via multiple channels, with websites, mobile applications, social media, and email being the most common. As a result, consumer requirements and expectations play an essential role in the implementation of e-business.

One of the driving forces behind the proliferation of e-business is the desire to accommodate consumers' ever-changing needs and expectations. To maintain a competitive advantage in the current market, businesses must be able to provide consumers with a consistent omnichannel experience. Customers anticipate the ability to access information and make purchases at any time, from any location, and on any device they choose. If these expectations are not met, consumer satisfaction and loyalty may decrease, eventually contributing to a decline in revenue. The proliferation of mobile technologies is another essential factor to consider. As more people access the Internet via mobile devices such as smartphones and tablets, businesses must adapt to meet their consumers' changing expectations. This necessitates the development of mobile-responsive websites and mobile applications that provide a streamlined and intuitive user experience.

Amazon and Uber are excellent examples of this concept's application in the real world. Amazon has revolutionized the retail industry by providing consumers with an extensive selection of items at affordable prices, the convenience of placing orders online, and the swiftness with which they are fulfilled. Uber's introduction of a user-friendly and uncomplicated smartphone application that allows users to request rides with a button has upended the transportation industry.

Companies need help to meet the changing needs of their target demographic and run the risk of losing market share to competitors who are better equipped to meet those needs. This is particularly true in retail, hospitality, and entertainment industries, where consumer experience quality is paramount. Implementing e-business may result in significant cost reductions for businesses, enabling them to meet consumer demands. For instance, e-commerce can reduce the costs of maintaining physical storefronts and managing inventory. Online marketing and advertising may be more cost-effective than traditional methods, such as print and television commercials. Companies must also consider the potential hazards associated with e-business adoption, such as the risk of cyberattacks and data breaches. Customers are increasingly concerned about their personal and financial data security, so businesses are pressured to secure this information.

Infrastructure and connectivity constraints: Infrastructure and connectivity constraints may be a significant barrier to adopting e-business, particularly in developing nations or rural areas with limited access to a dependable and economical Internet connection and energy. This is particularly true in rural areas of developing nations. Due to inadequate infrastructure, Internet speeds may be slow, frequent disruptions may occur, and bandwidth may be restricted, making it difficult for businesses to conduct online transactions and communicate with customers and suppliers. In Africa, for example, where Internet penetration is still relatively low compared to other regions, infrastructure and connection restrictions may impede the adoption of e-business. According to a study conducted by the International Telecommunication Union (ITU), only 28.2% of Africa's population had access to the Internet in 2021, whereas the global average was 59.5%. Due to the limited connection, businesses may need to help communicate with their consumers and suppliers, particularly in rural areas.

Infrastructure deficiencies may also affect the security and dependability of online transactions. Inadequate infrastructure may make it difficult for businesses to secure sensitive consumer data, leading to security breaches and a loss of customer confidence. This may result in financial losses and damage to the company's reputation. Infrastructure and connection constraints present additional obstacles, including obtaining the most recent versions of the hardware and software required for conducting e-business. Particularly difficult for small and medium-sized enterprises (SMEs), transitioning to new technology and software can take time and effort. This may result in a loss of competitiveness and an incapacity to keep up with the latest industry developments.

Nevertheless, despite these obstacles, initiatives and solutions are currently being developed to address infrastructure and connectivity limitations. For instance, developing nations' governments are investing in expanding access to electricity and enhancing Internet connectivity, which may increase e-commerce opportunities for businesses. Mobile payment systems, cloud-hosted software, and offline e-commerce solutions are just a few of the tools and platforms currently available to businesses to assist them in overcoming connection-related issues. For instance, the Rural Taobao program offered by Alibaba in China enables farmers and other rural residents without a reliable Internet connection to engage in e-commerce through authorized service centers.

The extensive use of electronic commerce is significantly influenced by the multitude of the above factors. E-business adoption can be facilitated or impeded by several factors, such as the complexity of the technology, the availability of IT skills, security concerns, legal and regulatory issues, organizational culture and resistance to change, market competition and pressure, potential benefits and return on investment, customer demand and expectations, and infrastructure and connectivity constraints. The quantity of money required to establish an e-business is one of the factors that is significantly affected by these other factors. Businesses with limited financial resources may find it challenging to adopt e-business, especially if implementing e-business requires a substantial investment in new technology or IT skills. Alternatively, businesses with more significant financial resources may be better positioned to invest more in e-business, thereby gaining a competitive advantage. These factors also affect the quantity of risk associated with e-business implementation. For example, businesses that store sensitive client information in an online repository may face a significant threat from data security concerns. Legal and regulatory issues may also pose risks to businesses, especially if they do not comply with the rules and regulations of the countries in which they operate. The resistance to change within an organization's culture can also be a significant barrier to e-business implementation. Some employees will resist new procedures or technologies, making it more challenging to implement e-business strategies. Companies must invest in change management programs to address this issue and ensure that employees are on board with the transition to e-business. The adoption of e-business may also be influenced by market competition and pressure. Companies that are slow to adopt e-business may be at a competitive disadvantage compared to their rivals, who have already taken measures to enhance their e-business capabilities.

In contrast, businesses that are early adopters of e-business may gain a competitive advantage by utilizing new technologies and processes. This is because they tend to be more inventive. During the implementation of an e-business strategy, the realization of potential benefits and a satisfactory return on investment may be crucial factors. Businesses demonstrating a positive return on investment (ROI) for their e-business activities are likelier to invest in these initiatives. For example, e-commerce platforms may help businesses increase sales while reducing expenses, resulting in a favorable return on investment (ROI). The quantity of consumer demand and expectations is another crucial aspect of e-business adoption. Customers expect businesses to provide online channels through which they can purchase and interact with the products and services offered by those businesses. Businesses that fail to meet these requirements risk losing customers to competitors with superior e-business capabilities. Lastly, infrastructure and connection limitations may be a significant barrier to the widespread adoption of e-business. E-business strategies may be challenging in regions with intermittent or unstable Internet access or infrastructure. Table 12.1 summarizes the factors influencing e-business adoption.

Table 12.1 Factors influencing e-business adoption

Factors	Positive impact	Negative impact
Complexity of technology	Enables companies to leverage innovative technologies and offer new capabilities	It can be challenging to implement and require significant investment
Availability of IT skills	Allows companies to implement and manage e-business initiatives successfully	Limited availability can hinder the adoption
Security concerns	Protects sensitive data and ensures customer trust	It can pose a significant risk if not managed properly
Legal and regulatory issues	Ensures compliance and protects companies from legal liability	It can be difficult and expensive to navigate
Organizational culture and resistance to change	Ensures buy-in from employees and promotes successful adoption	This can be a significant barrier to implementation
Market competition and pressure	Drives companies to innovate and stay competitive	This can create pressure to adopt e-business initiatives without proper planning or consideration
Potential benefits and return on investment	Offers significant cost savings and revenue opportunities	It can require significant investment and time to achieve ROI
Customer demand and expectations	Offers greater convenience and accessibility for customers	Failure to meet expectations can lead to loss of customers
Infrastructure and connectivity limitations	Hinders adoption in areas with limited Internet connectivity or unreliable infrastructure	Limits ability to leverage e-business capabilities

3 The Adoption Process

The implementation of e-business is a multistep process, with each step functioning independently. The incorporation of e-business into a business is described in the following paragraph (Fig. 12.2).

Awareness: Awareness is the initial stage in the process of e-business standardization. Companies must know the benefits of conducting business online and any potential disadvantages. They must know how conducting online business can help them expand their company, increase their customer base, and reduce operating expenses. During this phase, you will typically research and collect information about e-business and its potential effects on the organization.

The next stage in the adoption procedure is to express interest in the infant. When businesses know the prospective benefits of online transacting, they become interested in adopting the practice. At this stage, we will investigate the various types of e-business solutions available on the market and evaluate their potential effects on the organization.

Evaluation is the third and final stage in the infant adoption procedure. In order to proceed to the next phase of the process, businesses now begin analyzing the various e-business solutions they have identified as viable options. At this stage, we evaluate the features and functions of each solution and the costs associated with implementing and maintaining the solution.

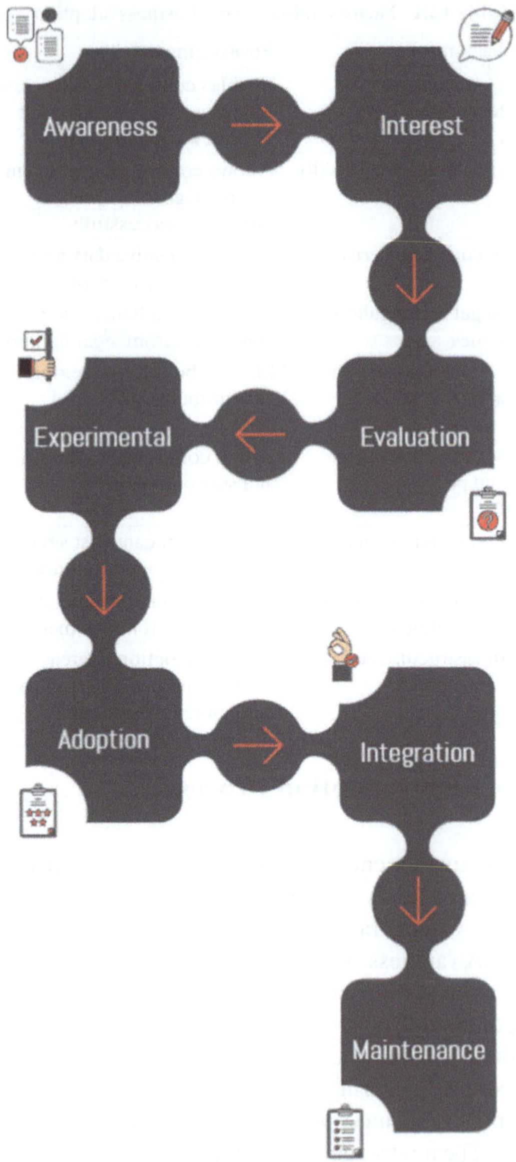

Fig. 12.2 The adoption process in an e-business

The adoption process consists of four steps, the fourth of which is the experimental phase. At this juncture, the selected e-business solution will undergo a series of experiments on a smaller scale to evaluate its viability and effectiveness. During the trial phase, businesses can identify any prospective problems or difficulties that must be resolved prior to the full implementation.

Adoption is the fifth and final stage in acquiring a child. At this juncture, the enterprise has fully implemented the e-business solution selected earlier in the procedure. During this phase, employees are instructed on how to use the new system, and infrastructure and support are evaluated to ensure they are in place.

The sixth and final stage in employing a new system is integration. Throughout this phase, the organization will work to integrate the new e-business solution with its existing processes and systems. This phase is crucial for ensuring the new solution is fully integrated and can be utilized effectively across the organization to achieve the desired results.

This is the final stage of the adoption process, and it is also the most crucial. It is now essential to provide ongoing support and maintenance for the e-business solution to ensure it continues operating effectively and efficiently. This phase includes monitoring performance, identifying issues or obstacles, and implementing necessary updates and adjustments.

It is essential to bear in mind that the procedure for implementing e-business can take various forms, determined not only by the scale and complexity of the organization but also by the specific e-business solution being adopted. In addition, the adoption process is not a one-time event but a continuous process that entails review, adaptation, and refinement.

The case of Amazon is an example taken from real life that demonstrates how the adoption process for an online company works. In 1995, Amazon launched its first website, which was a virtual bookshop. Since then, the company has expanded to become one of the most successful online retailers in the world. The success of the organization may be linked to many factors, including its use of e-business methods, its capacity for ongoing innovation, and its responsiveness to changing market circumstances. The establishment of Amazon's online platform was the first stage in the company's transition to doing business online. The company's founder, Jeff Bezos, saw the promise of the Internet as a medium through which to sell things directly to consumers. As a result, he made significant investments in the development of Amazon's online platform. This included the creation of a website that was easy to use, the implementation of safe payment processing systems, and the execution of solid supply chain and logistical operations. Following the completion of the platform, Amazon turned its attention to expanding its client base. In order to bring in new clients, the firm used several different marketing methods, such as running targeted advertising campaigns, giving discounts and specials, using social media, and forming connections with influential people. Amazon has been able to establish itself as a trustworthy and reputable source for online purchasing because it has developed a client base that is loyal to the company. Amazon continues to invent new products and broaden its product selection as customers' use of its services increases. In addition to selling books, Amazon started offering a wide variety of things, which included anything from apparel and beauty products to consumer electronics and home goods. In addition, the corporation established additional services such as Amazon Prime, which provided its users with free delivery and other perks. Amazon's ability to harness data to enhance its internal operations and the quality of the shopping experience for its customers significantly contributed to the company's overall success. Amazon was able to improve its suggestions and product selection by gathering and analyzing data from its consumers. As a result, the company made it simpler for customers to locate and buy the goods that met their needs. The business also used data to improve the effectiveness of its supply chain and logistics operations, which led to reduced expenses and increased productivity. Amazon's readiness to try new things and take calculated risks was another essential component of the company's success [2]. The firm did not avoid trying new initiatives, even if they were unorthodox or without prior evidence of success. For instance, Amazon was among the first corporations to experiment with drone delivery.

Additionally, the company has significantly invested in artificial intelligence and machine learning to enhance its business processes. Overall, Amazon's success may be credited to the company's adoption of best practices for doing business online and its ability to innovate and continually adapt to changing circumstances in the market. Amazon has established itself as a market leader in the e-commerce sector by creating a user-friendly online platform, cultivating a dedicated customer base, using data to enhance company operations and the overall consumer experience, and taking calculated risks to investigate emerging technologies and business models.

4 Barriers and Challenges to E-Business Adoption

The implementation of e-business presents companies and enterprises with a great deal of difficulty and obstruction. These obstacles can be categorized as technical, organizational, and environmental variables in that order. In this section, we will examine the most common obstacles that businesses face when attempting to implement e-business, as well as several strategies that can be used to overcome these obstacles (Fig. 12.3).

The complexity of the technology itself is one of the most significant obstacles preventing the widespread adoption of e-business. Online business systems are frequently complex and require substantial investments in technology and infrastructure. The cost of implementing systems for conducting online business may be prohibitive for many small and medium-sized businesses. In addition, implementing e-business systems requires a high level of information technology expertise, which may need to be more readily available within an organization. To surmount these obstacles, businesses should explore the possibility of establishing partnerships with external service providers to assist them with deploying and maintaining e-business systems.

Interoperability is an additional pervasive technical barrier to the widespread adoption of e-business. E-business systems frequently incorporate various types of systems and technologies, which can be problematic regarding data compatibility and system integration. Businesses must invest in data management tools and technology that will help them overcome interoperability challenges.

When it comes to the implementation of e-business, organizational considerations may also present substantial obstacles. Change resistance is one of the most significant organizational obstacles that may be encountered. Many employees need

Fig. 12.3 E-business adoption challenges and barriers

to refrain from employing new technology or procedures, which may slow the implementation process. Investing in change management programs that can assist employees in adapting to new technologies and procedures is a smart move for companies seeking to overcome this obstacle.

The absence of a clearly defined strategic vision is another significant organizational barrier to adopting e-business. Many businesses may need help understanding how e-business influences their larger strategic objectives. To overcome this barrier, businesses must develop e-business strategies that align with their larger organizational goals and objectives.

External variables, such as the environment, may also impede the implementation of e-business significantly. The lack of infrastructure and connection is one of the most significant environmental obstacles. Internet access and other technological infrastructure may need to be improved in many parts of the world, making it difficult for enterprises to implement e-business systems. Organizations should collaborate with municipal governments and numerous other stakeholders to overcome this barrier to constructing infrastructure and enhancing connectivity in unserved regions.

The legal and regulatory climate of the environment presents yet another significant environmental barrier to the widespread adoption of e-business. E-business systems are required to comply with a vast array of regulations. Laws about data privacy, consumer protection, and intellectual property are among these requirements. If these regulations are not followed, severe legal and monetary penalties could occur. Investing in legal and regulatory compliance processes may aid businesses in traversing the complex legal environment and surmounting this obstacle, so companies should consider doing so.

Regarding e-business implementation, businesses face these obstacles and several others. The operation of an e-business system, for example, requires substantial investments in the education and training of the workforce. In addition, implementing e-business systems may necessitate significant modifications to existing organizational protocols and procedures, which can be difficult.

Organizations should develop strategies for incorporating e-business that account for the numerous obstacles and barriers they must surmount to succeed. These plans must include crystal-clear objectives, timelines, budgets, and strategies for overcoming obstacles and challenges. Companies should investigate the possibility of partnering with third-party service providers to assist with deploying and maintaining e-business systems. Table 12.2 summarizes the barriers and challenges to e-business adoption.

To ensure the effective implementation of e-business initiatives and maximize the benefits they may provide, it is crucial to devise strategies for overcoming the obstacles and challenges that companies face when adopting e-business. In the following paragraphs, we will discuss a variety of strategies that can be used to surmount the difficulties described in the preceding section.

Investing in change management initiatives focusing on corporate culture and individuals' innate resistance to change is one of the most crucial strategies. With the aid of change management programs, the need for change can be conveyed to employees, their concerns and anxieties can be addressed, and training and support can be provided to facilitate the adoption of e-business initiatives by enterprises. When IBM shifted its focus to cloud computing, it made substantial investments in change management programs to ensure that employees were on board with the new direction and possessed the necessary skills to succeed in the new environment.

Implementing strict data protection protocols and ensuring compliance with all applicable laws and regulations are different methods for addressing security issues.

Table 12.2 Barriers and challenges to e-business adoption

Barrier/challenge	Impact on e-business adoption	Possible solutions
Lack of resources and budget	Limited investment in new technology or IT skills	Seek funding opportunities; implement low-cost solutions
Resistance to change	Employees may resist new technology or processes	Invest in change management programs; involve employees in decision-making
Lack of technical expertise	Limited knowledge and skills in implementing and managing e-business systems	Train and educate staff; hire experienced personnel
Security concerns	Risk of data breaches and loss of customer trust	Implement secure systems and protocols; educate staff on best practices
Legal and regulatory compliance	Risk of noncompliance and legal consequences	Ensure compliance with laws and regulations; seek legal advice if needed
Infrastructure and connectivity limitations	Limited Internet connectivity or unreliable infrastructure	Invest in reliable infrastructure and connectivity; implement offline solutions
Customer trust and adoption	Customers may be hesitant to share sensitive information or use online channels	Implement secure and user-friendly systems; educate customers on benefits and security measures
Integration with existing systems	Difficulty in integrating new systems with existing ones	Evaluate and select compatible systems; seek professional help if needed

This may include encrypting sensitive data, performing routine backups, and instituting access restrictions to limit the number of individuals who can view sensitive data. Regular audits and assessments must be conducted to identify and close any compliance breaches with the rules and regulations that must be followed. For example, in 2018, the European Union fined Facebook 1.63 billion dollars for failing to safeguard its users' personal information and violating data protection regulations.

Taking a phased approach to the implementation of e-business is another method businesses can use to circumvent difficulties caused by limited financial resources. To achieve this, the adoption process must first be divided into a series of more minor, more manageable phases. Then initiatives must be prioritized based on their potential return on investment. As more financial resources become accessible, a company may begin its online presence with a primary e-commerce platform that allows customers to purchase goods over the Internet and then progressively expand their operations to include more complex endeavors.

Effective communication is crucial for overcoming resistance to change and an organization's culture. Companies must effectively communicate the benefits of e-business initiatives to all parties involved, including employees, customers, and business partners. This may involve highlighting the potential cost reductions, revenue growth, and enhanced consumer experience resulting from implementing an e-business strategy. For example, when Amazon first introduced its one-click ordering system, the company emphasized the new system's convenience and speed, which ultimately led to an increase in the number of people using it.

Successfully overcoming problems caused by infrastructure and connection constraints may also require an effective strategy, which can be achieved through collaboration with partners and suppliers. When collaborating with these parties, organizations can improve their connectivity and infrastructure by utilizing their partners' and suppliers' expertise and resources. Microsoft, for example, established its Affordable Access Initiative in 2016, which aims to increase Internet connection in underprivileged regions worldwide by providing grants and other resources to organizations working to improve Internet access in such regions.

By being proactive and innovative in their e-business endeavors, businesses can surmount the challenges of market competition. This may involve monitoring the e-business activities of competitors to identify areas with space for improvement, as well as utilizing new technology and procedures to provide customers with a unique and distinguishable experience. For example, Walmart launched its InHome Delivery service in 2019. This service utilizes intelligent home technology to permit delivery personnel to enter customers' residences and transport food directly to their refrigerators. This offers consumers a unique and convenient service that meets their requirements.

5 Evaluation of E-Business Adoption

Adopting practices related to e-business may confer several advantages on a company, including enhanced competitiveness, heightened client accessibility, and better operational efficiencies. However, there are some disadvantages associated with the use of online business that need to be carefully considered. The following paragraphs will discuss the positive and negative aspects of adopting an e-business model (Table 12.3).

Enhanced productivity is one of the critical advantages of the proliferation of e-business. E-business can improve corporate operations' efficiency, reduce

Table 12.3 Benefits and drawbacks of e-business adoption

Benefits of e-business adoption	Drawbacks of e-business adoption
Increased efficiency and productivity through automation of business processes	Initial investment costs in technology, infrastructure, and training can be high
Expanded reach and market penetration through online sales and marketing	Security and privacy concerns may arise, and data breaches can be costly
Improved customer experience through 24/7 accessibility, personalized communication, and faster response times	Resistance to change and lack of employee buy-in can hinder adoption and implementation
Increased data collection and analysis capabilities for better decision-making	E-business can lead to losing personal touch and face-to-face interactions with customers

paperwork, and remove the need for human data input. The use of e-procurement technologies, for instance, may cut down on both the amount of time and the costs often involved with conventional purchasing procedures. One further illustration of this would be the implementation of e-commerce platforms, which enable consumers to make purchases of goods and services in a manner that is both expedient and time-saving. The increased consumer base that may be reached is yet another advantage of adopting an e-business model. Through e-business, companies can access customers outside their immediate geographic area, expanding their potential customer base. By way of illustration, a small company in the United States may utilize the medium of electronic commerce to offer its wares to consumers in Europe or Asia. E-business also enables companies to customize their marketing efforts, allowing them to focus on a more narrow subset of their overall consumer base. Adopting e-business can also improve competitiveness by giving organizations the tools and capabilities necessary to react to shifting market conditions. This enables organizations to meet customer needs better. E-business provides companies with the ability to swiftly adjust to changing client preferences and behaviors, as well as to react to the challenges of competition rapidly. For instance, a firm implementing e-business may more readily deploy new goods, services, or marketing tactics to react to market changes.

The implementation of e-business comes with several advantages, but some downsides need to be carefully examined. One of the most significant drawbacks is the financial commitment required to adopt a child. The adoption of e-business may call for large technological, infrastructure, and training expenditures. Putting in place an e-commerce platform, for instance, would need expenditures in site development, payment processing, and safety precautions. The increased risk of data privacy violations and security breaches is another disadvantage associated with the proliferation of online commerce. Collecting and storing sensitive consumer data, which leaves an organization open to cyber assaults and data breaches, is required for doing business online. For instance, in 2013, Target suffered a significant data breach that exposed the personal information of more than 110 million consumers. The elimination of face-to-face client communication is another possible downside of adopting an e-business model. E-commerce may result in a less personal and more impersonal experience for the client since there are fewer opportunities for face-to-face connection. This may be a problem for businesses that attempt to separate themselves from their competition via personal connections and excellent customer service. Adopting e-business may also lead to a loss of employment, especially those that entail manual procedures and data input. This can be a cause for worry for businesses that have a large labor force or that are situated in regions with a scarcity of other employment options. For instance, the use of automated production methods in the automobile sector has directly contributed to a significant reduction in employment opportunities.

The continued development of technology and the growing number of companies that see the value of e-business both point to a bright future for the widespread implementation of e-commerce. In the following paragraphs, we will talk about some of the technologies and trends expected to impact the future of e-business adoption significantly. The increased use of artificial intelligence (AI) and machine learning (ML) in online businesses is a significant development that should be addressed. These technologies may be used to analyze large quantities of data, customize customer experiences, and automate mundane operations, all of which can contribute to higher productivity and cost savings. For instance, Amazon's recommendation engine uses AI and ML algorithms to study consumer activity and develop individualized product suggestions. One such development to watch is the increasing significance of mobile devices in online commerce. Because smartphones and tablets are becoming more commonplace, companies must ensure that their e-commerce platforms are suitable for mobile devices. Developing mobile applications, making websites mobile-friendly, and implementing mobile payment systems are all potential steps in this direction. By way of illustration, the mobile app offered by Starbucks enables users to order and pay for their beverages using their cell phones. The proliferation of blockchain technology will have a significant effect on the expansion of e-business in the not-too-distant future. Increased transparency, decreased likelihood of fraudulent activity, and streamlined corporate operations are all potential benefits of using blockchain technology to construct secure, decentralized networks for conducting transactions and storing data.

For instance, Walmart uses blockchain technology in its supply chain to improve traceability and lower the risk of foodborne infections. Another development that is anticipated to impact the development of e-business in the future significantly is the Internet of Things (IoT). The Internet of Things refers to linking commonplace things to the Internet so that they may receive and send data. This may be used to construct intelligent homes, communities, and enterprises, which enable users to operate and monitor their gadgets and systems remotely from a distant location. IoT technology is used in products such as the Nest Learning Thermostat, which observes the actions and preferences of users and then adjusts the temperature setting appropriately. As the use of online business continues to increase, companies will need to ensure that they can give customers a consistent experience across all channels to compete effectively. This entails making it possible for consumers to communicate with the company across various channels, including social media, email, live chat, and the telephone, and ensuring that the quality of service provided is uniform across all channels. Customers of Sephora, for instance, may use the company's website, mobile app, or kiosks located within its stores to schedule appointments, get individualized product suggestions, and make purchases.

In conclusion, the necessity for companies to maintain their competitive edge and adapt to the ever-evolving demands of their customers is likely a driving force behind the adoption of e-business. This indicates that companies will need to consistently innovate and adapt to meet the expectations of their customers as well as new technological developments. For instance, Amazon was able to join the grocery sector and give clients a new kind of service when it purchased Whole Foods sector in 2017. This was a way for Amazon to distinguish itself from its competitors.

In conclusion, the future of the widespread adoption of e-business is bright, with themes such as artificial intelligence and machine learning, mobile optimization, blockchain, the Internet of Things, omnichannel customer experiences, and innovation driving growth and innovation. E-commerce will continue to play an increasingly significant part in the economy of the whole world as long as firms continue to understand the many advantages of doing business online and invest in cutting-edge technology and procedures.

Summary

Definition of E-Business Adoption

- E-business adoption refers to the process by which organizations integrate electronic technologies and digital platforms into their business operations and strategies.

Factors Influencing E-Business Adoption

- Internal factors: These include organizational culture, management support, resources, and capabilities.
- External factors: These involve market forces, competitive pressures, technological advancements, and customer demands.

The Adoption Process

- Awareness: Organizations become aware of the potential benefits of e-business and its relevance to their operations.
- Evaluation: Organizations assess the feasibility, costs, and benefits of adopting e-business practices.
- Decision: Organizations decide to adopt e-business based on the evaluation.
- Implementation: Organizations implement the necessary changes to integrate e-business into their operations.
- Monitoring: Organizations continuously monitor the performance and effectiveness of their e-business adoption.

Barriers and Challenges to E-Business Adoption

- Technological barriers include concerns about infrastructure, security, and compatibility with existing systems.
- Organizational barriers involve resistance to change, lack of digital skills, and a rigid organizational structure.
- Regulatory and legal barriers: These pertain to compliance with laws and regulations related to e-business activities.
- Market barriers include market saturation, intense competition, and changing customer preferences.

Evaluation of E-Business Adoption

- Organizations need to evaluate the effectiveness of their e-business adoption to determine if it aligns with their goals and objectives.
- Evaluation can involve measuring key performance indicators (KPIs), such as increased revenue, improved customer satisfaction, and enhanced operational efficiency.

- Various evaluation methods, including surveys, analytics, and benchmarking, can be employed to assess the impact of e-business adoption.

Case Study

The XYZ Corporation is a medium-sized manufacturing company that has been in operation for over 30 years. The company specializes in manufacturing industrial equipment and has a solid reputation in the market. To remain competitive and expand its customer base, XYZ Corporation is considering implementing e-business techniques due to rapid technological advancements and shifting consumer preferences.

1. The business operations of XYZ Corporation have been conducted using conventional paper-based methods. How can e-business technologies such as enterprise resource planning (ERP) systems and customer relationship management (CRM) software help an organization's operations become more efficient and effective?
2. XYZ Corporation's organizational structure is hierarchical, and its culture resists change. How can the organization surmount internal opposition to e-business adoption and foster a culture that embraces innovation and technology?
3. The XYZ Corporation is a well-established competitor in a highly competitive industry. How can the organization use e-business adoption to gain a competitive edge, acquire new customers, and differentiate itself from rivals?
4. XYZ Corporation has just begun to investigate e-business adoption. What steps should the organization take to increase employee awareness of the benefits of e-business, generate enthusiasm, and facilitate the evaluation and testing of e-business solutions?
5. XYZ Corporation is concerned about the potential risks and complications associated with adopting e-business, such as data security and privacy issues. How can the organization surmount these obstacles and ensure the secure and seamless implementation of e-business procedures?

Discussion

1. What is e-business adoption, and why is it important for modern businesses?
2. What are the technological factors that influence e-business adoption?
3. How do organizational factors impact the adoption of e-business?
4. What environmental factors can affect e-business adoption?
5. What are the stages of the adoption process for e-business?
6. How can organizations increase awareness of e-business to facilitate adoption?
7. What are some best practices for generating interest in e-business?

8. How can organizations evaluate the effectiveness of their e-business adoption efforts?
9. What are some common barriers to e-business adoption?
10. How can organizations overcome resistance to e-business adoption from employees or stakeholders?
11. How can organizations address security concerns related to e-business adoption?
12. How can organizations ensure that their e-business initiatives align with their overall business strategy?
13. What are some potential drawbacks of e-business adoption?
14. How can organizations measure the return on investment of their e-business initiatives?
15. What are some emerging trends in e-business adoption that organizations should be aware of?
16. How can organizations stay up to date with advancements in e-business technology?

Multiple-Choice Question

1. What is the definition of e-business adoption?

   (a) The use of electronic communication for internal business operations
   (b) The implementation of online marketing strategies
   (c) The process of integrating electronic technologies into business processes
   (d) The utilization of social media platforms for customer engagement

2. Why is e-business adoption important for modern businesses?

   (a) It reduces operational costs and improves efficiency.
   (b) It increases brand visibility and expands market reach.
   (c) It enables real-time data analysis and personalized customer experiences.
   (d) All of the above.

3. Which of the following is NOT a factor influencing e-business adoption?

   (a) Technological factors
   (b) Organizational factors
   (c) Financial factors
   (d) Environmental factors

4. How do technological factors influence e-business adoption?

   (a) They determine the availability and effectiveness of e-commerce platforms.
   (b) They influence the speed and reliability of Internet connections.
   (c) They impact the security and privacy of online transactions.
   (d) All of the above.

5. Which stage of the adoption process involves creating awareness about the benefits of e-business?
   (a) Interest
   (b) Evaluation
   (c) Trial
   (d) Awareness
6. What is a best practice for the evaluation stage of the adoption process?
   (a) Conducting a cost-benefit analysis of implementing e-business
   (b) Developing a detailed implementation plan
   (c) Establishing a pilot project to test the effectiveness of e-business
   (d) Communicating the benefits of e-business to employees and stakeholders
7. What are some common barriers to e-business adoption?
   (a) Resistance to change from employees
   (b) Limited technological infrastructure
   (c) Security concerns regarding online transactions
   (d) All of the above
8. Which strategy can help overcome the barrier of resistance to change?
   (a) Providing comprehensive training and support to employees
   (b) Investing in robust cybersecurity measures
   (c) Upgrading technological infrastructure
   (d) Offering financial incentives to employees for adopting e-business
9. What are some benefits of e-business adoption?
   (a) Increased customer reach and market expansion
   (b) Improved operational efficiency and cost reduction
   (c) Enhanced customer engagement and personalized experiences
   (d) All of the above
10. What is a drawback of e-business adoption?
    (a) Increased competition from online competitors
    (b) Potential security risks and data breaches
    (c) Dependency on Internet connectivity and technology
    (d) All of the above
11. What is the future outlook for e-business adoption?
    (a) It is expected to decline due to privacy concerns.
    (b) It will become the primary mode of business operations.
    (c) It will remain stagnant without significant advancements.
    (d) It will be replaced by traditional brick-and-mortar stores.

12. How do organizational factors influence e-business adoption?
    (a) They determine the willingness of employees to embrace new technologies.
    (b) They impact the decision-making process and resource allocation.
    (c) They influence the company culture and hierarchy.
    (d) All of the above.
13. Which stage of the adoption process involves testing the effectiveness of e-business on a smaller scale?
    (a) Evaluation
    (b) Trial
    (c) Interest
    (d) Adoption
14. Which strategy can help overcome the barrier of limited technological infrastructure?
    (a) Partnering with technology providers or outsourcing IT services
    (b) Ignoring the infrastructure limitations and proceeding with e-business
    (c) Reducing the reliance on technology and focusing on traditional methods
    (d) None of the above
15. What is a best practice for the awareness stage of the adoption process?
    (a) Conducting market research to identify customer needs and preferences
    (b) Creating informative and engaging content about e-business benefits
    (c) Establishing partnerships with industry leaders to promote e-business
    (d) Developing a detailed implementation plan

References

1. Rowley, J., E-Business: Principles and Practice. 2002: Palgrave Global Publishing
2. Rogers, E.M., Diffusion of Innovations. 2010: Simon and Schuster
3. Chen, L. and C.W. Holsapple, E-business adoption research: State of the art. Journal of Electronic Commerce Research, 2013. 14(3): p. 261

Chapter 13
Change Management

Abstract E-businesses are experiencing significant transformation in the digital era, leading to continuous change in today's fast-paced business environment. Effective change management is crucial for e-businesses to achieve their objectives and stay ahead of the competition. This involves planning, implementing, and monitoring modifications to e-business activities for a seamless transition. This chapter provides an overview of change management in e-business, emphasizing its importance and methodologies. The change management process consists of phases like planning, communication, and training, with efficient procedures implemented at each stage. Overcoming resistance to change is vital in e-businesses, and identifying and addressing sources of resistance is crucial. Leadership plays a key role in change management, and employing leadership-enhancing strategies significantly impacts the process's efficiency. The chapter explores the significance of leadership and evaluates effective online business strategies. The future of change management in e-companies is examined in the final section, considering the benefits and drawbacks of its implementation. The chapter's objectives are to provide a comprehensive understanding of change management in e-business and equip e-businesses with the knowledge and tools needed for effective implementation of changes.

1 Introduction to Change Management in E-Business

In the context of e-business, "change management" refers to efficiently administering changes made to an e-business platform or system. During this phase, adjustments to the e-business system are identified and implemented to enhance the system's performance, functionality, and overall effectiveness. Change management is a crucial aspect of e-business because the company can respond to new technological advancements, altering market conditions, and ever-changing consumer demands.

Change management in e-business is crucially dependent on determining whether or not a change is required. Assessing the current state of the e-business

H. Taherdoost, *E-Business Essentials*, EAI/Springer Innovations in Communication and Computing, https://doi.org/10.1007/978-3-031-39626-7_13

system and identifying improvement opportunities are necessary steps in this process. For example, suppose an e-commerce platform observes a decline in sales. In that case, the company may need to analyze the platform's user experience, payment methods, and product selections to determine the underlying cause of the decline in sales. After deciding that change is necessary, the next stage is to devise a plan for implementing the required modifications. This plan should include the specific adjustments needed, the required resources, and the timetable for implementing these adjustments. For instance, if an e-business platform wishes to introduce a new payment gateway, the company should prepare a strategy outlining the precise procedures required to integrate the payment gateway and the resources and timeframe needed to complete the integration. This plan should also describe the necessary completion timeframe for the integration. Modifications must be implemented in a controlled and systematic manner. This involves verifying the changes to ensure they function as intended and do not negatively impact the system's overall functionality used to administer online enterprises. For example, suppose a company will add a new function to its e-commerce platform. In that case, it must test that function to ensure that it functions properly and does not interfere with the operation of any other platform components.

Communication is crucial to the change management process within the domain of e-business. Everyone interested in the organization's e-business system, including employees, customers, and vendors, must be informed of the system's upcoming modifications. This ensures that everyone knows the changes and how they will impact them individually. For example, suppose an e-commerce platform decides to modify its shipping policy. In that case, the company that operates the platform must inform its customers so that they are aware of the new policy and how it will affect the products they purchase. As part of the change management process in an online company, employees must be trained to utilize any new features or systems being implemented. This ensures that personnel have the necessary skills and knowledge to utilize newly implemented features and systems effectively. For example, if a company is implementing a brand-new order management system, management must ensure that employees receive adequate training on using the new system [1].

In conclusion, it is essential to conduct continuous surveillance and analysis of the effectiveness of the e-business system's modifications. As part of this procedure, it is necessary to monitor the most critical performance indicators to determine whether or not the company's changes have had the intended effect on the organization's goals and objectives. For instance, if a company has just launched a new marketing campaign to increase the number of website visitors, the company must monitor the number of website visitors to determine the campaign's success.

Change management is an indispensable component for the successful adoption of e-business systems. In today's digital environment, businesses must endure a continuous process of adaptation in order to remain competitive and meet consumers' ever-changing needs. The acquisition and installation of software and hardware represent only a minor portion of implementing an e-business system. It is a challenging procedure that requires the participation of numerous stakeholders, such as employees, consumers, and suppliers. Effective change management ensures that all stakeholders are on board with the modifications and that the implementation of the e-business system proceeds without hiccups. Change management is required for the successful implementation of an e-business strategy in large part because it serves to reduce resistance to change. Everyone involved, including employees, consumers, and vendors, may find change difficult. Resistance to change can lead to delays, errors, and decreased productivity, negatively affecting the e-business system's operation. Businesses involving stakeholders in the change management process can better resolve concerns and identify potential roadblocks before they become significant issues.

Efficient change management ensures the current e-business system meets the company's goals and objectives. Involving stakeholders in the planning and implementing of an e-business system enables an organization to ensure that the system will meet the stakeholders' requirements and contribute to accomplishing the organization's strategic objectives. For instance, if an organization aims to increase sales, its e-business system should be designed to enhance the user experience, expedite the checkout process, and provide consumers with various payment options. Change management also ensures that implementing the e-business system disrupts the company's daily operations as little as possible. Effective change management requires developing a detailed implementation strategy that specifies the processes required to deploy the e-business system, the resources required, and the completion timeline. By devising a strategy that considers the organization's operations and protocols, businesses can ensure that the implementation of the e-business system will proceed smoothly and that the system will not cause unwarranted disruptions.

In addition, change management ensures that employees are adequately trained to use the e-business system, an additional advantage of the practice. Implementing a new system or function with sufficient training may result in clarity, errors, and decreased overall productivity. Effective change management requires developing a training plan that equips all employees with the necessary skills and knowledge to utilize the e-business system effectively. For instance, if a company is installing a brand-new order management system, the entire staff should be trained on how to use the new system, including placing orders, processing payments, and monitoring shipments. Effective change management requires effective communication. Communication is essential to ensure all parties know the changes and how they will affect them. The organization's employees, consumers, and vendors should be informed of any changes so that everyone understands how the modifications will affect their day-to-day work and other interactions with the business. For instance, if a business decides to transition to a different payment gateway, it must inform its customers of the impending changes so that they know how to take advantage of the increased number of payment options.

Last, effective change management necessitates ongoing monitoring and evaluation of the e-business system after its implementation. Monitoring and evaluating the performance of the e-business system enables businesses to identify potential problem areas and make the necessary adjustments. This ensures that the e-business system will continue to meet the company's needs and that the company will maintain its competitive position in the market. For instance, if a company adds a new feature to its website to increase website traffic, it must monitor website traffic to determine if the feature was successful and if any modifications are necessary.

In conclusion, to successfully integrate e-business systems, it is essential to have effective and efficient change management. Change management ensures that the e-business system aligns with the organization's goals and objectives, minimizes disruption to its operations, ensures that employees are adequately trained, communicates changes to stakeholders, and monitors and evaluates the e-business system's performance after its implementation. Change management reduces opposition to change. Organizations can enhance the performance and efficacy of their e-business systems, allowing them to maintain a competitive advantage in the ever-changing digital landscape. This is made possible by managing change effectively.

2 The Change Management Process

Change is unavoidable in the e-business era, and effectively managing change is imperative for success. Change management is the systematic and organized process of planning, implementing, and monitoring modifications to an organization's procedures, systems, and culture. It achieves this by minimizing the negative impact of change while maximizing its positive benefits, thereby assisting businesses in achieving their goals and objectives. The change management process can be initiated by following these six steps (Fig. 13.1).

1. *Identify the need for change*

In managing change for an e-business organization, the first and most crucial stage is recognizing the need for change. In the age of e-business, change is inevitable, and for businesses to maintain their competitive edge, they need to learn to adapt to this reality. However, recognizing the need for change is only sometimes a straightforward process. As a result, organizations may need to recognize or correctly interpret the indicators that change is required. In the following paragraphs, we will discuss the significance of recognizing the need for change inside an organization and the strategies that may be used. The ability of businesses to continue to be relevant and competitive in the e-business market is one of the primary benefits that can be gained from recognizing the need for change, which is one of the primary reasons why identifying the need for change is crucial. Take, for instance, an online retailer that is seeing a drop in sales due to its website becoming older and more challenging to browse. If this is the case, recognizing the need for change may create a new website that is friendlier to customers. This can enhance customers' shopping experience and lead to a revenue increase. Organizations can proactively address issues before they develop into significant problems if they first recognize the need for change. This results in enhanced competitiveness and success for the firm [1].

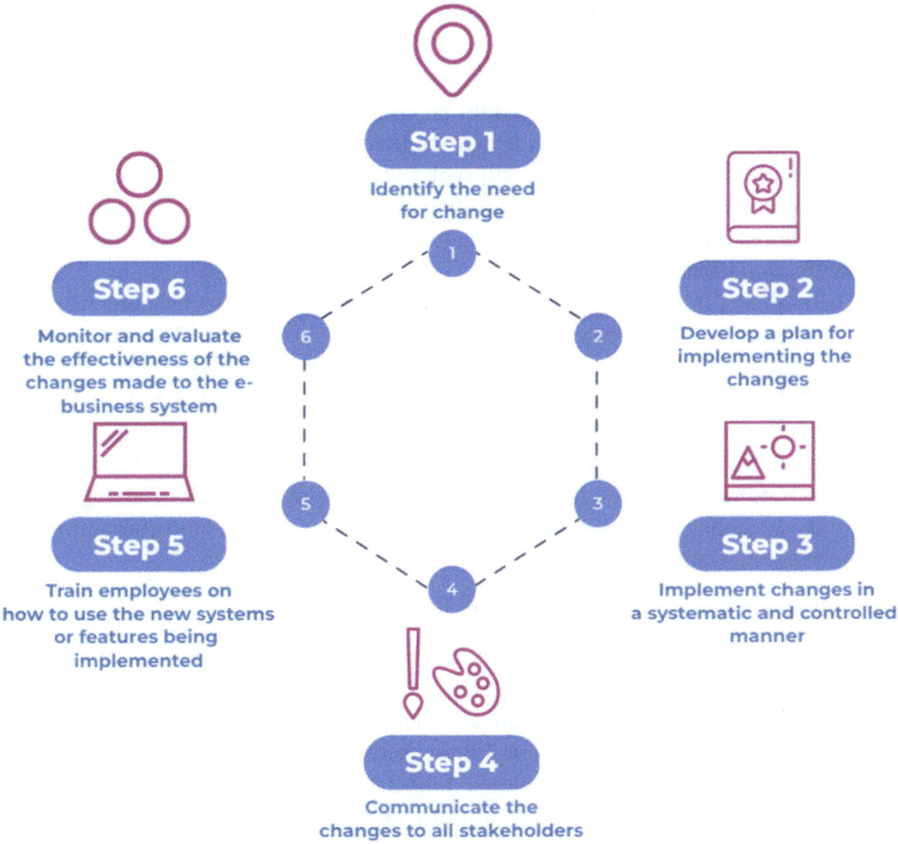

Fig. 13.1 Steps of the change management process

Online firms may use a few different approaches to determine whether or not there is a need for change. Conducting a SWOT analysis is one of the most productive approaches that can be taken. Performing a SWOT analysis on a company involves figuring out its strengths, weaknesses, opportunities, and threats. E-businesses may determine their opportunities and threats in the e-business market, as well as their strengths and weaknesses in contrast to those of their rivals, by completing a SWOT analysis and comparing their results to those of their competitors. An online company may find out via a SWOT analysis that it is falling behind its rivals in social media marketing, suggesting a need for change in that domain. Conducting market research is yet another method that online businesses can use to determine whether or not there is a need for change. Conducting market research entails acquiring and evaluating information about the e-business industry, such as the requirements and preferences of customers, the tactics used by competitors, and the developing trends. Online firms can uncover areas where the market could be improved and the innovation potential by conducting market research. For instance, an online firm may discover a rising demand for environmentally friendly items, which may signal a need to adjust its products. When determining whether or not there is a need for change, input should be solicited from all levels of an organization, which is another point that should be considered. People in customer-facing roles in an online business are frequently the first to recognize problems and opportunities, so it is essential to value and seek out their input. For instance, customer service representatives may detect an increase in the number of customers complaining about a particular feature of the e-business, which may signal the need for change in that particular area [2].

The realm of online business provides several real-life examples demonstrating the significance of correctly recognizing the need for change. The evolution of Netflix is a good illustration of this kind of phenomenon. When Netflix started, it was a DVD rental business, but as online streaming services became more popular, the firm had to change to remain competitive. Netflix recognized the need for change and pivoted its business model to concentrate on streaming services, leading to its success. Similarly, Amazon has consistently recognized the need for change, leading to the company's development into many sectors, such as cloud computing and traditional brick-and-mortar retail.

2. *Develop a plan for implementing the changes.*

After determining that change is necessary, the next step in the change management process for e-business is to develop a strategy for implementing the changes. At this stage, the objective is to create a road map outlining how the changes will be implemented and what resources will be required to carry out the plan effectively. When designing a strategy for implementing changes, the first step is to assemble a project team responsible for monitoring the change implementation process. This team must include members from all company departments with a vested interest in the outcomes of future modifications. Additionally, the group should have a project manager responsible for coordinating the group's activities. After that, the project team should conduct an in-depth analysis of the current state of the online business and identify the specific areas where adjustments are necessary. This research must assess the e-business's current technology, procedures, and personnel. During the transition process, this analysis will help identify the voids and obstacles that must be addressed to succeed. After completing the analysis, the project team must create a list of objectives and goals that will serve as a road map for the change process. These objectives and goals must be specific, measurable, and achievable within the established time frame.

Additionally, the group should collaborate to develop a set of performance metrics that can be used to evaluate the change implementation process. After establishing the project's objectives and goals, the team working on the project should collaborate to develop a comprehensive plan for implementing the changes. This plan must include a timetable that specifies the activities that must be completed, the required resources, and the roles and responsibilities of the project team members. The plan should also include a communication strategy that describes how the project team will communicate with stakeholders throughout the change process. This communication plan must include opportunities for stakeholders to provide input and ask questions and frequent updates on the status of the change process. As part of their efforts to ensure the successful implementation of the changes, the team working on the project should also develop a risk management strategy. This plan will identify potential hazards and detail strategies for mitigating them. A fallback strategy must be incorporated into the plan for risk management if the adjustments do not go as planned [2].

3. *Implement changes in a systematic and controlled manner.*

The next step in the change management process is the methodical and controlled implementation of the changes. This step follows identifying a need for change and formulating a strategy to resolve the need. During this phase, the plan will be implemented, and necessary adjustments will be made to ensure the changes are carried out effectively. This is especially true in e-business, where technology constantly advances, and businesses must stay one step ahead of the curve to maintain their competitive advantage. In order to effectively implement changes, a meticulous and organized strategy is essential. It is essential to ensure that implementing changes in e-business is well planned, well executed, and effectively communicated to all involved parties. This is a crucial aspect of implementing alterations. It is essential to establish specific timetables, milestones, and performance indicators to monitor the development of the implementation process effectively. In addition, it is crucial to maintain clear lines of communication with all relevant stakeholders to ensure that everyone is aware of the changes, their impact, and what is expected of them. For instance, to ensure a successful launch of its new Prime membership service, Amazon communicated the program's benefits to its customers via multiple channels, including email, social media, and advertising. This was done to ensure that the launch ran as efficiently as feasible. Suppose you want to implement changes in your e-business successfully. In that case, one of the essential steps you can take is to ensure that the implementation team has the necessary resources and experiences. This may involve hiring new employees, providing current employees with additional training, or outsourcing specific tasks to third-party vendors. To ensure that the modifications are implemented without hiccups, the team responsible must have access to the necessary tools, infrastructure, and technological resources. For example, prior to the release of iPhone X, Apple had to confirm that it had the necessary production capabilities and an established supply chain to meet the demand for the new product. It is also essential to ensure that the implementation process is controlled, with as little interference as feasible to the company's daily operations. To complete this task successfully, meticulous planning and cooperation with all relevant parties are required. For example, there may be better times to implement changes during prime business hours or a busy season, as they may negatively impact the customer experience and result in lost revenue. In order to minimize the amount of disruption, it is essential to schedule the installation process for off-peak hours or to carry it out progressively over time. In addition, monitoring and evaluating the implementation process regularly is crucial to ensure that it is proceeding according to plan and that any issues are resolved promptly.

Establishing a monitoring and feedback procedure, soliciting stakeholder input, and making any necessary adjustments to the plan are all potential actions associated with this step. Continuous monitoring and evaluation can keep the implementation process on track by identifying any potential obstacles or roadblocks that may arise [3].

In conclusion, it is essential to commemorate the successful implementation of changes and recognize the accomplishments of the implementation team and the involved stakeholders. If the achievement is celebrated, a culture of innovation and change can be nurtured, and a sense of ownership can be created among the stakeholders. This can aid in boosting morale. When Nike debuted its Nike+ app, for instance, the company acknowledged the accomplishments of the app's development team. It celebrated the app's successful launch via various internal and external communication channels.

4. *Communicate the changes to all stakeholders.*

One of the most essential stages in managing change for e-business is communicating any changes to all relevant stakeholders. It is essential to communicate explicitly so that everyone knows the changes and understands how they will affect them personally. Communication should be clear, concise, and tailored to the intended recipients to ensure that one's message is received and processed appropriately. Identifying who must be informed is the first step in effectively communicating changes to stakeholders. This includes shareholders, consumers, suppliers, employees, and other relevant parties.

Given that the requirements and concerns of each group may differ, it is essential to tailor the message accordingly. For example, employees may need to know how the changes will impact their responsibilities, and consumers may need to know how they will impact the orders or services they place. Once the stakeholders have been identified, formulating a communication strategy is the subsequent step. This should include the primary message, the channels that will be used to communicate the adjustments, and a schedule for when communication will occur. In addition, the plan must specify who will be responsible for conveying the message and who will be available to answer any inquiries or resolve any potential issues. In addition to developing a communication strategy, it is essential to consider how the message will be communicated. Avoid using technical jargon or terminology that is excessively complex, as this may confound or alienate the audience. The message should be clear and concise, with an optimistic and encouraging tone, highlighting the benefits of the changes and resolving any concerns or objections that stakeholders may have. A town hall meeting or all-hands assembly is one efficient and effective method for communicating changes to stakeholders. Due to this, the leadership team will have the opportunity to directly clarify the changes to the staff, address their questions and concerns, and answer their inquiries. In addition, it may aid in forming trust and cultivating a sense of shared purpose and commitment to the alterations. A document titled "Frequently Asked Questions" (FAQ) is another effective communication instrument. This may be posted on the company's website or distributed to employees, consumers, and vendors. The paper must address common issues and concerns and provide straightforward, plain, essential solutions for stakeholders to comprehend.

In conclusion, it is crucial to maintain communication and provide regular updates as the modifications are implemented. This may involve additional town hall or all-hands meetings, frequent progress updates, and implementing feedback systems. Building trust among stakeholders and maintaining their active participation throughout the change process enable an organization to reduce resistance to the change, establish credibility, and ensure a seamless transition.

5. *Train employees on how to use the new systems or features being implemented.*

Throughout any change management process for an e-business endeavor, employees must be made aware that their participation is crucial to achieving the objective. Employees may resist change, particularly when it involves learning new procedures or systems. Therefore, one of the essential stages in the change management process is training employees on utilizing the newly implemented features or systems. Identifying the required training is the first step in educating personnel. This can be accomplished by analyzing the knowledge and skill disparity between employees' actual knowledge and skills and the knowledge and skill set required to utilize the new features or systems. This evaluation may be conducted using questionnaires, individual interviews, or focus groups. The results of the analysis can be used to develop a training program. This plan may include the training's objectives, the content that must be addressed during the training, the training's duration, and the technique used to deliver the training. Introducing a brand-new e-commerce platform is a prime example of the need to train employees in e-business best practices. In this instance, employees must be instructed on using the platform to process orders, manage inventory, and manage consumer data. In addition, employees must be trained in customer data management. With the proper training, staff may be able to use the platform, which may result in errors and delays that can negatively impact the customer experience and, as a result, the company. Formulating training materials is the second step in the employee training procedure. This may involve creating user manuals, instruction films, and interactive simulations as examples. In addition to being tailored to the target audience's needs, the materials' content should be educational and engaging. It is essential to ensure that the information can be readily accessed and is simple to comprehend, particularly for employees unfamiliar with technology. The second phase, which follows the creation of training materials, entails conducting the training itself. The needs of the organization and the available resources will determine the distribution mode [3].

Training may be delivered in person, virtual, or a combination. It is essential to provide employees with ample advance notice of training sessions and to schedule them at convenient times. This will permit them to make the necessary preparations. In addition to conducting the training, evaluating how well the participants absorbed the training is essential. This can be accomplished through pre- and post-training assessments or questionnaires that assess the employees' knowledge and skill levels in the pertinent areas. This evaluation has the potential to help identify areas where additional training may be required and to provide valuable feedback on the training materials and delivery method. Amazon's training program for warehouse employees is a prime example of an exceptional employee training program in the context of an online business. Amazon has implemented a comprehensive employee training program that includes classroom instruction, on-the-job training, and e-learning opportunities. The curriculum is designed to help employees acquire the skills and knowledge necessary to succeed in the fast-paced environment of an Amazon warehouse, which is the program's focal point. Amazon's success in attaining high levels of warehouse productivity and efficiency is mainly attributable to the implementation of this training program.

6. *Monitor and evaluate the effectiveness of the changes made to the e-business system.*

In order to ensure the success of change management in e-business, it is essential to monitor and assess the efficacy of changes made to the e-business system. This is the only method to ensure the success of e-business transformation management. This is the final phase of the change management process. It entails monitoring the effects of the changes and making any necessary adjustments to ensure the desired outcomes are achieved. Metrics are essential to the surveillance and evaluation of the effectiveness of the change process. Before and after the implementation of the new changes, the efficacy of the system used for conducting online business can be evaluated using metrics. This category may include website traffic, customer interaction, conversion rates, and revenue, among other things. When these metrics are monitored, it is possible to determine whether the modifications have the desired effect and whether additional adjustments are necessary.

Real-life: For instance, Amazon, one of the largest online retailers in the world, consistently monitors and evaluates the effectiveness of its modifications. In 2018, Amazon modified its search algorithm, significantly changing the website's search results. After implementing these enhancements, the company followed up with customers for feedback, monitored website traffic, and analyzed conversion rates. Due to this feedback, Amazon made additional modifications to its search algorithm, increasing both consumer satisfaction and revenue.

In addition to monitoring and analyzing the efficacy of enhancements, requesting feedback from both consumers and employees is crucial. This feedback may be gathered through questionnaires, evaluations, or other direct channels. By collecting feedback from consumers and employees, businesses can identify areas for improvement and make the necessary adjustments to ensure the improvements' effectiveness. In addition to monitoring and analyzing the effectiveness of the modifications, it is crucial to disseminate the findings to all relevant parties. This applies to both the clients and the personnel and management. By disseminating the assessment's findings, the stakeholders can observe the effects of any necessary changes and comprehend the reasoning behind them.

In conclusion, it is crucial to continuously monitor and evaluate the effectiveness of changes made over time. Due to the swift evolution of e-business systems, what is effective today might be unsuccessful tomorrow. If businesses continue to collect input and monitor KPIs, they can ensure that the e-business systems they have in

Table 13.1 Best practices for each step of the change management process

Step in change management	Best practices for e-business
Identify the need for change	Conduct a thorough analysis of the current e-business system Involve stakeholders in identifying potential areas for improvement
Develop a plan for implementing the changes	Set clear goals and objectives for the changes Assign responsibilities to team members, and set timelines for the completion of tasks
Implement changes in a systematic and controlled manner	Use pilot testing to identify and address any issues before full implementation Monitor progress and adjust plans as necessary
Communicate the changes to all stakeholders	Develop a communication plan that includes clear and consistent messaging Use a variety of communication channels to reach all stakeholders
Train employees on how to use the new systems or features being implemented	Provide training in a variety of formats, including online and in-person options Ensure that training is tailored to the specific needs of each department or team
Monitor and evaluate the effectiveness of the changes made to the e-business system	Use metrics to track the success of the changes over time Use feedback from stakeholders to identify areas for continued improvement

place will continue to be efficient and competitive over time. Table 13.1 summarizes best practices for each step of the change management process.

3 Resistance to Change in E-Business

Change is inevitable in all businesses, but it is especially unavoidable in the world of e-business, where the digital environment is constantly evolving. On the other hand, change is not always welcomed with enthusiasm, and organizations must be aware of the numerous causes of resistance to change to navigate it successfully. In this section, we will examine how organizations can effectively manage familiar sources of resistance to change in e-business and identify and evaluate such sources (Table 13.2).

Fear of employment loss is one of the most prominent causes of resistance to change in the e-business industry. When changes are implemented, employees may worry that they will lose their jobs or cannot acclimate to the new technology or procedures due to a lack of the necessary skills. This fear may be a significant barrier to effective change management. Your concern about losing your employment may be warranted under certain conditions. For instance, if the implemented enhancements involve automation or artificial intelligence (AI), it is conceivable that some occupations will no longer require human interaction, resulting in job

Table 13.2 Sources of resistance to change in e-business systems

Resistance source	Definition	Examples	Impact on change	Mitigation strategies
Fear of job loss	Resistance due to fear of job loss or inability to acquire new skills	Employees may resist changes if they fear losing their jobs or lack confidence in learning new skills	Reduces the likelihood of successful change	Offer retraining and opportunities for skill development, emphasize the new system's benefits, and ensure employee job security
Lack of trust	Resistance due to a lack of trust in management or the change process	Employees may be skeptical about the motives behind the change or may have had negative experiences with past changes	Reduces the likelihood of successful change	Foster transparency and open communication, involve employees in the change process, address concerns, and provide clear explanations
Loss of control	Resistance due to perceived reduction in control over work or decision-making process	Employees may resist changes that will reduce their control over their work or ability to make decisions independently	Reduces the likelihood of successful change	Involve employees in the decision-making process, provide opportunities for feedback and suggestions, and clarify new roles and responsibilities
Uncertainty	Resistance due to anxiety caused by uncertainty about the future	Employees may worry about how the changes will affect their job security, workload, or work environment	Reduces the likelihood of successful change	Provide clear communication about the change process and its impact, offer reassurance and support to employees, address concerns, and provide information as it becomes available
Lack of knowledge/ skills	Resistance due to perceived lack of knowledge or skills required to adapt to the new system	Employees may resist changes if they lack the necessary knowledge or skills	Reduces the likelihood of successful change	Offer training and support for employees to develop necessary skills and provide clear expectations and guidance for adapting to the new system
Comfort with the current system	Resistance due to being comfortable with the current system	Employees may not see the need for change or may feel that the change is unnecessary	Reduces the likelihood of successful change	Emphasize the new system's benefits, provide clear explanations for the need for change, and offer incentives for adopting the new system

(continued)

Table 13.2 (continued)

Resistance source	Definition	Examples	Impact on change	Mitigation strategies
Inadequate communication	Resistance due to poor communication about the change process	Employees may feel they are not being adequately informed or that their concerns are not being heard	Reduces the likelihood of successful change	Foster transparency and open communication, provide regular updates and opportunities for feedback, address concerns, and provide clear explanations
Culture clash	Resistance due to conflict with company culture or values	Changes may conflict with the company culture or values, particularly in cross-border e-business ventures	Reduces the likelihood of successful change	Respect cultural differences and values, provide opportunities for cultural awareness and understanding, and involve employees from different cultures in the change process
Technical barriers	Resistance due to technical issues with the new system	Technical issues such as system downtime, bugs, or glitches can be frustrating and can lead to a lack of confidence in the new system	Reduces the likelihood of successful change	Address technical issues promptly, provide training and support for technical issues, and offer incentives for adopting the new system

losses. However, the fear of being laid off may be unwarranted in other situations [3]. Employees' reluctance to change may be due to a lack of understanding of new work methods or a sense of unease with these methods. In order to surmount this potential source of resistance, executives of e-businesses must communicate openly and plainly with their employees about the changes being implemented and the potential impact those changes will have on their employment. Because it enables workers to adapt to new technologies and assume new responsibilities within the organization, providing employees with opportunities for training and skill enhancement may also help alleviate concerns about potential job loss.

Deploying self-checkout systems in retail establishments is a real-world example of resistance to change in e-business based on the dread of employment loss. Employees are anxious because they do not want to risk losing their employment. When self-checkouts were first introduced, many retail employees feared for their jobs due to the possibility that the new technology would eliminate the need for human cashiers. Self-service checkout is an illustration of this possibility. However, many retail organizations could implement the changes without suffering significant job losses because they provided employees with training and support on using and maintaining the self-checkout systems and reallocated employees to other tasks, such as customer service and restocking.

A common cause of resistance to change in e-business is a need for more trust, which may also cause resistance. Trust is the foundation of all productive relationships, and the same holds for the relationship between an employer and their employees. Employees will likely only support change if they trust the organization's leadership and the proposed changes. A lack of trust may result from various factors, including ineffective communication, a history of broken promises, or a lack of transparency.

Yahoo is an example of how a lack of trust can result in resistance to change in an online business environment. In 2013, Marissa Mayer, then CEO of Yahoo, introduced a brand-new performance review system called the Quarterly Performance Review (QPR). Accountability and transparency were given top priority in the new system's design. On the other hand, the staff members lacked confidence in the new system and fought against the change. Many employees believed that the QPR was intended to encourage them to resign from their positions.

To address the lack of trust as the propelling force behind resistance to change in e-business, it is necessary to cultivate a culture of trust within an organization. This can be achieved by ensuring open and transparent communication regarding the changes and consequences. In addition, leaders are expected to keep their word and maintain behavioral consistency. In addition, involving employees in the change process and soliciting their feedback may help to establish trust and reduce resistance to the change. Providing employees with opportunities for training and advancement is yet another method for combating a lack of confidence. If a company invests in its employees' growth and development, employees are more likely to believe in the organization and its leaders. Additionally, the consistent deliverance of feedback and appreciation can facilitate the development of trust among colleagues and promote a positive work environment.

Another common factor contributing to resistance to change in e-business environments is losing control. When an organization modifies its e-business system, employees may be required to adjust their working methods or adopt new technologies. This may give employees the impression that their autonomy is being infringed upon, resulting in losing control over the work process. As a result of this loss of control, employees may experience anxiety and uncertainty, resulting in their resistance to change.

Introducing new project management software at a software development company is an example of loss of control as a reason for resistance to change in e-business in the real world. This exemplifies how individuals believe they are losing control of the situation. The program was created to expedite project reporting and monitoring and reduce the time spent on administrative duties. Even with this, some team members had the impression that the software limited their ability to make decisions and their level of project control. As a result, they resisted implementing the new software and continued to operate using the previous methods, resulting in difficulties with the team's ability to communicate and coordinate.

It is essential to include employees in the transition process and provide them with training and assistance to adapt to the new system. This will aid in reducing the resistance generated by the loss of control. Communication is also crucial because it enables employees to comprehend the factors propelling the change and how it will benefit them and the organization. Providing a detailed action plan can help employees feel more in control of the transition and reduce their anxiety. Involving employees in the planning and implementing of the new system is another method for regaining control of an out-of-control situation. This may give individuals a sense of ownership and control over the process and make them more invested in its success. This strategy was employed by a major retail organization introducing a brand-new point-of-sale system. In order to develop and implement the new system, the company convened a cross-functional team comprised of professionals from various departments. By involving employees in the process, the organization was able to assuage their concerns about losing control of the situation and cultivate a more collaborative work environment.

Uncertainty is another significant factor that contributes to resistance to change in e-business. Employees who need more clarity regarding the impact of the change on their work duties and responsibilities are more likely to experience feelings of uncertainty. Employees confronted with unpredictability may oppose changing out of a dread of the unknown or a belief that they cannot manage the changes if placed in an uncertain environment. Communicating with employees about the change, explaining its reasons, and providing as much information as possible about what will occur due to the change can reduce workplace uncertainty. It is possible to reduce uncertainty by providing training and assistance to staff members to facilitate their adjustment to the changes.

When a company decides to transition to a new e-commerce platform, for instance, the staff members may need clarification on how the new platform will operate and what their duties and responsibilities will be on the new platform. To reduce confusion in the workplace, the organization may organize information sessions to clarify the new platform, provide training on how to use it, and designate specific tasks and responsibilities for each employee.

In addition, employees may be concerned about how the change will affect the organization's overall operations. They may be interested in how the change will affect the future of the company or how it will affect the company's short-term success. Communicating with employees about the potential benefits of the change, such as enhanced productivity, increased customer satisfaction, and increased profitability, can help reduce their anxiety.

Inadequate knowledge of the subject is another common reason for resistance to change in e-business. It may be difficult for employees to embrace new technologies or procedures if they are unfamiliar or have not received adequate training. Employees who need more information to comprehend the reasons for the change and how to properly implement it may develop resistance or become overburdened by its implementation.

A company implementing a brand-new customer relationship management (CRM) system is a pertinent example of this concept in practice. If employees are adequately trained on how to use the new system, they may need help inputting customer information, monitoring sales prospects, and generating reports. Due to this information divide, people may experience frustration, make errors, and resist the new system.

It is crucial to provide employees with opportunities for training and open communication before and throughout the transition process to surmount this potential source of resistance. This category may include workshops, online courses, and individual sessions with a coach. When an employer invests in training and education for workers, those workers gain the knowledge and self-assurance required to embrace change and successfully implement new technologies or processes. Involving employees in the planning and implementation of the change process is another example of a recommended practice. This may make people feel more invested in the outcome and increase their motivation to acquire new skills and adapt to various methods and systems. Employees may feel more at ease and assured as they navigate the transition if the employer provides continuous assistance and feedback.

Another common cause of resistance to change in e-business is a company's familiarity with its current system. It is conceivable that employees are accustomed to the current system and do not see the need for change. They are content with the status quo and do not wish to make the necessary effort to learn anything new. This may be particularly challenging when the new system is significantly distinct from the old one. There is a possibility that employees are concerned that the new system will disrupt their routine and make it more difficult for them to do their jobs effectively. This foreboding may manifest more dramatically if the employees are already overburdened with responsibilities. They may be concerned that the time required to master the new system will interfere with their ability to carry out their current responsibilities. Workers who perceive they have invested a substantial amount of time and effort in the current system may also be reluctant to change. They may believe their experience and expertise will become less valuable if the system is modified. This may cause employees to believe their knowledge will become less valuable if a new system is implemented, contributing to resistance to change [4].

The financial industry serves as a concrete example of this principle in action. Because banks have already invested a substantial amount of time and money in their existing systems, they sometimes need more time to implement new technology. Despite the availability of more modern technology, several institutions continue to rely on antiquated mainframe systems for the execution of their core banking operations. This aversion to change may impede innovation and make it challenging for financial institutions to maintain their market leadership position.

In order to surmount this resistance, businesses must educate their employees on the benefits of the new system they are implementing. They should emphasize how the new system will increase productivity and assist employees in performing their duties more effectively. In addition, they should emphasize how the new system will assist employees in increasing their efficacy. It is also crucial to provide staff with training and support to become accustomed to and confident with the new system.

Inadequate communication is another prevalent factor contributing to resistance to change in e-business. Throughout the process of implementing changes, it is essential that all communication channels remain open and that all stakeholders are kept informed of the changes and their impact. In a given situation, the inability to communicate plainly may result in misunderstanding, irritation, and resistance to change. Establishing communication channels that are both open and transparent is one strategy for dealing with this type of resistance. This can be achieved by holding frequent meetings, distributing regular updates and newsletters, and creating a feedback system where stakeholders can discuss their concerns and pose questions. It is crucial to provide information that is not only clear and concise but also pertinent to each stakeholder group, and it is also crucial to tailor the communication to the specific needs of each stakeholder group.

For instance, when Amazon first began offering the Prime membership program, the company ensured that the many benefits of becoming a member were prominently advertised via email, advertisements, and the Prime section of Amazon's website. The organization was very forthcoming with information regarding the nature of the program, including its objectives, the benefits available to clients, and the enrollment procedure. This resulted in a decline in opposition and an increase in the number of individuals adopting the program. Another illustration of this would be a company's decision to move to a new e-commerce platform. Inadequate communication among those tasked with administering the new platform could result in uncertainty among those personnel. This may result in implementation delays, discontent, and organizational change resistance. To prevent this from occurring, it is crucial to provide this staff with exhaustive training and support and ensure that they have access to all the data they require to administer the new platform effectively.

In e-business, cultural conflicts are among the most frequently encountered causes of resistance to change. When a company decides to install a new e-business system or process, it frequently requires a substantial shift in the way its employees operate and communicate. Alternatively, if the company's culture does not embrace such changes, this could result in resistance and hinder the e-business transformation's effectiveness. When there is a significant disparity between the values, beliefs, and practices of the established organizational culture and those required for the effective adoption of e-business, a convergence of cultures is conceivable within an organization. This may result in a negative outlook, resistance to change, and an unwillingness to adopt new methods of operation.

For instance, a typical brick-and-mortar retail company could implement an e-commerce platform to expand its customer base and increase sales. Even with the company's culture being hierarchical and resistant to change, employees may view the new e-commerce platform as threatening how they have traditionally conducted business. They may be unwilling to implement new technologies or methods of interacting with clients, resulting in a protracted adoption process and a loss of potential revenue. This may be the situation.

The initial step in resolving the culture dispute is for businesses to comprehend their current culture and identify any potential conflicts that may arise during the e-business transformation process. They must ensure that all employees know the intended outcomes and objectives of the transformation and encourage participation and collaboration throughout the process. Additionally, businesses can cultivate a more adaptable culture to change by adopting a culture of continuous improvement and employing a more agile method of working. Employees may discover that they are better able to acclimate to change due to this, and it may also encourage them to develop more adaptable and flexible work practices.

Regarding e-business, the prevalence of technical obstacles is one of the most prevalent causes of resistance to change. This may occur when employees have grown accustomed to the organization's current technology and systems and are hesitant to adopt newer ones. Some employees may be unable to use the new systems because they need more technical skills, which could cause confusion and annoyance. In addition, the new system may not be compatible with the existing one, resulting in additional employee resistance. A corporation's implementation of a new customer relationship management (CRM) system is an example of a real-world scenario illustrating the presence of technological obstacles in e-business. Even though the new CRM system was designed to be more efficient and user-friendly, some employees resisted the change due to the technical obstacles they encountered. They were accustomed to the previous system and had no desire to master the new one. Due to this reluctance, the installation was delayed, and additional training was required to familiarize employees with the new system. It is crucial to provide staff with the necessary technical training and assistance to conquer the new systems and surmount any technological obstacles. This category may include training sessions, access to online seminars, and other technical assistance resources. Participating in selecting and implementing new processes may help employees feel more invested in the change and increase their likelihood of adjusting. Providing consistent communication and feedback throughout the transition process also helps alleviate anxieties and concerns regarding the existence of technological obstacles [4].

Emotional attachment to the old system: In e-business, an emotional attachment to the old system may be one of the primary sources of resistance to change. People may have become accustomed to and familiar with the previous system, making it difficult to let go of what they know. They may not be motivated to learn a new system because they devoted so much time and effort learning the previous system, and they may not be willing to put in the same amount of effort to learn the new system.

Blockbuster, a once-thriving chain of video rental stores, is an excellent example of how a passionate attachment to the status quo can result in resistance to change. Blockbuster illustrates how emotional attachment to the old system can result in opposition to change. Blockbuster was emotionally invested in its brick-and-mortar storefronts and physical rental business; consequently, it was reluctant to adapt to the shift toward online streaming and digital downloads. This refusal to adapt ultimately led to the company's decline, as competitors such as Netflix embraced the digital age and obtained a competitive advantage.

In order to resolve the issue of emotional attachment as a cause of resistance, it is essential to acknowledge and affirm people's feelings about the previous system. They may require assurance that their prior expertise and abilities will continue to be valued in the new system and that their efforts to learn the new system will be supported. Individuals may feel less emotionally attached to the old system if they can develop a sense of comfort and familiarity with the new system, which can be facilitated by training and assistance. Involving people in the change process and giving them a sense of ownership and control over the transition is another approach that can be taken. Consequently, they may experience less emotional attachment to the previous system, making them feel more invested in the new system's success. Individuals may have an easier time recognizing the value of the change if the benefits of the new system and how it will improve their jobs are highlighted. This may also reduce their resistance to the change.

Humans are accustomed to resisting change, which may pose a formidable obstacle when attempting to implement enhancements to e-business systems. However, there are strategies that businesses can employ to surmount opposition and facilitate a more seamless transition to the new system. Companies can use these strategies.

1. Communication and participation: Effective communication is one of the essential factors in e-business for overcoming resistance to change. The management is responsible for effectively communicating the reasons for the change, its anticipated benefits, and its impact on the employees. They should also solicit and include the employees' input in developing and implementing the strategy. In this way, employees will experience being acknowledged and respected, which will help alleviate their anxieties and concerns.
2. Instruction and assistance: Providing training and assistance to employees is another effective method for overcoming resistance to change in an e-business environment. When employees are provided with adequate training and support, they are more likely to feel confident in the new system, which may facilitate their adaptation to the change. Employees are more likely to feel confident using the new system when provided with adequate training and support. Businesses owe it to their employees to provide custom-tailored training on the new system and user manuals, support centers, and other resources that employees can use while navigating the new system.
3. Offering employees incentives and rewards: In an e-business setting, providing employees with incentives and rewards may also help surmount resistance to change. If employees comprehend how a new policy or procedure would benefit them, they are more likely to accept it. Those who effectively demonstrate mastery of the new system may be eligible for rewards and incentives such as bonuses, promotions, and recognition. This could be a motivating factor for employees to learn and adapt to the new system.
4. Responding to employee concerns: To surmount resistance to change in an e-business environment, it is essential to resolve the concerns of employees. Employees may be concerned about their employment, the quantity of work they will be required to complete, or the change's impact on the workplace environment. The management must address these issues and provide those concerned as much reassurance as possible. They should also be forthright and honest about the change's impact and how it will affect the duties and responsibilities of the employees.

5. Testing the alteration: To surmount resistance to change in e-business, businesses can also implement a second strategy, namely, change piloting. This involves implementing the change on a more manageable scale and putting it through its trials before deploying it across the entire organization. Before implementing the change on a larger scale, conducting a pilot test of it may expose any potential problems or obstacles and provide the opportunity to make necessary modifications.

Setting the standard for others to follow: In e-business, a powerful method for overcoming resistance to change is to serve as a model for others. When management sets an example for employees to follow, it sends a clear message that the change being implemented is essential and being taken seriously. To demonstrate their commitment to the change, managers must maintain a visible presence throughout the change process and actively partake in its execution. One of the most important strategies for overcoming resistance to change in e-business is recognizing past achievements. After the change has been successfully implemented, it is essential to give credit where credit is due and commemorate the achievement. This may help employees develop a more positive attitude toward change and encourage them to embrace future modifications.

In order to be successful in e-business, businesses must address the issue of overcoming resistance to change. On the other hand, implementing the proper techniques may make all the difference in how smoothly the transition is accomplished. Businesses can surmount change resistance through various strategies, including effective communication, training and assistance, incentives and rewards, addressing concerns, piloting the change, leading by example, and celebrating success. Businesses can effectively incorporate enhancements to their e-business systems and ensure their continued competitiveness in today's rapidly evolving digital world by putting these strategies into action and following through with their implementation.

4 The Role of Leadership in Change Management

Leadership is one of the most significant determinants of the success of change management efforts in an e-business. Leaders are responsible for providing employees with direction, motivation, and support throughout the change process. They are also responsible for ensuring the change is consistent with the company's overarching strategy and vision. The function of leadership in effectively administrating change in e-business will be discussed in the following paragraphs.

To begin, the change's leaders must devise a vision for it. They must clarify the rationale for the change and its benefits to the company. Jeff Bezos, the founder and CEO of Amazon, had a crystal-clear mental image of how the company would transform from a conventional bookstore into a global leader in e-commerce. Bezos communicated this vision to the company's employees and led the company's restructuring by investing in technological and logistical infrastructure.

The second responsibility of leaders is to impart a sense of urgency. They must convey the significance of change and the dangers associated with maintaining the status quo. To encourage people to embrace change, leaders must provide convincing evidence as to why the change is required. In the 1990s, when Lou Gerstner became the new CEO of IBM, he promptly recognized the urgent need for the company to endure a significant transformation. He was the primary force behind IBM's successful transition from a hardware-centric company to a services-focused one, ultimately saving the company from bankruptcy.

Thirdly, leaders are charged with assembling a coalition of support. They must identify the key influencers and constituents who can support the change within the organization. Transparency with employees and involving them in the change process are two of the most essential things leaders can do to gain their trust and credibility. This fosters a sense of ownership and responsibility for the change. Steve Jobs, the CEO of Apple, formed a coalition of support for the company's transition from computer manufacturer to consumer electronics corporation. He ensured that key stakeholders were involved in developing new products and that employees were aware of the benefits of the change.

Fourthly, efficient transition communication is crucial for leaders. They must utilize multiple communication channels to ensure employees comprehend the change and its consequences. In addition, leaders are responsible for listening to the comments and concerns of their employees and responding appropriately. This contributes to the growth of trust while simultaneously reducing change resistance. For instance, when Microsoft CEO Steve Ballmer announced a significant restructuring in 2013, he emailed all employees and produced a video to convey the news. In addition, he organized town hall meetings to resolve the staff's concerns and inquiries.

Fifthly, leaders are required to provide resources and assistance to the change. To ensure the transition's success, they must commit the necessary resources, such as money, technology, and personnel. Additionally, executives are liable for providing training and professional development opportunities to employees in order to facilitate their transition to the new system. For example, when Dell undertook a significant reorganization in 2007, CEO Michael Dell ensured that employees had adequate opportunities for training and advancement to help them adapt to the new environment.

Sixthly, it is the responsibility of leaders to monitor and evaluate the change. They should assess the degree to which the change has occurred and identify the areas lacking. In order to maintain momentum and maintain support for the change, leaders must emphasize both the successes and the obstacles that have been encountered. This increases the likelihood that the change will be permanent and integrated into the company's culture. Larry Page, the CEO of Google, is one individual who does this. Regularly, he verifies that the firm's numerous ongoing initiatives are consistent with its mission and values. Figure 13.2 illustrates the leadership role in successful change management in e-business [3, 4].

Effective leadership is crucial for successful change management in e-business. Leaders who can guide their teams through the change process and inspire them to

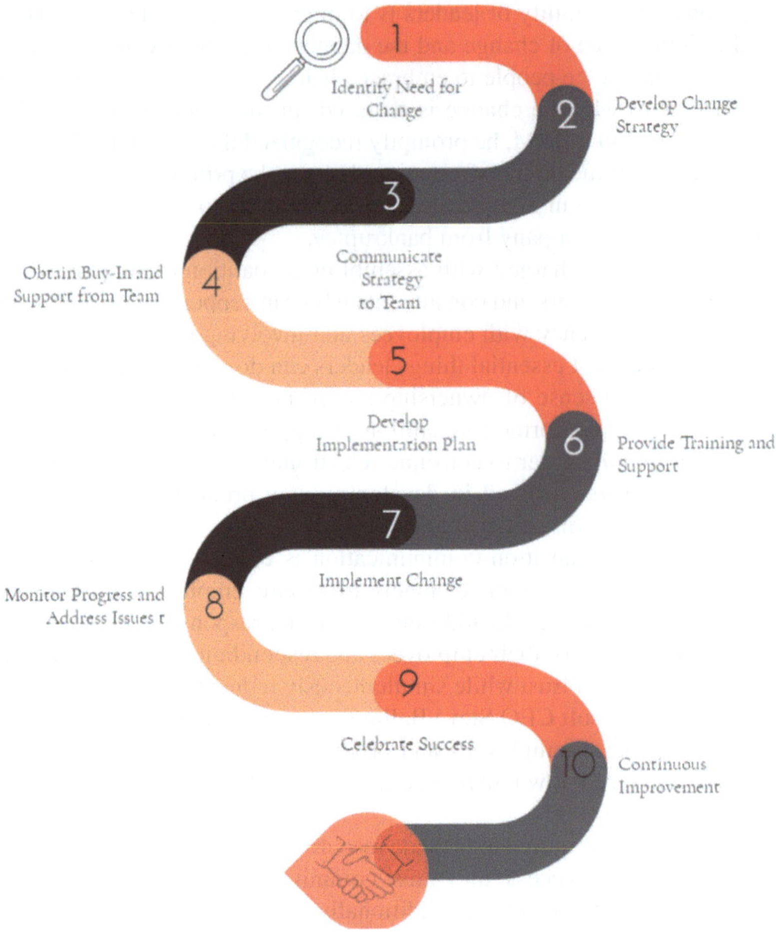

Fig. 13.2 Leadership role in successful change management in e-business

adapt to new technologies and processes can help businesses stay competitive in the rapidly evolving digital landscape. This section analyzes some of the most effective leadership strategies for change management in e-business (Fig. 13.3).

Effective e-business change management requires a clear vision. Leaders must articulate a vision of what the change will accomplish and how it will benefit the organization. They should convey this vision to inspire and motivate their team members to accept the change. When Steve Jobs returned to Apple in the late 1990s, he had a clear vision of what the company could become, and he inspired his team to work toward that vision, resulting in the company's revival.

Effective communication is another essential leadership strategy for e-business change management. Leaders must communicate the change's rationale, anticipated benefits, and impact on team members. They should be open and honest about the impact of the change and address any team members' concerns or inquiries. For

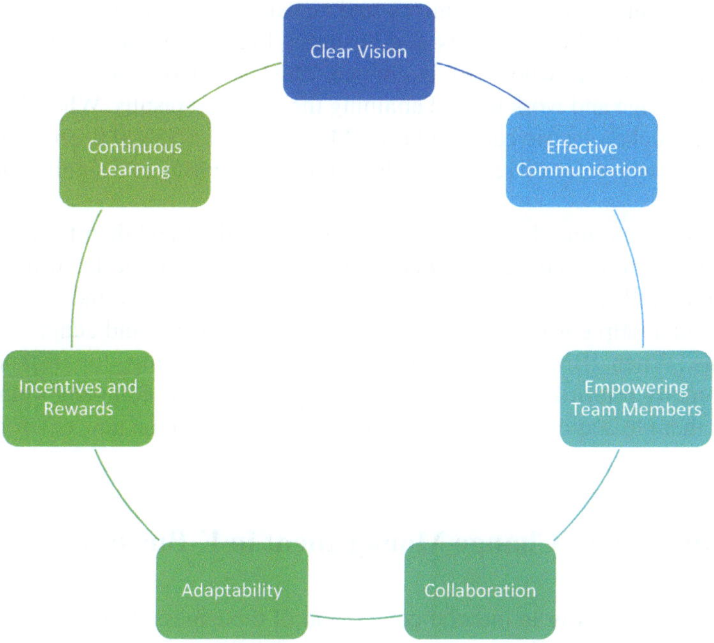

Fig. 13.3 Requirements of e-business change management

instance, when Amazon introduced its Prime membership program, Jeff Bezos informed his team of its benefits and encouraged them to contribute to its success.

A further effective leadership strategy for change management in e-business is empowering team members to assume responsibility for the change. Leaders should encourage team members to take the initiative and support the change with their decisions. They should provide the resources and assistance required for team members to assume change ownership. For instance, when Facebook acquired Instagram, Mark Zuckerberg gave the Instagram team the autonomy to continue operating the platform independently, enabling them to make decisions by their vision.

Collaboration is an additional crucial leadership strategy for e-business change management. Leaders should encourage team members to collaborate and work toward a common objective. They should foster an atmosphere that encourages frank dialogue and exchanging ideas. Google collaborated with other companies to incorporate their data into the Google Maps platform, resulting in a more comprehensive mapping tool.

Adaptability: Leaders must be able to modify their strategy as necessary to accomplish the intended results. They should be receptive to feedback and willing to adapt their approach to shifting conditions. When eBay acquired PayPal, for instance, the company shifted its strategy to concentrate on its primary marketplace business, allowing it to remain competitive in the ever-changing e-commerce landscape.

Incentives and rewards: Incentives and rewards can also be effective change management leadership strategies in e-business. Leaders should acknowledge and reward team members who adapt to change. This can encourage team members to accept the change and work toward attaining the intended results. When Tesla introduced its Model S electric car, CEO Elon Musk offered a $5000 financial incentive to employees who referred recruits to the company, resulting in a significant referral increase.

Continuous learning: Leaders must also prioritize their and their team members' continuous learning and development. They should encourage team members to acquire new skills and remain current with the latest e-business trends and technologies. This can help ensure the business remains competitive and adapts to altering conditions. When IBM shifted its focus to cloud computing, for instance, CEO Ginni Rometty prioritized learning and development for the company's employees, allowing them to acquire the skills necessary to succeed in the new environment.

5 Evaluation of Change Management in E-Business

Change management is an integral component of e-business. It involves modifying existing business procedures and computerized information systems to increase productivity and profitability. Change management may provide excellent value to organizations, but several disadvantages must be considered (Table 13.3).

Utilizing change management in e-business has several benefits, including increased productivity. Businesses can increase productivity by modifying their processes and systems, streamlining operations, and achieving greater efficiency. For instance, an e-commerce company may modify its website to make it easier for users to navigate and purchase, increasing its customers and revenue.

Enhanced competitiveness is an additional benefit of effective change management. E-businesses operate in a highly competitive environment, and organizations that fail to respond to market changes risk losing customers and revenue. This is particularly true for online businesses. Change management enables organizations

Table 13.3 Benefits and drawbacks of change management in e-business

Benefits of change management	Drawbacks of change management
Reduces resistance to change	Costly and time-consuming
Helps maintain business competitiveness	May disrupt business operations
Ensures effective implementation of changes	It may require additional resources and expertise
Encourages employee participation and buy-in	This may lead to a temporary decrease in productivity
Improves communication and collaboration	It may require a shift in organizational culture
Enhances overall business performance	It may not guarantee the success of the change
Enables businesses to adapt to the digital landscape	It may require ongoing maintenance and updates

to rapidly adapt to fluctuating client demands and market conditions, thereby preserving their competitive advantage. For instance, a company that sells retail products may modify the functionality of its inventory management system to accommodate changing consumer demands and market trends.

On the other hand, change management has disadvantages that must be considered. Among the disadvantages of making modifications is the associated cost. Implementing changes to corporate processes and systems can be costly, requiring organizations to commit resources and invest in new technology and training. If enhancements are made to an e-commerce platform, engage additional personnel, purchase new software, and train employees to use the new system.

Another disadvantage associated with change management is the possibility of business operations interruptions. Implementing changes to existing business procedures and computer systems may result in interruptions, leading to decreased productivity and outages. For instance, implementing changes to a website could make the site inaccessible for some time, resulting in a decrease in revenue and sales.

Despite the potential drawbacks of change management, organizations must routinely analyze their operations and implement changes to maintain a competitive advantage in the digital age. In order to minimize the adverse effects of change management and maximize its positive effects, businesses should meticulously plan and implement changes, ensuring that they are aligned with the overall company strategy and goals. In addition, they must ensure a seamless transition by effectively communicating changes to employees and customers and providing adequate training and support.

The domain of e-business is perpetually evolving, and as a result, change management must also evolve. Although the methods presented earlier in this book effectively overcome resistance to change in e-business, it is essential to consider the future of change management and how emerging technologies and altering business practices will impact it.

The increasing use of artificial intelligence (AI) and machine learning is a significant trend that will significantly impact the future of change management in e-business. As firms increasingly rely on artificial intelligence to automate processes and make decisions, change management will need to evolve to accommodate the specific issues posed by introducing new technologies. Workers may be hesitant to collaborate with AI, for instance, out of concern that their jobs will eventually be taken over by robotics and other forms of automation. Change management should focus on educating people about the positive aspects of AI and how it can make their jobs more efficient rather than focusing on finding methods to replace workers. The need to maintain data privacy and security will only increase in the coming years, another trend that will influence the future of change management in e-business. Due to the escalating number of high-profile data breaches and the increased scrutiny from regulatory authorities, businesses are compelled to take preventative measures to safeguard consumers' personal information. Change management must focus on implementing new security measures, educating employees on the importance of data privacy, and ensuring that any new e-business systems or technologies are developed with security in mind. In addition to assuring that any new e-business systems or technologies are developed with security in mind, these tasks will be performed. The increasing prevalence of telecommuting and distributed work environments is an additional factor that will impact the evolution of change management in e-business. To ensure that employees can collaborate effectively from a distance, change management will need to concentrate on ensuring that they are equipped with the necessary tools and technologies to work remotely and provide adequate support and training. This is because an increasing number of companies are employing flexible work arrangements [1].

The present COVID-19 pandemic has also significantly impacted e-business change management. As a result of the pandemic, businesses have been compelled to rapidly adopt new ways of functioning and responding to consumer demands. The epidemic has demonstrated the significance of adaptability and resilience in change management. The pandemic has demonstrated the significance of businesses' ability to pivot and adapt rapidly to new circumstances. With an eye toward the future, businesses must implement a culture of continuous development in which change is viewed as an essential element of remaining competitive in the ever-changing e-business landscape. To ensure the successful implementation of these changes, change management must prioritize the development of a culture in which employees are actively encouraged to innovate and experiment and provide the necessary support and resources.

By analyzing instances of effective change management in e-business from the real world, it is possible to gain valuable insights into the future potential of change management. Amazon, for example, can maintain its position as one of the most successful online businesses in the world by constantly experimenting with new forms of technology and business strategies. They quickly transitioned and modified their business model in response to changing market conditions, such as the emergence of mobile commerce and the growing importance of AI, because their culture encourages innovation and is open to change.

The conclusion is that the future of change management in e-business will be influenced by advancing technology, shifting business practices, and the need to be flexible and competitive. Change management will need to focus on training people, incorporating new technology, emphasizing data privacy and security, facilitating remote work, and cultivating a culture that values continuous improvement to adjust to these changes. If businesses adopt these measures, they will have a greater chance of successfully navigating the challenges of change management and remaining competitive in the rapidly evolving world of e-business.

Summary

The Change Management Process

- The change management process refers to the structured approach organizations adopt to transition from the current state to a desired future state in e-business. It involves planning, implementing, and monitoring changes to achieve organizational goals and objectives.

Resistance to Change in E-Business

- Resistance to change in e-business refers to the opposition or reluctance displayed by individuals or groups within an organization toward implementing new strategies, technologies, or processes. It is a natural response to change and can hinder the successful adoption of new initiatives.

The Role of Leadership in Change Management

- The role of leadership in change management encompasses the actions and behaviors exhibited by leaders to guide and support employees through the process of change in e-business. Effective leadership is crucial in fostering a positive change culture, addressing resistance, and ensuring the successful implementation of change initiatives.

Evaluation of Change Management in E-Business

- Evaluation of change management in e-business refers to the systematic assessment of the effectiveness and impact of change initiatives on the organization's performance and overall goals. It involves collecting and analyzing data to determine the extent to which the intended outcomes of change have been achieved and identifying areas for improvement.

Case Study

Company X, a traditional brick-and-mortar enterprise, has chosen to go digital by deploying an e-commerce platform. The company aims to expand its client base, increase its revenue, and improve its operational efficiency through Internet platforms. During the transition to e-business, the organization's procedures, technology, and culture must undergo significant changes. The management recognizes the importance of change management in assuring the successful adoption of e-business.

1. Identify and describe three potential causes of resistance to change that Company X employees may exhibit during the transition to e-business. How will the organization surmount each of these obstacles?
2. Discuss the steps Company X must take during the change management planning phase to ensure a smooth transition to e-business. How can the organization effectively identify and include key stakeholders in the planning process?
3. Company X recognizes the importance of effective communication during the change management process. Outline three recommended practices for communicating the e-business deployment-related changes to the organization's employees. How can the organization ensure its communications are understandable, expeditious, and pertinent?
4. Leadership is indispensable for the success of change management. Examine two effective leadership strategies that the management team of Company X

should employ to assist and guide employees through the transition to e-business. How can leaders cultivate trust and instill confidence during this revolutionary period?

5. Consider the benefits and drawbacks of implementing change management throughout the e-business transition at Company X. How will the efficacy of the organization's change management initiatives be measured and evaluated? Discuss the long-term effects of effective change management on Company X's future growth and competitiveness in the e-business environment.

Discussion

1. What is change management in e-business, and why is it important?
2. How can change management impact the success of e-business implementation?
3. What are some common challenges associated with change management in e-business?
4. How can organizations prepare for change management in e-business?
5. What is the change management process, and what are its key components?
6. How can effective planning contribute to successful change management in e-business?
7. What are some best practices for communication during the change management process in e-business?
8. How can training and education be used to support change management in e-business?
9. What are some common sources of resistance to change in e-business?
10. How can organizations identify and address resistance to change in e-business?
11. What strategies can organizations use to overcome resistance to change in e-business?
12. How can leadership support successful change management in e-business?
13. What are some effective leadership strategies for managing change in e-business?
14. How can organizations evaluate the effectiveness of their change management efforts in e-business?
15. What are the benefits of effective change management in e-business?
16. What are some potential drawbacks of change management in e-business?
17. How can organizations balance the need for change with the potential risks associated with change management in e-business?
18. What is the future of change management in e-business, and how might it evolve over time?
19. How can organizations stay up to date with best practices and emerging trends in change management for e-business?
20. What are some key considerations for organizations looking to implement effective change management processes for e-business?

Multiple-Choice Question

1. What is change management in the context of e-business?
   (a) A process of changing business models in e-business
   (b) A process of managing organizational change in the implementation of e-business
   (c) A process of managing technical changes in e-business systems
   (d) A process of managing financial changes in e-business projects
2. Why is change management important for successful e-business implementation?
   (a) It helps reduce costs in the implementation process.
   (b) It ensures smooth transitions and minimizes disruptions.
   (c) It increases profit margins for e-business organizations.
   (d) It automates business processes in e-business systems.
3. Which of the following is a step in the change management process?
   (a) Product development
   (b) Customer acquisition
   (c) Training and development
   (d) Market research
4. Which of the following is NOT a best practice for planning in change management?
   (a) Setting clear objectives and goals
   (b) Identifying key stakeholders
   (c) Communicating the change to employees
   (d) Implementing the change immediately without testing
5. What is a common source of resistance to change in e-business?
   (a) Lack of technological infrastructure
   (b) Employee motivation and engagement
   (c) Inadequate financial resources
   (d) Customer demand and preferences
6. Which strategy can help overcome resistance to change in e-business?
   (a) Ignoring employee concerns
   (b) Providing proper training and support
   (c) Implementing changes abruptly
   (d) Not involving employees in the change process

7. What is the role of leadership in successful change management in e-business?
   (a) To resist change and maintain the status quo
   (b) To communicate and inspire a shared vision
   (c) To delegate all change-related tasks to employees
   (d) To solely focus on financial outcomes of the change
8. Which of the following is an effective leadership strategy for change management in e-business?
   (a) Micromanaging employees during the change process
   (b) Promoting open communication and collaboration
   (c) Limiting employee involvement in decision-making
   (d) Implementing top-down directives without explanation
9. What are the benefits of change management in e-business?
   (a) Increased customer loyalty and retention
   (b) Higher employee turnover rates
   (c) Reduced efficiency in business processes
   (d) Limited market expansion opportunities
10. Which of the following is a drawback of change management in e-business?
   (a) Increased innovation and creativity
   (b) Resistance and pushback from employees
   (c) Improved customer satisfaction and loyalty
   (d) Streamlined decision-making processes
11. What does the future of change management in e-business look like?
   (a) Decreased importance and relevance
   (b) Greater reliance on traditional business models
   (c) Continued evolution and adaptation to technological advancements
   (d) Complete elimination due to automation and artificial intelligence
12. Which step of the change management process involves identifying and analyzing resistance?
   (a) Planning
   (b) Communication
   (c) Training
   (d) Evaluation
13. How can resistance to change in e-business be overcome?
   (a) By enforcing strict disciplinary measures
   (b) By providing ongoing training and support
   (c) By ignoring employee concerns and proceeding with the change
   (d) By avoiding communication with employees during the change process

14. What is one characteristic of effective leadership in change management?
    (a) Lack of transparency and communication
    (b) Top-down decision-making without employee input
    (c) Ability to inspire and motivate employees during the change process
    (d) Micro-management and excessive control over employees
15. What is one benefit of evaluating change management in e-business?
    (a) Identifying areas for improvement and making necessary adjustments
    (b) Eliminating the need for ongoing change initiatives
    (c) Minimizing employee involvement and feedback
    (d) Ignoring the impact of change on customer satisfaction

References

1. Beer, M. and N. Nohria, *Cracking the code of change*. Harvard business review, 2000. **78**(3): p. 133–141.
2. Hurley, R., *Transformational Change in Sales Organizations: A Special Case of Managing Change*. 2015. p. 237–242.
3. Tan, C.C., *The Theory and Practice of Change Management*. Asian Business & Management, 2006. **5**(1): p. 153–155.
4. Tidd, J., J.R. Bessant, and K. Pavitt, *Managing innovation: integrating technological, market and organizational change*. 1997, J. Wiley: Chichester [etc].

Chapter 14
Analysis and Performance Measurement in E-Business

Abstract In online business, conducting research and monitoring performance are crucial for achieving objectives. Understanding how a company operates facilitates informed decisions, operational improvements, and overall development. This chapter provides an overview of essential metrics and technologies for analysis and performance measurement, including web analytics, social media metrics, customer satisfaction evaluation, financial analysis, and performance measurement tools. Web analytics is indispensable for e-business analysis and performance evaluation. It provides valuable insights into website traffic, conversion rates, exit rates, and other critical metrics, enabling data-driven decision-making and operational enhancements. Social media metrics are equally important, offering information on engagement rates, reach, impressions, and other factors affecting social media platform performance. Measuring customer satisfaction is crucial for e-business performance evaluation. Feedback collection through surveys and other means allows businesses to identify areas for refinement and enhance the customer experience. Financial analysis is essential, providing insights into parameters like revenue, profit margins, and return on investment. By utilizing financial analysis tools such as ratio analysis and benchmarking, businesses can identify development opportunities and make informed decisions for enhanced financial performance. Effective utilization of resources increases the chances of success in the competitive e-business landscape.

1 Introduction to Analysis and Performance Measurement in E-Business

In today's rapidly evolving digital market, e-commerce companies must regularly evaluate their performance to maintain a competitive advantage. Analysis and performance measurement are crucial in e-business because they provide insight into how well the company operates and where it can be improved. These insights are indispensable for making prudent business decisions. By analyzing data, online

H. Taherdoost, *E-Business Essentials*, EAI/Springer Innovations in Communication and Computing, https://doi.org/10.1007/978-3-031-39626-7_14

businesses can identify patterns, trends, and correlations to help them make informed business decisions. Analysis can reveal, for example, which products or services are the most popular, which marketing strategies are the most effective, and which client categories generate the most revenue. This information may aid online businesses in modifying their strategies and allocating their resources more effectively.

Performance measurement is also an essential element of e-business because it enables businesses to track their progress toward achieving their objectives. When they have clearly defined, measurable objectives, e-businesses can determine if they are on the correct path and make any necessary adjustments. A company could, for instance, determine that they want to see a 10% increase in website traffic over the next 30 days. By measuring website traffic and comparing it to the goal, the company can determine if it needs to modify its marketing strategy or make other adjustments. This enables the company to determine whether or not the objective is achievable. Performance can be measured using a vast array of metrics, all of which are accessible to online businesses. Popular metrics include website traffic, conversion rates, customer acquisition cost, lifetime value, and return on investment (ROI). With the help of these metrics, which provide crucial data on numerous aspects of the company, e-businesses can identify areas where they can make improvements.

Amazon is an excellent example of the importance of conducting research and monitoring one's progress when managing an online business. Amazon is renowned for the importance of data analysis in the company's decision-making process. The company collects massive amounts of data regarding the actions of its customers, the number of website visitors, and the results of its sales and then uses this information to enhance its plans perpetually. For example, Amazon's recommendation engine, which makes product recommendations to users based on browsing and purchase history, is powered by complex algorithms that analyze vast amounts of client data. Another example would be eBay, which analyzes performance metrics to monitor the success of its vendor community. eBay considers a variety of metrics, such as sales volume, feedback ratings, and shipping timeframes, when determining the efficacy of its vendors. This data determines the most successful vendors, provides constructive feedback to sellers whose performance could be improved, and develops policies and initiatives to aid seller communities [1].

In addition to providing insights into performance, analysis and performance evaluation can assist e-businesses in identifying areas where they can improve efficiency and reduce costs. By identifying inefficiencies in their processes, e-businesses can increase productivity and decrease waste. For instance, an analysis may reveal that the production of a particular item takes significantly longer than anticipated, resulting in additional costs. The e-commerce company will be able to modify its production procedure, allowing it to become more cost-effective while simultaneously increasing productivity.

In the domain of e-business, performance evaluations and measurements are indispensable. In this era of digital advancement, a company has access to many indicators and technologies that can be used to monitor, analyze, and improve its performance. One of the most essential metrics in e-commerce is website traffic, which can be analyzed using programs like Google Analytics. This technology gives businesses access to vital information regarding the traffic generated by their websites, such as the total number of visitors, their location, and the pages they view.

In e-business, the conversion rate is an additional statistic that plays an important function. This metric measures the percentage of website visitors who complete a desired action, such as purchasing or filling out a form. Utilizing A/B testing tools, a company can increase its conversion rate by placing different website designs, layouts, and calls to action through its trials. E-business relies significantly not only on website-related data but also on social media-related data. Businesses may evaluate the efficacy of their social media marketing strategies by tracking engagement rate, reach, and click-through rate, among other metrics. With the assistance of software applications such as Hootsuite and Sprout Social, it is possible to monitor the metrics of a company's social media presence and make informed decisions regarding its social media strategy. Utilizing measurement and analysis tools is crucial for most significant facets of e-business, including email marketing. Tracking data such as the percentage of messages opened, the percentage of messages viewed, and the conversion rate can be used to determine the efficacy of an organization's email marketing efforts. Email marketing systems, such as Mailchimp and Constant Contact, equip businesses with the means to evaluate and enhance the efficacy of their email marketing campaigns.

In e-business performance evaluation, data visualization software is an indispensable tool. Using these tools, businesses can generate graphical representations of their data, such as graphs and charts, which can assist them in quickly identifying recurring patterns and trends. Tableau and Power BI are two extensively utilized examples of data visualization technologies. E-business performance monitoring also uses customer relationship management software or CRM software. These technologies enable businesses to manage consumer contacts and monitor client behavior, such as purchase history and preferences. Customer relationship management (CRM) tools, such as Salesforce and HubSpot, may aid businesses in enhancing their interactions with customers and targeting their marketing efforts more precisely toward specific customer segments.

Last but not least, search engine optimization (SEO) tools are crucial to the success of any online business. These tools aid businesses in optimizing their websites for search engines, which may lead to higher search engine rankings and increased website traffic. Moz and SEMrush are examples of well-known search engine optimization (SEO) applications. These provide organizations with valuable information regarding the SEO performance of their websites and development suggestions.

In summary, the surveillance and evaluation of performance are essential in the world of online business. Website traffic, conversion rate, social media metrics, email marketing metrics, data visualization software, customer relationship management software, and search engine optimization tools are among the metrics and tools available to businesses. Businesses can make informed decisions regarding their e-business strategy and remain competitive in today's swiftly evolving digital marketplace if they effectively measure and analyze their performance using the available tools. Table 14.1 compares critical metrics for website analysis and performance measurement.

Table 14.1 Comparison of key metrics for website analysis and performance measurement

Metric	Description	Tool
Conversion rate	The percentage of website visitors who complete a desired action, such as purchasing or filling out a form	Google Analytics
Click-through rate (CTR)	The percentage of people who click on a specific link compared to the number of people who view the page or email it is on	Google Analytics, email marketing software
Bounce rate	The percentage of website visitors who navigate away from the site after only viewing one page	Google Analytics
Customer lifetime value (CLTV)	The estimated value a customer will bring to a business over their lifetime	CRM software, analytics tools
Average order value (AOV)	The average amount of money customers spend per order	The e-commerce platform, analytics tools
Return on investment (ROI)	The ratio of profit or loss generated by an investment compared to the amount invested	Analytics tools, financial software
Churn rate	The percentage of customers who stop using a product or service within a given period	CRM software, analytics tools
Net promoter score (NPS)	A customer satisfaction metric measures the likelihood of customers recommending a business to others	Surveys, CRM software

2 Web Analytics

Web analytics is accumulating, measuring, and analyzing data associated with website traffic and user activity. Web analytics is a crucial component of online business and plays a role in e-commerce. This data is then used to make decisions based on the data and optimize websites to improve their performance and ultimately achieve corporate objectives. This section will discuss the significance of web analytics in e-business and how it can promote growth and success (Table 14.2).

One of the primary benefits of web analytics is that it provides companies with information about the traffic generated by their websites and user actions. Businesses can better understand how their website functions and identify areas where it can be improved by measuring the number of page visits, the exit rate, and the time spent. For instance, if a company observes that a particular page has a high exit rate, it can conduct additional research to determine the cause and modify the page to increase user engagement. Monitoring and assessing website visitors' actions is another essential aspect of web analytics. This category includes click-through rates, conversion rates, and the path visitors take through a website. Through the analysis of this data, businesses can gain insights into the preferences and behavior of users and make data-driven decisions to optimize their websites for enhanced user experience and increased conversions [2].

Table 14.2 Comparison of key web analytics metrics for website

Metric	Definition	Importance	Calculation	Example	Actionable insight
Website traffic	Total number of visitors to a website during a specific period	It helps businesses understand the size of their audience and how it changes over time	Number of unique visitors during a specific period	10,000 unique visitors in March	Increase website traffic by improving SEO and social media promotion
Pageviews	Number of times a specific page on a website is viewed	It helps businesses understand how users engage with a website's content	Number of pageviews on a specific page during a specific period	5000 pageviews on the homepage in March	Optimize popular pages for better engagement and user experience
Bounce rate	Percentage of visitors who leave a website after viewing only one page	Indicates the website's content or user experience is not engaging	Number of single-page visits divided by the total number of visits	40% bounce rate in March	Improve website usability and relevance to reduce bounce rate
Average session duration	The average length of time visitors spend on a website	It helps businesses understand the level of engagement visitors have with the website's content	The total duration of all sessions divided by the number of sessions	Average session duration of 2 min in March	Create more engaging and relevant content to increase session duration
Conversion rate	Percentage of visitors who complete a desired action on a website	Measures the effectiveness of the website in converting visitors into customers or leads	The number of conversions divided by the total number of visits	5% conversion rate for purchases in March	Optimize website design and user experience to increase conversion rate
Exit rate	Percentage of visitors who leave a website after viewing a specific page	It helps businesses identify pages that are causing visitors to leave	Number of exits from a specific page divided by the total number of pageviews	20% exit rate on the product page in March	Improve product page content and user experience to reduce exit rate

(continued)

Table 14.2 (continued)

Metric	Definition	Importance	Calculation	Example	Actionable insight
Click-through rate (CTR)	Percentage of users who click on a link or call to action (CTA) on a website	Measures the effectiveness of CTAs and website design in encouraging user engagement	The number of clicks on a link or CTA divided by the total number of impressions	3% CTR on a "learn more" CTA in March	Improve CTA design and placement to increase CTR
Referral traffic	Number of visitors who come to a website from another website	It helps businesses understand which external sources are driving traffic to their website	Number of visitors from a specific external source during a specific period	500 visitors from a blog post in March	Invest more in external sources that generate significant referral traffic
Social media engagement	Measures the level of interaction users have with a business's social media content	It helps businesses understand the effectiveness of their social media strategy in engaging their audience	Number of likes, comments, and shares on a social media post	100 likes, 50 comments, and 20 shares on a Facebook post in March	Create more relevant and exciting social media content to increase engagement

Web analytics can also be used to measure the effectiveness of marketing strategies. Through conversion monitoring and subsequent analysis of the collected data, businesses can determine which campaigns produce the most conversions and adjust their marketing strategy accordingly. Using a pay-per-click (PPC) advertising campaign as an example, a company can monitor the conversion rate for each advertisement and make any necessary adjustments to increase the campaign's effectiveness. In addition to monitoring website traffic and user behavior, web analytics can also be used to track and evaluate the metrics of social media platforms. Suppose businesses measure metrics such as engagement rates, click-through rates, and follower growth. In that case, they can gain insights into the efficacy of their social media strategy and make data-driven decisions to maximize their social media presence.

Google and Amazon are concrete examples of the significance of web analytics to the operation of an online business. For instance, Amazon uses web analytics to track user activity and customize each customer's shopping experience based on their preferences. Amazon can recommend products tailored to each user's preferences because the company analyzes user data, including search and purchase history. This increases the likelihood that a user will purchase something from Amazon. Similarly, Google employs web analytics to monitor and analyze user activity across numerous platforms, such as YouTube and its search engine. This information is used to improve the relevance of search results and provide users with an overall better experience. Using Google's web analytics tools, businesses can measure and analyze the traffic to their websites as well as the behavior of their users. Google Analytics is an example of such a tool.

Web analytics is collecting data from websites and monitoring, analyzing, and reporting on that data to understand better and enhance website performance. Web analytics provides organizations with the data to make informed decisions regarding their online presence and marketing strategies. This section will discuss ten of the most critical metrics in web analytics and how these indicators can help organizations improve their online performance.

Traffic on a website refers to the total number of users who have visited a website during a given period. It is a crucial metric because it enables businesses to understand the magnitude of their audience and how it varies over time. Monitoring website traffic can help organizations identify patterns and trends in user behavior, which can then be used to inform website optimization and marketing strategies. For example, if the number of visitors to a website decreases after it has been redesigned, this may indicate that the new design could be more user-friendly.

Pageviews are the number of times a website's visitors have viewed a page. This metric is beneficial for determining how consumers interact with the content presented on a website. A high number of visits to a page may indicate that the content on that page is well liked. Businesses can use this information to optimize their website's content and improve the user experience.

The percentage of website visitors who view only one page before exiting the site is known as the "bounce rate." A high exit rate may indicate that the website's content or user experience could be more engaging, which may explain why visitors leave the site quickly. Companies may use this metric to determine whether certain aspects of their websites, such as page loading performance or content relevance, require refinement.

The average duration of a visitor's session (average session duration) refers to the average quantity of time a visitor spends on a website. Using this metric, businesses can better comprehend how engaged website visitors are with the website's content. For instance, if the average duration of a session is short, this may indicate that the information could be more exciting or relevant to the visitors' requirements.

The website's conversion rate is the percentage of site visitors who complete a desired action, such as purchasing or filling out a questionnaire. This metric is essential for businesses, as it measures the website's ability to convert visitors into customers or prospects. If a website's conversion rate is low, it could indicate that the site's design, content, or user experience needs improvement.

The exit rate is the proportion of website visitors who leave after viewing a page, also referred to as the rebound rate. This indicator helps businesses determine which pages on their website are causing users to leave. For instance, if the exit rate on a product page is high, this may indicate that the page is ineffective at convincing visitors to purchase.

Click-through rate (CTR) is the percentage of users who click on a website's links or calls to action (CTA). This metric is necessary for determining whether a website's calls-to-action (CTAs) and layout effectively drive user interaction. This measure enables businesses to enhance the design of their websites and the calls-to-action (CTAs) that appear on those websites.

Referral traffic refers to the number of visitors arriving at a website by selecting a link on another website. This measure helps businesses identify the numerous external sources responsible for sending website visitors to their domain. For instance, if a company discovers that a particular blog or social media platform generates a large amount of referral traffic, it can invest more in that platform to increase brand exposure.

Engagement on social media pertains to the quantity of interaction with a company's social media content. This participation may manifest as favorites, remarks, and shares. This indicator can help businesses determine if their social media engagement strategy is effectively attracting the attention of their target audience. For example, a high engagement rate may indicate that the audience finds the information pertinent and exciting.

CPA is the expense of acquiring a new consumer or prospect through a specific marketing channel, such as pay-per-click (PPC) advertising or email marketing. The cost of acquiring a new consumer or prospect is denoted by CPA. This indicator can assist businesses in gauging the efficacy and efficiency of their marketing initiatives and identifying which channels offer the highest return on investment.

3 Social Media Metrics

Social media platforms are essential for any e-business to build brand awareness, connect with its audience, and drive sales. However, measuring the success of social media campaigns can be challenging. This section will explain ten critical metrics used to measure social media performance in e-business and how they can help businesses improve their social media strategy [2].

1. Reach: Reach is the number of unique users who see a post or piece of content on social media. This metric is crucial as it helps businesses understand the size of their audience and how many people they can reach with their content. A high reach can indicate that the content is engaging and shareable, and businesses can use this information to optimize their content and improve their social media strategy.
2. Engagement: Engagement is the number of interactions users have with a business's social media content, such as likes, comments, shares, and follows. This metric is essential as it helps businesses understand how their content resonates with their audience. A high engagement can indicate that the content is relevant and exciting to the audience, and businesses can use this information to create more engaging content.
3. Impressions: Impressions are the total number of times a post or piece of content is viewed on social media. This metric helps understand how often the audience sees the content. Strong impressions indicate that many people share and view

the content. Businesses can use this information to improve their content strategy and reach a larger audience.

4. Click-through rate (CTR): The click-through rate is the percentage of users who click on a link or call to action (CTA) on a social media post. This metric is essential for measuring the effectiveness of CTAs and content in encouraging user engagement. Businesses can use this metric to optimize their content and CTAs for higher click-through rates, leading to increased website traffic and conversions.
5. Conversion rate: Conversion rate is the percentage of users who complete a desired action, such as making a purchase or filling out a form, after clicking on a social media post. This metric is essential for businesses as it measures the effectiveness of their social media campaigns in converting users into customers or leads. A low conversion rate can indicate that the content or landing page needs improvement. In contrast, a high conversion rate indicates that the content and landing page effectively drive conversions.
6. Audience demographics: Audience demographics refer to the audience's characteristics, such as age, gender, location, and interests. This metric is essential for businesses as it helps them understand their audience and create relevant and engaging content. For example, if a business's audience is primarily female and interested in fashion, they can create content that appeals to those interests and demographics.
7. Sentiment analysis: Sentiment analysis is the process of analyzing social media mentions to determine the overall sentiment of the audience toward a brand or product. This metric is essential for businesses as it helps them understand how their audience perceives their brand. For example, if a business notices a high volume of negative sentiment, it can take steps to address any issues and improve its brand reputation.
8. Share of voice (SOV): Share of voice is the percentage of mentions or conversations about a brand compared to its competitors. This metric is crucial as it helps businesses understand their position in the market and how they are perceived relative to their competitors. A higher share of voice can indicate that a business is more visible and top-of-mind with its audience, while a lower share of voice can indicate a need for improved brand awareness and visibility.

Every online business must have a social media presence to increase brand awareness, engage its audience, and generate sales. However, measuring the efficacy of social media campaigns can be challenging. In the following section, we will discuss the ten main metrics used to evaluate the effectiveness of social media in e-business, as well as how these metrics can help businesses improve their social media strategy (Table 14.3).

1. Reach: The platform's reach is the number of unique individuals who view a particular social media post or content. This statistic is crucial because it enables businesses to comprehend better the size of their audience and the potential number of people their content can reach. Businesses can use this data to opti-

Table 14.3 Comparison of key social media metrics

Metric	Definition	Importance	Calculation	Example	Actionable insight
Reach	The number of unique users who see a particular post or piece of content on social media	Indicates the size of the potential audience that the content can reach	Count of unique users who has viewed a post or content	A post reaches 10,000 unique users	Optimize content and improve social media strategy to increase the reach of the audience
Engagement	The number of interactions users have with a business's social media content	Indicates how well the content resonates with the audience and its relevance	Likes, comments, shares, and follows received by posts or content	A post receives 500 likes, 100 comments, 200 shares, and 50 follows	Use the insights to create engaging content that resonates well with the audience
Impressions	The total number of times a particular post or content is viewed on social media	It helps businesses understand how often the audience sees the content	Total views of a post or content	A post has 20,000 views	Identify the type of content that receives more impressions and optimize the content strategy
Click-through rate	The percentage of users who click on a link or call to action (CTA) on a social media post	Measures the effectiveness of CTAs and content in encouraging user engagement	(Number of clicks/number of impressions) × 100	A post receives 2000 impressions and 100 clicks	Optimize content and CTAs to improve the click-through rate and drive website traffic and conversions
Conversion rate	The percentage of users who complete a desired action after clicking on a social media post	Measures the effectiveness of social media campaigns in converting users into customers or leads	(Number of conversions/ number of clicks) × 100	A post receives 100 clicks, and 10 users complete the desired action	Analyze the content and landing page to improve the conversion rate and drive more sales or leads

(continued)

Table 14.3 (continued)

Metric	Definition	Importance	Calculation	Example	Actionable insight
Audience demographics	The characteristics of the audience, such as age, gender, location, and interests	It helps businesses understand their audience and create content that is relevant and engaging	Analysis of demographic data of the audience	50% of the audience is female and interested in fashion	Create relevant and engaging content for the audience based on their demographics and interests
Sentiment analysis	The process of analyzing social media mentions determines the overall sentiment of the audience toward a brand or product	It helps businesses understand how their audience perceives their brand	Analysis of the tone of social media mentions related to the brand or product	70% of social media mentions related to a brand are positive	Take steps to address negative sentiment and improve brand reputation

mize their content and improve their social media strategy by using high reach to indicate engaging and shareable content.

2. Engagement: Engagement is the number of times a user interacts with a company's social media content. Likes, comments, shares, and followers are examples of possible interactions. This statistic is significant because it helps businesses determine how well their content resonates with their intended audience. A high level of interaction may indicate that the content is engaging and pertinent to the audience, and companies can use this information to create even more engaging content.
3. Impressions: An impression is the cumulative number of times a piece of content or a post has been viewed on a social media platform. Using this metric, one can obtain a deeper comprehension of how frequently the audience views the content. Many impressions indicate that many individuals are sharing and viewing the content. This information can be utilized by businesses to improve their content strategy and expand their audience reach.
4. CTR: Click-through rate (CTR) is the percentage of users who click on a link or call to action (CTA) in a social media post. This metric determines whether calls to action (CTAs) and content successfully encourage user interaction. This metric enables businesses to enhance their content and calls to action (CTAs) to achieve higher click-through rates, increasing website traffic and conversions.
5. Rate of conversion: After engaging in a social media post, the conversion rate is the proportion of users who take the desired action, such as making a purchase or filling out a form. This category includes activities such as completing a form or making a purchase. This metric is crucial for businesses, as it measures the effectiveness of their social media marketing in converting consumers into customers or prospects. A high conversion rate may indicate that the content and

landing page generate conversions effectively. In contrast, a low conversion rate may indicate that the content and landing page need improvement.

6. Audience demographics: "Audience demographics" refers to the audience's characteristics, such as age, gender, location, and interests. In particular, a target audience may be defined by age, gender, location, or interests. This statistic is essential for businesses because it enables them to comprehend their audience better and provide relevant and engaging content. For example, if most of a company's audience consists of females interested in fashion, the company may create content that appeals to both demographics and interests.
7. Sentiment analysis is the process of monitoring social media mentions to determine the general attitude of a target audience toward a brand or product. This may be accomplished by examining how individuals feel about the brand or product. This statistic is crucial for businesses because it allows them to comprehend better how their target audience perceives their brand. For instance, a company can improve its brand reputation by addressing any issues brought to its attention when they discover significant negative sentiment.
8. Share of voice (SOV): Share of voice refers to the ratio of the number of times a brand is mentioned or discussed relative to the number of times its competitors are mentioned or discussed. This statistic is essential because it enables businesses to comprehend better their position in the market and how they are perceived compared to their competitors. There may be room for development regarding brand recognition and visibility if the proportion of discourse is minimal. On the other hand, a larger share of discourse may indicate that a company is more visible and prominent in the minds of its target audience.

Social media platforms have progressed to the point where they are integral to every business's online presence. Because there are so many social media platforms to choose from, businesses must monitor the performance of each one before making informed decisions regarding their social media marketing strategy. Metrics for social media platforms provide businesses with insightful information about their performance on these platforms, allowing them to improve their online presence and overall success. One of the essential factors to consider when using social media is reach. "Reach" refers to the total number of people who have viewed a company's social media post. This indicator is crucial because it demonstrates how far a company's message has extended and how many people it has reached, as it signifies the message's extent of dissemination. For instance, if a company has a low reach, it may indicate that the content they share on social media must resonate with the target audience. Engagement is another essential social media metric to monitor. Engagement is a metric that measures how many people interact with a company's social media content. This interaction could consist of favorites, remarks, and shares. This indicator is significant because it demonstrates how interested people are in the content a company creates and how likely they are to interact with it. For example, a high engagement rate may indicate that the social media content shared by a company is engaging with its target audience, which may lead to increased brand exposure and potential sales.

Despite being equally important, sentiment is a social media metric often overlooked. Based on social media interactions, sentiment analysis determines how individuals feel about a company. This metric is crucial because it reveals how people perceive a company and whether they have positively interacted with it. For example, if a company's sentiment is consistently negative, it may indicate they need to modify their social media strategy to improve their brand image. If the sentiment is consistently optimistic, this may be the same. In addition to these data, businesses must track the number of individuals who follow them. The number of individuals who follow a company's social media accounts is a reliable indicator of the company's overall popularity and reach. For example, if a company has many followers, this may indicate that the content they share on social media intrigues the target audience. This may result in enhanced brand recognition and increased sales potential. Another essential social media metric to monitor is the click-through rate. The click-through rate is the proportion of people who view an online advertisement or social media post from a company and then click through to that company's website or landing page. This number is significant because it indicates how effectively a company's social media content attracts website visitors. For example, suppose a company's click-through rate is low. In that case, the content it shares on social media must be more engaging and intriguing to encourage users to visit the company's website. Metrics derived from social media platforms also provide businesses valuable insight into the consumers they are attempting to reach. By analyzing demographic data, such as age, gender, location, and interests, businesses can gain a better understanding of their target audience. These factors can help businesses determine the interests of their intended audience. With this information, businesses can engage their target audience more effectively and increase their overall success by tailoring the content they post on social media. Lastly, businesses must track their social media efforts' ROI (return on investment). ROI is a metric that measures how much a company earns back from its social media investments. This metric considers both financial and nonfinancial benefits. Suppose a company decides to invest in social media advertising, for example. In that case, it can calculate the return on its investment by tallying the sales or leads generated by the advertisement. A greater comprehension of the return on investment (ROI) afforded by social media may assist businesses in making more informed decisions regarding their social media investment and strategy.

4 Customer Satisfaction Measurement

The level of consumer satisfaction with a company's products or services can be measured using a technique known as customer satisfaction measurement. This metric is essential for online businesses because it enables them to understand their clientele's needs and preferences better, thereby allowing them to improve the quality of their products or services and establish stronger customer relationships. In this

Table 14.4 Methods for measuring customer satisfaction in e-business

	Method	Description	Pros	Cons
1.	Surveys	Collecting feedback through questionnaires or surveys	Provides specific feedback, can be tailored to measure specific aspects of the customer experience	Response rates may be low; survey bias can occur
2.	Net promoter score (NPS)	Measures customer loyalty by asking customers how likely they are to recommend a company to others	Simple and easy to understand, it provides a single score for benchmarking	Limited in scope, does not provide in-depth feedback
3.	Customer effort score (CES)	Measures the ease of a customer's experience with a company, product, or service	Focuses on the customer experience, helps identify pain points in the customer journey	Limited to specific touchpoints or interactions
4.	Online reviews and ratings	Collecting feedback from online platforms such as Google reviews or Yelp ratings	Provides public feedback for potential customers to view, allows for prompt response to negative feedback	Biased toward extreme ratings, potential for fake reviews
5.	Social media monitoring	Monitoring social media platforms for feedback about a company, product, or service	Provides real-time feedback, allows for prompt response to negative feedback	Limited in scope, does not provide in-depth feedback
6.	Customer interviews	Conducting one-on-one interviews with customers	Provides in-depth feedback and understanding of the customer experience, allows for follow-up questions	Time-consuming and expensive
7.	Focus groups	Bringing together a group of customers to discuss their experiences	Provides a variety of perspectives, allows for follow-up questions	Time-consuming and expensive, group dynamics can bias results
8.	Complaint tracking	Tracking and analyzing customer complaints and feedback	Identifies recurring issues, allows for prompt response to negative feedback	Limited to negative feedback, does not measure overall satisfaction
9.	Customer churn rate	Measures the percentage of customers who stop using a company's products or services	Provides a measure of overall satisfaction and loyalty	Limited in scope, does not provide in-depth feedback

section, we will discuss the various methods for measuring consumer satisfaction and the importance of doing so in the context of online enterprises (Table 14.4).

Measuring consumer satisfaction with a company's products or services is crucial to the success of an online business for various reasons. However, one of the most important is that it enables businesses to identify weak points in their offerings. Businesses may better understand which aspects of their products or services fall short of consumers' expectations if they collect customer feedback and use it to make the necessary adjustments. For instance, if customers complain that an e-commerce website's checkout procedure is too complicated, the website proprietor may decide to modify the procedure to simplify it for customers and thus improve the overall quality of the customer experience. Measuring consumer satisfaction is essential in e-business for various reasons, including enabling businesses to develop more meaningful interactions with their clientele. Businesses can increase customer loyalty and retention by demonstrating their customers' feedback is valued. In addition, delighted consumers are more likely to tell others about their positive experiences with the company, which may increase brand recognition and sales. For instance, a satisfied online consumer may leave a positive review on the company's website or one of its social media platforms. These assessments have the potential to attract new customers [3].

When it comes to e-business, there are a variety of methods for measuring the level of consumer satisfaction. The use of questionnaires that can be completed online or via email is one of the most common methods. Typically, a survey consists of queries designed to evaluate various aspects of the consumer experience. These elements may include product quality, customer service, and website usability. The responses can then be analyzed to identify patterns and areas that require improvement. Social media monitoring tools are yet another method online businesses use to gauge customer satisfaction. These technologies enable businesses to monitor and analyze customer comments and sentiments on social media platforms like Twitter, Facebook, and Instagram. Observing the conversations on social media platforms enables businesses to quickly identify any issues or concerns their customers may have and take immediate action to address them. Utilizing website analytics is a third method employed in e-business to determine consumer satisfaction. Website analytics solutions may provide organizations with information regarding consumer behavior, such as the time customers spend on a website and the most frequently visited pages. By analyzing this data, businesses can determine which aspects of their websites may be causing consumers to become frustrated or perplexed and then make the necessary modifications to improve the overall quality of the user experience.

Customer feedback collection is one of the most crucial aspects of operating a prosperous online business. Customer satisfaction metrics provide organizations with a means of gauging how effectively they are meeting the needs and expectations of their clientele. The following paragraphs will discuss ten of the most common methods for measuring consumer satisfaction.

1. Surveys: Surveys are a popular method for measuring customer satisfaction. They can be conducted online or offline and are designed to gather specific information about customers' experiences and opinions. For example, a business may rate customers' satisfaction with a product or service on a scale of 1–10.
2. Net promoter score (NPS): The net promoter score is a metric used to measure customer loyalty. Customers are asked to rate on a scale of 0–10 how likely they are to recommend the business to others. The NPS score is calculated by sub-

tracting the percentage of detractors (customers who give a rating of 0–6) from the percentage of promoters (customers who give a rating of 9–10).

3. Customer feedback forms: Customer feedback forms are a simple and effective way to gather customer feedback. These forms can be placed on a business's website or handed out in-store. They can gather specific information about customers' experiences with a product or service.
4. Social media listening: Social media platforms like Facebook, Twitter, and Instagram allow businesses to listen to customer feedback in real time. By monitoring social media conversations, businesses can identify areas where they excel and need to improve.
5. Customer reviews: Customer reviews are an essential source of feedback for e-businesses. These reviews can be collected through online platforms like Google, Yelp, or Amazon. Reviews can provide businesses with valuable insights into customer preferences and opinions.
6. Online analytics: Online analytics tools like Google Analytics can provide businesses with valuable insights into customer behavior on their website. This information can be used to identify areas of the website that need improvement, such as slow load times or confusing navigation.
7. Customer retention rate: Customer retention rate measures the percentage of customers who continue to do business with a company over a specific period. A high retention rate indicates customers are satisfied with the products or services offered.
8. Complaints and resolutions: Complaints and resolution data can provide businesses with valuable information about areas where they need to improve. By tracking customer complaints and how they are resolved, businesses can identify areas where they are falling short and take steps to improve.
9. Focus groups: Focus groups involve gathering a group of customers together to discuss their experiences with a business's products or services. These groups can provide businesses with in-depth insights into customer behaviors and preferences.
10. Customer churn rate: Customer churn rate measures the percentage of customers who stop doing business with a company over a specific period. A high churn rate indicates customers are unsatisfied with the products or services offered.

In conclusion, measuring customer satisfaction is a critical aspect of any e-business. By combining these ten methods, businesses can gain valuable insights into customer preferences and behaviors. This information can be used to improve products and services, increase customer loyalty, and ultimately drive business success.

5 Financial Analysis in E-Business

In the world of e-business, financial metrics are vital in measuring the performance and success of a business. E-businesses can make informed decisions about their financial strategies, investments, and growth opportunities by analyzing these metrics. In this section, we will discuss nine critical financial metrics that are commonly used to measure e-business performance (Table 14.5).

1. Gross revenue: Gross revenue refers to the total revenue generated by an e-business from all sources before any deductions or expenses. It is a critical metric that overviews a business's overall performance. For example, if an e-business generates $1 million in gross revenue, it shows a significant market share and generates substantial sales.
2. Net revenue: Net revenue is an e-business's revenue after all deductions and expenses have been accounted for. This metric gives a more accurate picture of a business's profitability. For example, if an e-business generates $1 million in gross revenue but has $200,000 in expenses, its net revenue would be $800,000.
3. Gross profit margin: Gross profit margin is the percentage of revenue that a business retains after deducting the cost of goods sold (COGS). It measures how efficiently an e-business produces and sells its products or services. For example, if an e-business generates $100,000 in revenue and has $60,000 in COGS, its gross profit margin would be 40%.
4. Net profit margin: Net profit margin is the percentage of revenue that a business retains after all expenses have been accounted for, including taxes, interests, and other deductions. It measures how efficiently an e-business manages its overall costs. For example, if an e-business generates $1 million in revenue and has $600,000 in expenses, its net profit margin would be 40%.
5. Return on investment (ROI): ROI measures the return on the investment made by the business. It is the ratio of the profit or loss made on an investment relative to the amount invested. For example, if an e-business invests $50,000 in a marketing campaign and generates $150,000 in sales, its ROI would be 200%.
6. Customer acquisition cost (CAC): Acquiring a new customer or lead through a specific marketing channel, such as pay-per-click (PPC) advertising or email marketing. It is a critical metric that measures the efficiency of an e-business's marketing efforts. For example, if an e-business spends $5000 on a PPC advertising campaign and acquires 500 new customers, its CAC would be $10.
7. Lifetime value of a customer (LTV): LTV is the total revenue a business can expect to generate from a customer throughout their relationship. It is a critical metric that helps businesses determine the long-term value of their customers. For example, if an e-business's average customer spends $1000 annually and stays with the business for five years, their LTV would be $5000.
8. Average order value (AOV): AOV is the average value of an order or purchase made by a customer. It is a critical metric that measures the revenue generated per customer order. For example, if an e-business generates $100,000 in revenue from 1000 orders, its AOV would be $100.

Table 14.5 Key financial metrics used to measure e-business performance

Metric	Definition	Importance	Calculation	Example	Actionable insight
Gross revenue	Total revenue generated from sales before accounting for any expenses	Measures the overall financial performance of the business	Gross revenue = units sold × price per unit	An e-commerce business generated $100,000 in revenue by selling 1000 units of a product for $100 each	Increase sales volume or price per unit to increase revenue
Net revenue	Total revenue generated from sales after accounting for any returns or discounts	Measures the actual revenue the business earns from sales	Net revenue = gross revenue – returns/discounts	A business had $100,000 in gross revenue, but $10,000 in returns and discounts, resulting in $90,000 in net revenue	Monitor returns and discounts to ensure they do not significantly impact revenue
Gross profit margin	The percentage of revenue that is left after accounting for the cost of goods sold	Measures the profitability of the business's sales	Gross profit margin = (gross revenue - cost of goods sold) / gross revenue × 100	A business had $100,000 in gross revenue and $60,000 in cost of goods sold, resulting in a gross profit margin of 40%	Increase prices or decrease costs of goods sold to improve gross profit margin
Net profit margin	The percentage of revenue that is left after accounting for all expenses, including operating expenses	Measures the overall profitability of the business	Net profit margin = net income/total revenue × 100	A business had $100,000 in revenue, $70,000 in expenses, and $10,000 in taxes, resulting in a net income of $20,000 and a net profit margin of 20%	Monitor and control expenses to maintain or increase net profit margin
Return on investment (ROI)	The percentage of return earned on the investment	Measures the profitability of an investment relative to its cost	ROI = (gain from investment – cost of investment)/ cost of investment × 100	A business invested $10,000 in a marketing campaign and generated $30,000 in sales, resulting in an ROI of 200%	Monitor the ROI of investments to determine the most effective use of resources

(continued)

Table 14.5 (continued)

Metric	Definition	Importance	Calculation	Example	Actionable insight
Customer acquisition cost (CAC)	The cost of acquiring a new customer	Measures the cost-effectiveness of marketing and sales efforts	CAC = total marketing and sales costs/ number of new customers acquired	A business spent \$20,000 on marketing and sales efforts and acquired 100 new customers, resulting in a CAC of \$200	Monitor CAC to ensure it does not exceed customer lifetime value
Customer lifetime value (CLV)	The total value a customer brings to a business over their lifetime	Measures the long-term profitability of acquiring and retaining customers	CLV = average value of Sale × number of repeat transactions × average retention time	A business has an average sale value of \$50, 5 repeat transactions per year, and an average customer retention time of 3 years, resulting in a CLV of \$750	Increase CLV by increasing the value of each sale, increasing repeat transactions, or increasing customer retention time

9. Conversion rate: Conversion rate is the percentage of website visitors who complete a desired action, such as purchasing or filling out a form. It measures the effectiveness of an e-business's website and marketing strategies in converting visitors into customers.

Financial analysis instruments are required for determining an online company's financial condition. They aid business proprietors and administrators in making informed decisions, evaluating performance, and identifying areas for improvement. This section will cover online enterprises' most common financial analysis instruments (Table 14.6).

Income statement: An income statement is a financial statement that depicts a company's sales and expenses over a specified period, such as a quarter or a year. A profit and loss statement is another name for an income statement. It assists businesses in determining their net income and profitability. For instance, a business may use an income statement to determine if introducing a new product was successful.

Balance sheet: A balance sheet is a snapshot of a company's financial status at a given instant and is used to assess the viability of a business. It depicts the assets, liabilities, and equity of a business at a specific period in time. This instrument can help businesses better understand their liquidity, solvency, and financial stability. A balance sheet is one instrument a company can use to determine whether or not it has adequate assets to meet its obligations.

Table 14.6 Standard financial analysis tools used in e-business

Tool	Description	Benefits	Limitations
Financial ratios	A comparison of financial data that can be used to assess a company's performance over time or against competitors	Easy to calculate and understand. Can identify strengths and weaknesses in a company's financial performance	Accounting practices can distort ratios. Comparing ratios between companies in different industries can be difficult
Trend analysis	Examines changes in a company's financial data over time to identify patterns and trends	It can help identify patterns and trends in a company's financial performance. It can help predict future performance	It only shows a company's financial history, not its current situation. Predictions are only sometimes accurate
Cash flow analysis	A tool that measures a business's cash inflows and outflows to determine its liquidity	It helps identify cash flow problems before they become critical. It can help identify areas where cash inflows can be increased or outflows can be reduced	Does not consider noncash items like depreciation. It can be challenging to interpret if the business has multiple revenue streams
Break-even analysis	A tool used to determine the point at which a company's revenue will cover its expenses	It can be used to evaluate the impact of changes in revenue or costs on the company's profitability. It can help identify the minimum level of sales necessary to break even	Assumes that all costs are fixed or variable. Does not consider the impact of changes in market conditions or competition
Return on investment (ROI)	A measure of the profit earned from an investment, expressed as a percentage of the investment	Useful for evaluating the profitability of investment opportunities. It can help identify the most profitable investments	Assumes that all costs and revenues are accurately measured. Does not consider nonfinancial benefits
Net present value (NPV)	A tool evaluates an investment's profitability by comparing the present value of expected cash inflows to the present value of expected cash outflows	Takes into account the time value of money. It helps evaluate the profitability of long-term investments	Requires accurate forecasting of future cash flows. Assumes that all cash flows can be accurately estimated
Payback period	A tool determines the time it takes for an investment to generate enough cash inflows to recover its initial cost	Helpful in evaluating the time it takes to recover an investment	Does not consider the time value of money. Assumes that all cash flows occur at the end of each period

Cash flow statement: A cash flow statement is a financial document that summarizes the capital inflows and disbursements of a business over a specified period. This instrument helps businesses monitor their current financial situation and better understand their ability to generate cash from their activities. Using a cash flow statement, for instance, a business can determine if it has sufficient cash to cover its operational expenses.

Ratio analysis is a technique that involves comparing two or more financial measures in order to gain insight into an organization's financial health. Ratios can facilitate the evaluation of a company's liquidity, profitability, and efficiency. For instance, a business may use a gross margin ratio to determine the profit it generates from its sales. This would permit them to determine their income.

Profitability analysis: A break-even analysis is a technique that helps businesses determine when they will begin to generate a profit. This can be determined by comparing the company's current condition to its historical averages. It facilitates businesses in determining the level of revenue required to cover their expenses. For example, a business may conduct a break-even analysis to determine how many units of a product it must sell to recoup its fixed and variable expenses.

Return on investment (ROI) is a metric that measures the profitability of an investment by comparing the amount of money acquired to the amount of money invested. This instrument facilitates businesses in determining the performance of their investments and identifying those investments with the maximum return on investment. For instance, a company may use ROI to determine whether or not to invest in a brand-new marketing campaign.

Payback period: The amount of time it takes for a business to recoup its initial investment in a specific undertaking. This utility helps businesses determine whether a project is financially feasible and how long it will take to generate a return on investment. Before deciding whether or not to invest in a brand-new production facility, for instance, a business may consider the amount of time it would take to recoup its initial investment.

The term "net present value," or "NPV," refers to a method that helps organizations evaluate the prospective profitability of an investment by comparing the present value of the venture's cash inflows and cash disbursements. NPV is also referred to as "net present worth." For instance, a company might use NPV to determine if it should invest in a brand-new technological platform.

IRR: Internal rate of return IRR is a technique that aids businesses in determining the prospective profitability of an investment by calculating the rate at which cash inflows equal cash disbursements. This assists businesses in determining whether an investment is likely to be profitable. For instance, a business may use IRR to determine whether it should invest in a new product line.

Sensitivity analysis involves evaluating multiple circumstances to determine the potential impact that changes in significant factors, such as sales volume or product cost, may have on a company's financial performance. In order to better plan future pricing decisions, a company may use sensitivity analysis to determine how variations in the cost of a primary material would impact its profit margin.

6 Performance Measurement Tools

In today's fast-paced business world, the success of e-businesses is heavily dependent on their ability to measure their performance and make data-driven decisions effectively. Using performance measurement tools and best practices to gain valuable insights and improve business outcomes is essential. This section will discuss the best practices for using performance measurement tools to drive e-business success (Table 14.7).

1. Web analytics: Web analytics is a tool used to measure and analyze website visitors' behavior. It can provide insight into how visitors interact with a website, how long they stay on a page, and which pages are most popular. Google Analytics is a popular web analytics tool that many businesses use to track their website performance.
2. Conversion rate optimization (CRO) tools: CRO tools optimize a website's conversion rate. They provide insight into how visitors engage with a website and suggest changes that can be made to improve the conversion rate. Examples of CRO tools include Optimizely and Unbounce.
3. Heat maps: Heat maps are a visualization tool showing where website visitors click, scroll, and spend the most time. They can be used to identify areas of a website that are popular and areas that may need improvement. Examples of heat map tools include Crazy Egg and Hotjar.
4. Social media analytics: Tools measure and analyze social media performance. They can provide insight into the effectiveness of social media campaigns and information on engagement and audience demographics. Examples of social media analytics tools include Hootsuite Insights and Sprout Social.
5. Customer feedback tools: Customer feedback tools gather customer feedback about their experience with a product or service. They can provide valuable insights into customer needs and preferences and areas for improvement. Examples of customer feedback tools include SurveyMonkey and Qualtrics.
6. Key performance indicators (KPIs): KPIs are metrics used to measure the success of a business or specific marketing campaign. Examples of KPIs include revenue growth, customer acquisition cost, and return on investment (ROI).
7. Search engine optimization (SEO) tools: SEO tools are used to optimize a website's content for search engines. They can provide insight into keyword rankings, backlinks, and other factors that affect search engine rankings. Examples of SEO tools include Ahrefs and SEMrush.
8. A/B testing tools: A/B testing tools are used to test different website or marketing campaign versions to determine the most effective. They can provide valuable insights into which elements of a website or campaign are most effective at driving conversions. Examples of A/B testing tools include Google Optimize and VWO [1].

These tools allow businesses to understand their online performance better and make data-driven decisions to improve their e-business strategy.

Table 14.7 Overview of performance measurement tools in e-business

Tool	Description	Metric(s)	Advantages	Limitations	Example use cases
Google Analytics	A web analytics tool that tracks website traffic and user behavior	Website traffic, bounce rate, conversion rate, time on site, etc.	User-friendly interface, free version available, integrates with other Google tools	Limited ability to track offline activity may only accurately capture some website interactions	Understanding customer behavior on the website, optimizing website design and user experience
Customer surveys	Surveys to gather feedback from customers about their experience with a product or service	Customer satisfaction, net promoter score (NPS), customer effort score	Provides direct feedback from customers, can uncover areas for improvement	Limited sample size may not be representative of the entire customer base	Measuring customer satisfaction, identifying areas for improvement
Social media analytics	Analyzing data from social media platforms to understand user engagement and sentiment	Number of followers, engagement rate, sentiment analysis	Can help understand customer preferences and sentiments, track brand reputation	Limited to social media interactions, may not capture all customer sentiment	Measuring social media effectiveness, understanding customer sentiment
A/B testing	Testing different versions of a webpage or marketing campaign to determine which performs better	Conversion rate, click-through rate	It helps identify the most effective version and provides data for informed decision-making	It can be time-consuming and expensive to set up and run tests	Optimizing website and marketing campaign performance
Sales forecasting	Predicting future sales based on historical data and market trends	Sales revenue, sales growth rate, market share	Provides insight for budgeting and resource allocation, helps identify growth opportunities	Predictions may only sometimes be accurate and can be impacted by unforeseen events or changes in the market	Planning and budgeting for future growth
Key performance indicators (KPIs)	Specific metrics used to measure progress toward business goals	It depends on business goals and industry	Provides focus for business strategy, helps track progress toward goals	It can be challenging to identify relevant KPIs and may not capture all aspects of business performance	Measuring overall business performance, tracking progress toward goals

Summary

Analysis and Performance Measurement

- Analysis and performance measurement refer to the systematic process of gathering, examining, and evaluating data to assess the effectiveness and efficiency of business operations in e-business. It involves using various metrics and techniques to track and measure key performance indicators (KPIs) to inform decision-making and improve overall performance.

Web Analytics

- Web analytics collects, analyzes, and interprets data related to website usage and user behavior. It involves tracking and measuring website traffic, pageviews, bounce, conversion, and click-through rates to gain insights into website performance, user experience, and marketing effectiveness. Web analytics helps businesses optimize their websites, enhance user engagement, and increase conversions.

Social Media Metrics

- Social media metrics refer to the quantifiable measurements used to evaluate the performance and impact of social media activities and campaigns. These metrics include the number of followers, likes, shares, and comments, reach, engagement rate, and sentiment analysis. By analyzing social media metrics, businesses can assess the effectiveness of their social media strategies, understand customer sentiment, and identify opportunities for improvement.

Customer Satisfaction Measurement

- Customer satisfaction measurement involves gathering and evaluating data to assess customers' satisfaction with a company's products, services, or overall experience. This can be done through surveys, feedback forms, online reviews, and customer support interactions. By measuring customer satisfaction, businesses can identify areas for improvement, enhance customer loyalty, and drive long-term success.

Financial Analysis in E-Business

- Financial analysis in e-business involves examining financial data and performance metrics to assess an online business's financial health and viability. It includes analyzing financial statements, key financial ratios, cash flow, profitability, and return on investment (ROI). Financial analysis helps businesses understand their financial position, make informed financial decisions, and ensure long-term sustainability.

Performance Measurement Tools

- Performance measurement tools refer to the various software applications, platforms, and technologies used to collect, analyze, and present data for performance measurement purposes. These tools may include web analytics tools, social media analytics platforms, customer satisfaction survey tools, financial analysis software, and dashboard reporting systems. Performance measurement tools facilitate the collection and interpretation of data, enabling businesses to make data-driven decisions and track progress toward their goals.

Case Study

XYZ Electronics is an online store that sells electrical devices and goods. They offer various products, including smartphones, desktops, entertainment consoles, and smart home devices. The organization has offices in numerous nations and serves customers worldwide. They have a significant online presence through their website and numerous social media channels. XYZ Electronics endeavors to improve its e-business performance and increase customer satisfaction.

1. Why is it crucial for XYZ Electronics to assess and monitor their e-business performance? How could analysis and performance evaluation aid them in identifying and optimizing areas for operational improvement?
2. Explain the importance of web analytics to XYZ Electronics. How can web analytics be used to monitor and measure website traffic, conversion rates, and exit rates? How can these measurements enhance the user experience and increase sales?
3. As the social media manager for XYZ Electronics, describe the significance of evaluating social media performance. How can essential metrics such as engagement rates, reach, and impressions be monitored and analyzed? How can these metrics enhance brand recognition and consumer engagement by providing crucial data on the effectiveness of social media marketing strategies?
4. Discuss the importance of evaluating customer satisfaction for XYZ Electronics. How can they evaluate client satisfaction using surveys and feedback forms? What actions can be taken from consumer feedback to enhance the overall customer experience and cultivate long-term customer loyalty?
5. As a financial analyst for XYZ Electronics, describe the essential financial metrics that should be examined to evaluate the company's e-business success. How can sales, profit margins, and return on investment provide insight into a company's financial viability? How can ratio analysis and benchmarking be employed to compare the company's performance to industry norms?

Discussion

1. Why is analysis and performance measurement important in e-business?
2. What are the key metrics used for analysis and performance measurement in e-business?
3. How does web analytics contribute to e-business success?

4. What are the key metrics used in web analytics?
5. How can social media metrics be used to measure e-business performance?
6. What are the key social media metrics used in e-business?
7. Why is customer satisfaction measurement important in e-business?
8. What are some common methods for measuring customer satisfaction in e-business?
9. How can financial analysis be used to measure e-business performance?
10. What are the key financial metrics used in e-business?
11. What are some financial analysis tools used in e-business?
12. How can performance measurement tools be used to drive e-business success?
13. What are some examples of performance measurement tools used in e-business?
14. How can dashboards and scorecards be used to measure e-business performance?
15. What are some best practices for using performance measurement tools in e-business?
16. How can organizations use analysis and performance measurement to improve their e-business operations?
17. What are some challenges associated with analysis and performance measurement in e-business?
18. How can organizations overcome these challenges to effectively measure their e-business performance?
19. How can organizations stay up to date with emerging trends and best practices in analysis and performance measurement for e-business?
20. What are some key considerations for organizations looking to implement effective analysis and performance measurement processes for their e-business?

Multiple-Choice Question

1. What is the primary purpose of analysis and performance measurement in e-business?

   (a) To track website traffic
   (b) To improve customer satisfaction
   (c) To measure financial performance
   (d) To optimize e-business operations

2. Which of the following is not a key metric used for analysis and performance measurement in e-business?

   (a) Conversion rates
   (b) Bounce rates
   (c) Impressions
   (d) Website design

3. Web analytics is important in e-business because it helps:
   (a) Increase social media engagement
   (b) Optimize website performance
   (c) Improve customer satisfaction
   (d) Measure financial performance
4. Which of the following is a key metric used in web analytics?
   (a) Return on investment
   (b) Revenue growth
   (c) Customer lifetime value
   (d) Website traffic
5. Social media metrics are used to measure:
   (a) Customer satisfaction
   (b) Revenue growth
   (c) Engagement rates
   (d) Profit margins
6. The importance of social media metrics in e-business is mainly due to their ability to:
   (a) Drive website traffic
   (b) Increase conversion rates
   (c) Improve customer loyalty
   (d) Evaluate social media marketing efforts
7. Customer satisfaction measurement is essential in e-business because it helps:
   (a) Increase website traffic
   (b) Enhance product quality
   (c) Optimize marketing campaigns
   (d) Improve customer retention
8. Which of the following is a common method for measuring customer satisfaction?
   (a) Ratio analysis
   (b) Benchmarking
   (c) Surveys
   (d) Financial analysis
9. Financial metrics used to measure e-business performance include:
   (a) Impressions
   (b) Reach
   (c) Profit margins
   (d) Engagement rates

10. Financial analysis tools used in e-business include:
    (a) Dashboards
    (b) Scorecards
    (c) Surveys
    (d) Feedback forms
11. Performance measurement tools in e-business can help:
    (a) Improve website design
    (b) Drive customer engagement
    (c) Optimize marketing strategies
    (d) Increase website traffic
12. Which of the following is not a performance measurement tool used in e-business?
    (a) Dashboards
    (b) Scorecards
    (c) Surveys
    (d) Feedback forms
13. The primary purpose of using performance measurement tools in e-business is to:
    (a) Track website traffic
    (b) Optimize financial performance
    (c) Improve customer satisfaction
    (d) Drive e-business success
14. Best practices for using performance measurement tools in e-business include:
    (a) Setting realistic goals and targets
    (b) Ignoring customer feedback
    (c) Focusing only on financial metrics
    (d) Using multiple tools simultaneously
15. Which of the following is not a key metric used for social media performance measurement in e-business?
    (a) Engagement rates
    (b) Reach
    (c) Impressions
    (d) Conversion rates

References

1. Chaffey, D. and F. Ellis-Chadwick, Digital Marketing. 2019: Pearson uk
2. Stair, R., & Reynolds, G, Principles of information systems. 2020: Cengage Learning
3. Combe, C., Chapter 12 – E-business: the future, Introduction to e-Business, C. Combe. 2006, Butterworth-Heinemann: Oxford. p. 325–344

Chapter 15
Legal, Regulatory, and Ethical Considerations in E-Business

Abstract The rise of the digital economy has necessitated e-commerce companies to give significant attention to ethical, legal, and regulatory concerns. To ensure responsible and sustainable business practices, companies must navigate a complex landscape of rules and regulations while adhering to ethical principles and standards. This chapter provides an overview of the key legal, regulatory, and ethical considerations for e-businesses. It covers topics such as patents, copyrights, trademarks, and the protection of intellectual property rights in online commerce. It also delves into privacy and data protection laws, their impact on online commerce, and the importance of safeguarding consumers' personal information and privacy. Furthermore, the chapter examines the laws and regulations governing advertising and marketing in e-commerce, as well as prevalent cybersecurity threats and fraud prevention techniques. It analyzes the potential for unethical behavior in online commerce, focusing on privacy, security, and social responsibility, and offers suggestions for overcoming these challenges. By thoroughly understanding the legal, regulatory, and ethical aspects of e-business, companies can earn the trust of their customers, operate transparently and accountably, and avoid potential legal and reputational risks.

1 Introduction to Legal, Regulatory, and Ethical Considerations in E-Business

In recent years, e-business has completely transformed how businesses operate. The development of new technologies has enabled organizations to conduct their operations in a more prosperous and efficient manner. This applies to all aspects of business, including marketing, sales, communication, and financial transactions. Due to the increased reliance on technology, companies must consider the legal, regulatory, and ethical considerations associated with online business.

Any e-business operation must always pay close attention to legal considerations. Many rules and regulations govern electronic commerce transactions,

H. Taherdoost, *E-Business Essentials*, EAI/Springer Innovations in Communication and Computing, https://doi.org/10.1007/978-3-031-39626-7_15

including data protection, consumer protection, intellectual property, and privacy laws. If the company fails to comply with these rules and regulations, they may face severe penalties in addition to having their reputation affected. Due to this, businesses that conduct their operations online must thoroughly understand the legal framework that governs their activities.

Regulatory issues are also a crucial factor for online businesses. There has been a recent increase in governments worldwide enacting e-commerce-related legislation. These policies include data privacy, cybersecurity, and tax policies. In order to avoid legal repercussions and maintain their clients' trust, online businesses must comply with the applicable regulations. In addition to issues pertinent to the law and regulations, ethical concerns are crucial in online commerce. Ethical considerations are the concepts and principles that guide the actions of individuals and businesses. It is the responsibility of e-commerce businesses to ensure that their business practices adhere to ethical standards such as transparency, accountability, and social responsibility. If you fail to adhere to these standards, you risk losing the trust of your customers, which could result in negative publicity and possible legal action.

When conducting business online, it is impossible to overstate the importance of adhering to all applicable laws, regulations, and ethical standards. For instance, the European Union data protection regulations such as the General Data Protection Regulation (GDPR) require businesses to safeguard their customers' personal information. The company could face substantial fines and suffer reputational damage if these regulations are violated. Similarly, ethical considerations such as social responsibility have become necessary after the COVID-19 outbreak. Consumers demand that businesses place a greater emphasis on the health and safety of their employees and customers. Online businesses should implement stringent rules and procedures to ensure compliance with all applicable legal, regulatory, and ethical requirements. This includes developing a transparent code of conduct, instructing employees on applicable laws and regulations, and routinely evaluating and revising policies to reflect changes in the environmental regulatory landscape. In order to ensure that they are adhering to all applicable rules and regulations, online businesses must seek legal counsel when required [1].

In conclusion, it is impossible to overstate the significance of ethical, legal, and regulatory concerns in online commerce. In addition to adhering to fundamental ethical principles such as accountability, transparency, and social responsibility, online businesses must also comply with a complex and constantly changing regulatory environment. If you do not comply, you risk legal consequences, losing the trust of your consumers, and harming your organization's reputation. Therefore, online businesses must prioritize these factors and implement stringent rules and procedures to ensure compliance.

In recent years, e-commerce has become an increasingly common business transaction method. This is partially due to the widespread accessibility of technology and the Internet. Numerous legal, regulatory, and ethical considerations must be adhered to when conducting business in this manner. If these requirements are met, the company may face severe penalties, legal action, or a damaged reputation.

Companies involved in e-commerce are responsible for analyzing and adhering to all legal concerns and requirements. This category includes compliance with regulations regarding data protection, intellectual property, and consumer protection. Examples of such regulations include the General Data Protection Regulation (GDPR) in the European Union and the Digital Music Copy Act (DMCA) in the United States. Before collecting users' personal information, online businesses must obtain the users' consent and then take the necessary precautions to safeguard the data. Similarly, the DMCA regulates the use of copyrighted content on the Internet and requires online businesses to take the necessary precautions to avoid infringing on the rights of others. In addition to legal considerations, e-commerce organizations must also consider regulatory issues. These are related to many fields, including privacy, taxation, and cybersecurity. For example, online businesses must adhere to the Payment Card Industry Data Security Standards (PCI DSS) to protect the integrity of their payment processing systems and their customers' sensitive information. E-commerce companies must comply with the applicable laws and regulations of each jurisdiction in which they conduct business; consequently, the complexity of the tax legislation in several countries may pose a significant obstacle [2].

Ethical issues are as essential as any other factor when managing an e-business. This category of issues contains criteria such as transparency, accountability, and social responsibility. Before acquiring users' personal information, online businesses must be transparent and truthful about their data collection practices and obtain their consent. In addition, they are required to assume responsibility for any damage caused by their actions and to take steps to mitigate the severity of such injury. E-commerce businesses are required to consider their social responsibilities and the impact of their actions on society and the environment.

Regarding intellectual property issues, online businesses must utilize copyright-, trademark-, and patent-protected materials with extreme caution and foresight. Online businesses must acquire the licenses and permissions to use the content above and take precautions to avoid violating intellectual property rights. In addition to protecting their intellectual property, such as trademarks and patents, e-commerce companies must also take legal action to prevent others from infringing on those rights. Disagreements are always possible during an e-business, just as with any other business form. Using an online forum for conflict resolution may be an expedient and cost-effective method for resolving disagreements. Online businesses need clear policies and procedures for handling consumer complaints and resolving issues transparently and equitably.

In conclusion, online businesses must operate in a challenging legal, regulatory, and ethical environment. Failure to comply with applicable rules, regulations, and ethical considerations may result in severe consequences, including potential legal action, monetary penalties, and a damaged reputation. Suppose they thoroughly understand and adhere consistently to the factors above. In that case, e-commerce businesses can ensure continued success while fulfilling their responsibilities to society and customers. Figure 15.1 summarizes critical considerations of e-business.

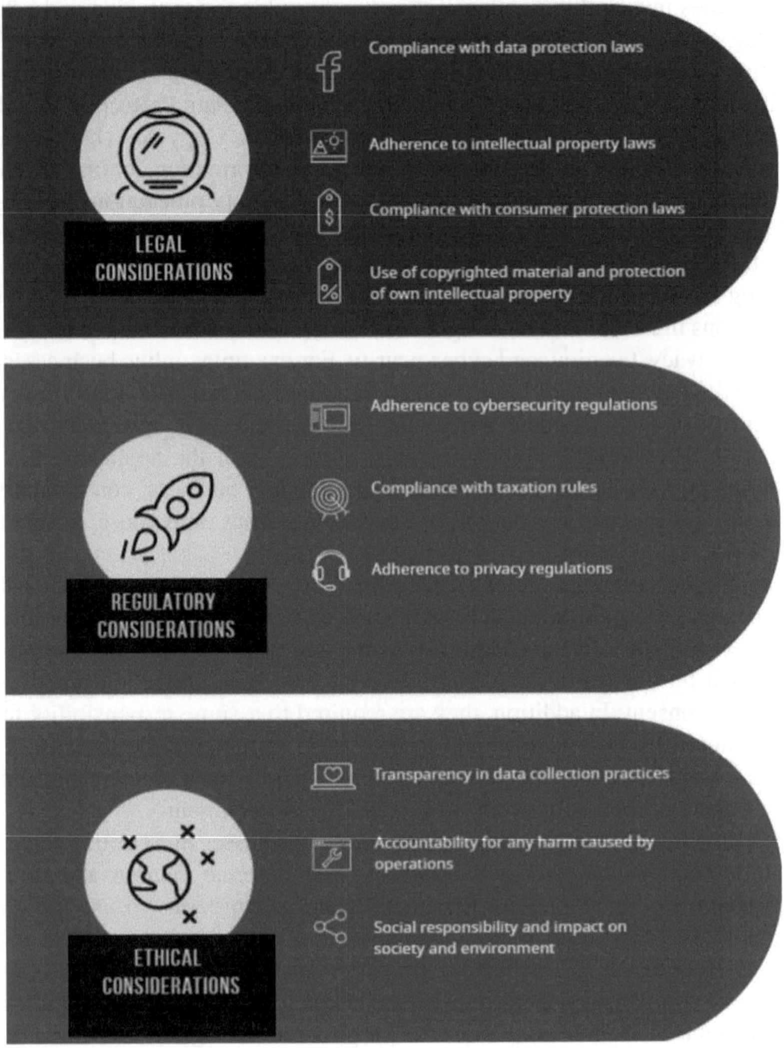

Fig. 15.1 Critical considerations for e-business: legal, regulatory, and ethical

2 Intellectual Property Rights

Protection of intellectual property rights (IPRs) is essential to online commerce. Intellectual property rights (IPRs) are the legal protections accorded to original invention works, such as innovations, designs, and creative works. IPRs are crucial in e-business because they safeguard companies from intellectual property infringement and guarantee they can profit from their ideas. This tutorial will discuss the significance of intellectual property rights (IPRs) in e-business and how businesses can protect their intellectual property.

Providing exclusive rights to a company's inventions is one of the most significant advantages of intellectual property rights for online businesses. Companies can exercise control over how their intellectual property is utilized and profit from the fruits of their labor due to these rights. Without the protection of intellectual property rights, firms would have no incentive to create new products, designs, or content because others could copy their work without their permission and profit from it. Online businesses require IPRs to defend themselves against infringement, another reason for their significance. Due to the proliferation of digital technology, it is now significantly more accessible for unauthorized parties to acquire and distribute digital property. This includes images, videos, and written content. IPRs provide corporations with the legal recourse necessary to prevent unauthorized use of their intellectual property. This assists businesses in avoiding financial losses and safeguards their brand reputations. IPRs are essential for e-businesses because they help build a company's reputation and brand identity, another critical aspect of IPRs. E-businesses rely heavily on their brand identities to attract new customers and retain existing ones. Using intellectual property rights (IPRs) in the form of trademarks is necessary to safeguard a company's brand identity. Registering trademarks prevent competitors from employing confusingly similar marks but also aids online businesses in constructing and preserving their reputations. When businesses own exclusive rights to their inventions, they are incentivized to invest in research and development, resulting in the creation of new and innovative products and services. This is yet another crucial role that intellectual property rights play in fostering innovation and originality. IPRs provide businesses with a framework for safeguarding their intellectual property, incentivizing them to invest in innovative ideas and technology that can generate success in the e-business environment. IPRs are developed by companies [3].

Patents are a type of intellectual property right that confers exclusive legal protection on the inventor or assignee of an invention for a specific period, typically 20 years. One of the rights granted to a patent holder is the authority to prevent others from manufacturing, using, selling, or importing an invention without authorization. Patents are indispensable for protecting novel computer programs, digital technologies, and algorithmic processes in online commerce. Patents provide novel digital technology developed by online businesses with legal protection, which is one of the primary benefits they offer. Patents can be utilized by e-commerce businesses to protect their intellectual property, thereby preventing competitors from using their technology without permission. This protection is crucial for online businesses that rely heavily on cutting-edge technology and software to create unique and highly valued products and services for their customers.

Patents may provide e-commerce businesses with a competitive advantage in the market. By acquiring patents for the digital technology that powers their operations, e-commerce companies can differentiate themselves from their competitors and strengthen their position as market leaders. This may be of the utmost importance for new and burgeoning small businesses working hard to stake out a market niche. Patents offer online businesses some benefits, including the ability to profit from patents. E-businesses can generate revenue by licensing or selling their patents to other companies or investors. This generates revenue and provides a crucial source of funds for further innovation and company growth. In addition, patents are integral to fostering innovation in the e-business sector. Patents incentivize firms to invest in research and development by granting exclusive rights to inventors, which ultimately results in the development of innovative products and services. This may facilitate the growth and success of the e-business sector, which is advantageous for both businesses and consumers.

However, patent procuring can be difficult and costly, particularly for online enterprises with limited resources. Companies involved in e-commerce must provide evidence that their technology is original and not readily surmised, which frequently requires in-depth legal knowledge. In addition, filing patent applications and participating in legal proceedings can be quite expensive, which can be a barrier to entry for startups and lesser businesses.

A trademark is a form of intellectual property that protects its proprietors' unique brand identities and logos. E-commerce trademarks are essential for developing brand recognition, establishing consumer trust, and protecting the company's reputation. Additionally, trademarks can be used to defend the business's reputation. A trademark is a term, phrase, symbol, or design—or any combination of these elements—that distinguishes one company's products or services from those of other companies in the same industry.

Establishing and sustaining a consistent brand identity is facilitated by trademarks, which is one of the most significant benefits trademarks offer to online businesses. By registering a trademark prohibiting the use of confusingly similar names or logos, e-commerce companies can prevent their competitors from using names or logos that are similar to their own. This can help safeguard the company's reputation. This security is of the uttermost importance for online businesses, which frequently operate in environments marked by intense competition and rapid market change. The use of trademarks may also increase customers' trust. Customers will choose an organization's products or services over its competitors if the organization has a strong trademark synonymous with quality and dependability. This may be of the uttermost importance for online businesses, many of which have no physical location and rely on Internet communication to reach their customers. The ability to generate revenue through trademarks is another benefit to online businesses. By licensing or selling their trademarks to other companies or investors, e-businesses may provide a vital source of revenue and funding for future development and expansion. This may be accomplished directly or indirectly through a third party.

Moreover, trademarks serve to shield online businesses from legal liability. If they register their trademarks, e-commerce companies can prevent third parties from using their name or logo dishonestly or misleadingly. This security may benefit Internet businesses susceptible to cyberspace-based swindling artists and fraudsters. In contrast, acquiring a trademark can be complicated and time-consuming. To affirm that their chosen trademark still needs to be used, online businesses must conduct specific queries, which may be challenging given the digital landscape's rapid pace and constant evolution. In addition, the registration of a trademark can be costly, which can be a barrier to entry for new businesses and lesser enterprises.

Copyrights are a type of intellectual property right whose primary function is the preservation of original works of authorship, such as literary, artistic, and musical works. Copyrights are essential to e-commerce when preserving digital information, such as the text on a website, photographs, videos, and software. Protecting intellectual property rights is crucial for online businesses that rely on digital content to attract customers, maintain their interest, and generate revenue.

The fact that copyrights grant exclusive rights to the author of the work for which they were granted is one of the most significant benefits they offer to online businesses. This indicates that the proprietor has the right to regulate the work's use, reproduction, and distribution, among other things. This control is crucial for online businesses that create and distribute digital content because it enables these businesses to protect the financial investment they made in the creation of the content and prevents others from pirating or otherwise exploiting their labor. Additionally, copyrights contribute to developing and protecting online businesses' reputations. By generating content that is both unique and of the highest possible quality, e-commerce businesses can gain their clients' trust and position themselves at the vanguard of their industry. The protection provided by copyright laws ensures that only the e-business may use and distribute the content, thereby aiding in preserving the content's originality and quality. Copyrights may also be a source of revenue for online businesses, which is another benefit of possessing them. By licensing their intellectual works to third-party organizations or individuals, e-commerce companies can earn additional income. For example, an online company that develops software and sells it to other businesses or individuals for a fee could be designated a software licensee. Copyrights are an additional method by which online businesses can avoid legal liability. By registering copyright to safeguard their intellectual property, e-businesses can prevent others from using their content for fraudulent or deceptive purposes and claiming it as their own. This security may benefit Internet businesses susceptible to cyberspace-based swindling artists and fraudsters.

On the other hand, acquiring copyright protection can be a challenging and time-consuming process. Since originality and creativity are two requirements for obtaining copyright protection, e-commerce companies must ensure that the content they distribute meets these standards. In addition, online businesses must register their copyrighted works with the appropriate authorities, which can be lengthy and costly.

Using a trade secret, a type of intellectual property right, to protect confidential information that gives a company a competitive advantage is crucial for businesses. The term "trade secret" in online commerce can refer to a wide range of nonpublic information, including but not limited to company strategies, customer databases, algorithms, and software source code. Trade secrets are an essential intellectual property right for e-businesses because they allow companies to secure their valuable and confidential information from being viewed by competitors and other third parties.

The fact that trade secrets provide continuous security for sensitive information is one of their most important benefits for online businesses. In contrast to patents, which only offer protection for a limited period, trade secrets can be protected indefinitely so long as the relevant information is kept private. This may be especially significant for online businesses, which rely heavily on technology and private information to maintain a competitive advantage. By utilizing trade secrets, e-commerce businesses have greater flexibility in protecting sensitive data. In contrast, trade secrets can be protected even if they are not disclosed, whereas patents require that the innovation be made public to be protected. Due to this, online businesses can maintain the privacy of their sensitive data while still enjoying the benefits of legal protection. Another benefit of trade secrets for online businesses is that they effectively deter the misappropriation of confidential information. This is especially useful for businesses that handle sensitive data. E-businesses can prohibit competitors and other third parties from using or disclosing sensitive information without establishing a trade secret, which safeguards the information against unauthorized use or disclosure. This is of utmost importance in the fast-paced and competitive environment of e-business, where valuable confidential information becomes outdated and irrelevant. In addition to providing a substantial economic advantage, trade secrets may be especially beneficial for online businesses. E-businesses can continue innovating and developing new products and services and maintain their competitive advantage if they protect sensitive information. This can be advantageous for e-commerce companies seeking to establish themselves as industry leaders and attract new consumers and clients.

However, several obstacles exist to surmount when preserving trade secrets in online enterprises. E-commerce businesses are obligated to take the necessary precautions to protect the privacy of their customer's personal information and prevent unauthorized parties from viewing, copying, or distributing it. This may involve implementing stringent security measures such as firewalls and encryption and developing clear rules and protocols for managing sensitive data. In addition, online businesses have a duty to safeguard the confidentiality of their customer's sensitive information. This may be a particularly challenging endeavor in the current era when information can be rapidly shared and disseminated across numerous Internet platforms. E-businesses are required to maintain vigilance in the surveillance and control of access to their sensitive data, as well as the prompt implementation of corrective measures in response to any unauthorized access or disclosure.

Protecting intellectual property rights is crucial for businesses that engage in e-commerce, as it enables them to maintain a competitive advantage in the digital marketplace. Infringement of intellectual property rights or unauthorized use of IPRs can result in substantial monetary losses for businesses. In addition, there is a possibility that the company's reputation will be harmed, resulting in diminished consumer trust and loyalty. If a company maintains its intellectual property rights (IPRs), it can prevent others from profiting from its creativity and innovation.

Registering with the appropriate government entity is one of the essential strategies for protecting intellectual property rights in online enterprises. Companies can obtain patents, trademarks, and copyrights through the United States Patent and Trademark Office (USPTO) or the United States Copyright Office. If they register their IPRs, businesses can assert dominion over their intellectual property rights and defend them against infringement. Contracts, such as nondisclosure agreements (NDAs) and licensing agreements, offer an additional means of protecting intellectual property rights. A person or business that signs an NDA is prohibited from disclosing confidential information to a third party, such as a trade secret. By utilizing licensing agreements, businesses can retain custody of their intellectual property while granting third parties the right to use it for a fee.

Enforcement of intellectual property rights in electronic commerce may be challenging due to the global nature of the digital marketplace and the wide variety of applicable rules and regulations. Enforcement of intellectual property rights (IPRs) may involve litigation, such as a lawsuit for intellectual property infringement. However, legal action can be extremely costly and time-consuming, so it may not be an option for some businesses. Digital rights management (DRM) and watermarking are alternative approaches to enforcing intellectual property rights (IPRs) based on technological solutions. While digital rights management technologies prevent the unauthorized duplication and distribution of digital content, watermarking allows businesses to identify their content and track its usage. These technologies have the potential to aid in preventing intellectual property infringement and protecting intellectual property rights. Table 15.1 summarizes the types of intellectual property in e-business.

3 Privacy and Data Protection

Due to the development of the digital era, online businesses are becoming increasingly concerned with privacy and data protection. Online businesses' collection, use, and dissemination of personally identifiable information have contributed to a growing concern for protecting individual privacy rights. This section investigates how privacy and data protection regulations affect online businesses. After providing an overview of privacy and data protection laws in the introduction, the discussion will continue to the impact of these laws on online enterprises. This section will conclude with recommendations for online businesses to ensure privacy and data protection legislation compliance. Privacy and data protection statutes govern individuals' and organizations' collection, processing, storage, and distribution of personal information. Both private individuals and government institutions may conduct these activities. In several nations, the right to one's privacy and the protection of one's data are regarded as fundamental human rights, and their defense is frequently enshrined in national constitutions and laws. For instance, the General Data Protection Regulation (GDPR) is a comprehensive privacy regulation that provides a framework for protecting personal data. This law applies to all organizations operating within the European Union (EU) and all businesses operating within the EU [1].

The implementation of privacy and data protection laws could have a significant impact on online businesses. Online businesses collect and process personally identifiable information, including names, addresses, credit card numbers, and browsing histories. These data are utilized for various purposes, including marketing, customer service, and research. Nonetheless, collecting and processing personal data

Table 15.1 Types of intellectual property in e-business

Type of intellectual property	Examples in e-business	Description
Patents	Software, algorithms, e-commerce platforms, business processes	Patents protect inventions or discoveries, such as new products or processes. In e-business, patents may protect software, algorithms, e-commerce platforms, and business processes. These patents give the owner the exclusive right to use and license the invention for a certain period
Trademarks	Domain names, logos, brand names, slogans	Trademarks protect words, symbols, logos, or designs that identify and distinguish the source of goods or services. In e-business, trademarks may protect domain names, logos, brand names, and slogans. These trademarks give the owner the exclusive right to use and license the trademark for a certain period
Copyrights	Website content, digital media, software code	Copyrights protect original works of authorship, such as books, music, plays, paintings, photographs, sound recordings, architectural works, and software. In e-business, copyrights may be used to protect website content, digital media, and software code. These copyrights give the owner the exclusive right to reproduce, distribute, and display the work for a certain period
Trade secrets	Customer lists, marketing strategies, business plans, algorithms	Trade secrets are confidential information that gives a business a competitive advantage. In e-business, trade secrets may include customer lists, marketing strategies, business plans, and algorithms. These trade secrets are protected through secrecy and nondisclosure agreements, giving the owner the exclusive right to use and license the information for as long as it remains confidential

involves several significant dangers, such as the possibility of data breaches, identity theft, and unauthorized access. Privacy and data protection laws require e-businesses to safeguard personal data appropriately. These measures include instituting data security measures, obtaining individuals' consent for data acquisition and processing, and granting individuals the right to access and amend their data. In addition, e-businesses must give individuals the right to access and correct their personal information. In order to comply with privacy and data protection laws, online businesses must disclose all information regarding their data collection and processing practices. E-commerce businesses are required by law to inform customers clearly and concisely about the types of personally identifiable information (PII) they collect, the reasons for doing so, and the recipients of the data. In addition, e-commerce businesses must obtain the explicit consent of individuals before engaging in data processing operations. In addition, privacy and data protection regulations require online businesses to implement adequate security measures to prevent unauthorized access, use, and disclosure of their customers' personally identifiable information.

Serious consequences may result if an online business fails to comply with applicable privacy and data protection regulations. If an online business is discovered to violate privacy and data protection regulations, it may face hefty fines, incur legal liability, and suffer reputational damage. In 2018, for instance, the United Kingdom's Information Commissioner's Office fined Facebook half a million pounds for failing to prevent unauthorized access to users' personal information by third-party applications. Due to this occurrence, Facebook's user trust and reputation significantly declined. In order to remain in compliance with the various privacy and data protection regulations, e-commerce businesses must take the necessary precautions to safeguard their consumers' personal information. E-businesses should conduct a privacy impact assessment to identify the risks associated with data collection and processing and then implement appropriate security measures to mitigate these risks once the results have been compiled. E-commerce companies are obliged to ensure compliance with applicable privacy and data protection regulations by implementing the necessary rules, procedures, and training.

As a result of the proliferation of online businesses, consumers now have access to a broader range of goods and services than ever before. However, as a result of this, an environment in which private information is routinely traded has been created, raising concerns about protecting personal privacy and data. It is impossible to exaggerate the importance of protecting customer privacy and maintaining customer confidentiality when conducting e-business. Maintaining consumer confidence is one of the most crucial reasons why online businesses should prioritize data security and privacy. Customers are more likely to conduct business with a company if they have confidence in its ability to safeguard their personal information. Customers who do not trust a company may be unwilling to disclose their confidential information, resulting in a decline in sales and a loss of revenue [4].

Online businesses must protect customers' personal information and privacy for +various fundamental reasons, including legal requirements. Governments worldwide have enacted laws and regulations requiring businesses to protect their customers' information. You risk accruing severe penalties, being brought to court, and harming your reputation if you fail to comply with these requirements. Data intrusions are an additional significant concern for online businesses. Theft of sensitive consumer information is typically the result of increasingly pervasive cyberattacks on businesses, which have become more prevalent in recent years. The fact that the stolen data could be used for identity theft and other unlawful activities makes this situation potentially hazardous for the company and its customers. Therefore, it is imperative to take precautions to prevent data breaches to safeguard personal data and consumers' privacy. E-commerce businesses must consider the ethical ramifications of administering customers' confidential information. Customers have the right to expect that their personal information will be handled responsibly and ethically by enterprises. Companies that do not comply risk adverse press coverage and damage to their brand reputation. Regarding client information administration, e-commerce businesses must prioritize ethics above all else. This may facilitate the development of trust and contribute to a positive reputation [4].

Numerous measures can be taken in online commerce to protect the confidentiality of sensitive consumer information and customer data. Encryption-based technologies can be used to protect and prevent unauthorized access to data. In order to restrict access to sensitive data, it is also possible to implement password policies and access controls. The protection of consumers' personal information is another crucial aspect of transparency. Online businesses need to have privacy policies that are not only comprehensible but also readily accessible. These policies should detail the collection, use, and security of personally identifiable information. In addition, they should allow consumers to opt out of any data collection or sharing.

Last but not least, online businesses must ensure that their employees are adequately trained on the most effective methods for protecting the confidentiality of client information and personal data. This requires educating personnel on access control mechanisms, data security, and privacy regulations. E-businesses that place a high priority on staff education can ensure that every employee understands the significance of protecting the privacy of client information and personal data.

In conclusion, online businesses must protect consumers' personal information and privacy. In addition to ensuring legal compliance, preventing data breaches, and upholding ethical standards, it helps build consumer trust. Encryption technology, access restrictions, explicit privacy rules, and staff education are ways online businesses can protect their customers' personal information and privacy. Implementing data security and privacy measures as a top priority can assist an online business in establishing a positive reputation and ensuring its sustained success.

4 Cybersecurity and Fraud Prevention

As a result of the constant evolution of the business world, more and more individuals are gravitating toward conducting their business transactions online. To address the new cybersecurity risks and fraud concerns that arise as a result of this transition, it is necessary first to identify them. In this part, we will provide an overview of the threats posed by cybercrime and the precautions taken to prevent online deception (Table 15.2).

Hacking is one of the most common types of cybercrime in online enterprises. Hackers use various methods to access classified information, such as customer data, financial records, and intellectual property. Using malware, luring schemes, or other forms of social engineering, they could deceive employees into providing them with access to private data. To prevent hackers from breaching their systems, e-commerce businesses should employ firewalls, antivirus software, and staff training on cybersecurity best practices. Data breaches represent an additional grave threat to the cybersecurity of online enterprises. A data breach occurs when sensitive data is accessed, misappropriated, or disclosed without authorization. A data intrusion may have significant consequences, including the loss of financial resources, damage to one's reputation, and legal consequences. Encryption, two-factor authentication, and secure data storage techniques are among the most effective methods for online businesses to prevent data intrusions.

Malware is one of the most common types of cybersecurity threats. Malware refers to any software specifically designed to cause damage to a computer system,

Table 15.2 Common types of cybersecurity threats and prevention measures in e-business

Type of threat	Description	Prevention tools/techniques
Phishing	An attempt to obtain sensitive information such as usernames, passwords, and credit card details by disguising as a trustworthy entity in electronic communication	Anti-phishing software, two-factor authentication, employee training, and awareness programs
Malware	Software designed to harm a computer system, such as viruses, Trojan horses, and spyware	Antivirus software, firewalls, software updates, and patches
DDoS attacks	An attempt to overwhelm a system by flooding it with traffic from multiple sources	DDoS protection services, firewalls, load balancers
Ransomware	A type of malware that encrypts a victim's data, making it inaccessible until a ransom is paid	Backup and recovery plans, antivirus software, employee training, and awareness programs
Insider threats	An employee or contractor who deliberately or accidentally exposes sensitive information or causes damage to a company's systems	Access control policies, employee monitoring software, employee training, and awareness programs
Social engineering	An attempt to manipulate individuals into divulging sensitive information or performing actions that compromise a system's security	Employee training and awareness programs, two-factor authentication, background checks

network, or device. Malware can access a computer system through phishing emails, malware-infected downloads, and compromised websites. Utilizing firewalls, keeping all systems and software up to date, and implementing antivirus software can defend businesses from malware attacks. Phishing is yet another prevalent form of cyberattack. Phishing is a form of social engineering in which a perpetrator uses deception and social engineering techniques to trick a target into divulging sensitive information such as login credentials or financial information. Companies can prevent phishing attacks by instructing employees to recognize and avoid spoofing emails, implementing two-factor authentication, and deploying anti-phishing software. DDoS attacks, or distributed denial of service attacks, are an additional common cybersecurity threat. An adversary employs a distributed denial of service attack when they intend to make a system inaccessible to its users by overwhelming it with traffic. Utilizing DDoS mitigation services, deploying firewalls and intrusion detection/prevention systems, and utilizing content delivery networks are all viable DDoS protection strategies for businesses [1].

Additionally, ransomware attacks are becoming more prevalent in the online commercial world. Ransomware is a type of malicious software that encrypts the data of its victim and then demands payment in exchange for the decryption key. Businesses can defend themselves from ransomware attacks in various ways, such as regularly backing up their data, employing antivirus software, and providing security awareness training to their employees. Businesses may also be the target of social engineering attacks such as business email compromise (BEC) and CEO deception. These forms of attacks are susceptible to targeting businesses. Business email compromise (BEC) is an attack in which the perpetrator poses as a high-level executive to trick an employee into transmitting money or disclosing sensitive information. Businesses can protect themselves from BEC attacks by adopting email authentication methods and anti-spoofing technologies and establishing explicit communication routes. Identity theft is among the most common forms of deception affecting online businesses. The perpetrator of identity theft impersonates the victim to perpetrate fraudulent acts such as making purchases or opening bank accounts using the victim's personal information. Businesses can combat identity theft by implementing two-factor authentication, employing encryption and secure data transmission protocols, and training employees on best practices for protecting sensitive information.

In conclusion, online businesses are vulnerable to various cybersecurity threats and fraud, but numerous tools and strategies are available to defend against these threats. Businesses can protect themselves and their consumers from potential harm if they stay current on the latest cybersecurity threats and implement the most effective fraud prevention and data protection policies.

5 Consumer Protection and Advertising Standards

E-commerce may be a convenient method for businesses to communicate with clients and customers, but there are inherent risks for end consumers. In order to protect the interests of consumers, a variety of rules and regulations have been implemented in the online marketplace.

The Electronic Commerce (EC Directive) Regulations of 2002 are frequently cited as one of the laws with the most significant impact on assuring the protection of consumers conducting business online. To comply with these requirements, businesses must provide customers with certain information, such as the company's name, address, and other contact information. In addition, they mandate that businesses make their policies, terms, and conditions readily accessible and understandable to consumers. In addition to the European Community Directive, several additional regulations are designed to protect online shoppers. For instance, the Consumer Rights Act of 2015 protects consumers when purchasing a product or service online. These rights include the right to rescind an order within 14 days and the right to receive a refund if the products or services are defective or do not match the supplied description. Due to the prevalence of advertising as a marketing instrument for companies' products and services, e-business also necessitates the establishment of advertising standards. The Advertising Standards Authority (ASA) is responsible for advertising regulation in the United Kingdom. The ASA-published Code of Advertising contains several rules businesses must follow when conducting Internet advertising. For instance, businesses are obligated to ensure that their advertising complies with all applicable laws and is ethical, truthful, and accurate. The Advertising Standards Authority (ASA) also has guidelines for social media advertising. When businesses advertise on social media, they must provide accurate information and refrain from making false or unsupported claims. Influencers compensated to promote products on social media must disclose their relationship with the companies they promote. In addition to rules and regulations, companies may take additional precautions to protect consumers' financial and personal information when conducting e-business. For instance, companies can protect the financial information of their consumers by utilizing encrypted payment processing systems. They can also encrypt sensitive information, such as the names and addresses of customers, for protection. A simple and accessible method for submitting complaints is another essential step in ensuring online shoppers' safety. This allows consumers to file complaints if dissatisfied with the purchased goods or services. Businesses must make it crystal clear to consumers how to file a complaint, and they must respond to customer concerns quickly and effectively [4].

Numerous laws and regulations are in place to protect online shoppers (also known as e-commerce consumers). Companies are responsible for ensuring compliance with these regulations and, if necessary, taking additional measures to protect their consumers. By acting this way, businesses can acquire customers' trust and ensure their online operations' long-term success.

It is essential to advertise and promote products and services in e-business to attract new customers and generate revenue. However, there are regulations and guidelines that online businesses must adhere to in order to ensure that their advertising practices are honest and ethical. Regarding advertising and marketing their products or services, this book will elucidate the regulations and standards that e-businesses are required to follow. The Federal Trade Commission (FTC) Act is one of the most critical laws online enterprises must adhere to. The Federal Trade Commission (FTC) is a government agency combating unethical and deceptive advertising practices. The Federal Trade Commission requires that all advertising be truthful and not deceptive. In addition, advertising claims must be supported by evidence that is both credible and pertinent to the product or service being sold. The Lanham Act is yet another piece of legislation that must be followed by online businesses. The Lanham Act is a federal statute that prohibits deceptive advertising practices. False advertising is defined as advertising that is likely to mislead a consumer who makes reasonable decisions. The Lanham Act gives competitors the right to sue for damages caused by misleading advertising, significantly threatening an online business's credibility and financial stability. E-businesses are required to comply not only with federal regulations but also with industry group recommendations.

The Digital Advertising Alliance (DAA) is an example of an organization that provides self-regulatory standards for Internet advertising. These regulations require e-businesses to give customers a plain and prominent notice regarding collecting and using their data for advertising purposes. In addition, the DAA requires online businesses to allow consumers to opt out of having their data collected and used for marketing purposes. Additionally, e-commerce businesses must adhere to the principles established by various social media sites. For instance, Facebook and Instagram have particular rules regarding the advertisements permitted on their respective platforms. Under these regulations, false, deceptive, or offensive advertising is prohibited. In addition, these platforms require that all advertising adheres to all currently applicable rules and regulations. Online enterprises must comply with all federal, state, and local advertising regulations. These principles are subject to change based on the online business's geographic location and the target audience's demographics. For instance, certain jurisdictions prohibit certain types of advertising to juveniles, such as advertising for tobacco products. Other jurisdictions prohibit specific forms of pornographic advertising. In order to comply with these rules and regulations, e-businesses must have a thorough understanding of their legal responsibilities and implement internal policies and procedures to ensure compliance. This will ensure that they comply with these regulations and recommendations. E-commerce companies should work closely with their legal and marketing departments to ensure that all of their advertising is truthful, does not mislead consumers, and adheres to all industry-specific rules and regulations.

E-businesses must comply with many laws and standards when advertising and promoting their products and services. These standards and guidelines were developed to protect consumers from misleading and erroneous advertising practices. E-businesses should clearly understand their legal obligations and collaborate closely with their legal and marketing teams to ensure compliance with all applicable laws and regulations. By adhering to these standards and guidelines, online businesses can establish a solid reputation for engaging in fair and ethical advertising, which may increase consumer trust and revenue. Table 15.3 summarizes laws, regulations, consumer protection, and advertising guidelines in e-business.

6 Ethical Issues in E-Business

E-commerce has become a fundamental component of today's economic landscape, raising several ethical concerns that require in-depth examination. In this piece, we will discuss some of the ethical challenges of online business and potential solutions to these problems.

Invading one's privacy is one of the most often encountered ethical dilemmas in online commerce. Concerns about customer privacy might be prompted by acquiring, using, and storing personal information. Many organizations use cookies and other tracking technologies to monitor the activity of customers, which might result in an invasion of privacy in some instances. E-commerce companies are responsible for solving this problem by developing and enforcing stringent privacy rules, obtaining consumer agreements before collecting personal information, and ensuring the safety of client data. Cybercrime presents yet another difficulty for companies doing business online. Phishing, identity theft, and hacking are all examples of fraudulent activities that may be carried out with the help of the Internet, which offers a platform for these sorts of activities. Both organizations and individual customers might suffer significant financial losses due to cybercrime. E-commerce companies may reduce their risk of becoming victims of cybercrime by implementing stringent security measures, including firewalls, antivirus software, and encryption

Table 15.3 laws, regulations, and guidelines for consumer protection and advertising in e-business

Regulation/law/ guideline	Purpose	Examples
Electronic Commerce (EC Directive) Regulations 2002	Provide consumers with certain information, make terms and conditions clear and accessible	Business's name, address, and contact details; terms and conditions made clear and easily accessible
Consumer Rights Act 2015	Provide consumers with rights when buying goods/services online	Right to a refund if goods/services are faulty or not as described; right to cancel an order within 14 days
Advertising Standards Authority (ASA) Code of Advertising	Regulate advertising in the United Kingdom, and set rules for businesses to follow when advertising online	Advertising must be legal, decent, honest, and truthful; businesses must make it clear when they are advertising on social media; businesses must not make claims that are misleading or unsubstantiated; influencers must disclose their relationship with the business
Federal Trade Commission (FTC) Act	Prevent unfair and deceptive advertising practices	Advertising must be truthful and not misleading; advertising claims must be backed by reliable and relevant evidence
Lanham Act	Prohibit false advertising	False advertising is defined as advertising likely to deceive a reasonable consumer; competitors can sue for damages caused by false advertising
Digital Advertising Alliance (DAA) guidelines	Provide self-regulatory guidelines for online advertising	Businesses must provide clear and prominent notice to consumers about data collection and use for advertising purposes; businesses must provide consumers with a way to opt out of data collection and use for advertising purposes
Social media platform rules	Set rules for what types of advertising are allowed on their platforms	Advertising must comply with all applicable laws and regulations; advertising must not be false, misleading, or offensive
State and local laws	Regulate advertising practices within specific locations	Some states have laws that prohibit certain types of advertising to children, such as advertising for tobacco products

technology. Another area of ethical concern in e-business is the protection of intellectual property. People can more easily duplicate and distribute content protected by intellectual property rights because of the proliferation of the Internet. E-commerce companies are responsible for avoiding infringing on the intellectual property rights of other parties and safeguarding their unique intellectual property. This may be accomplished by securing exclusive rights to their goods and services using intellectual property such as copyrights, patents, and trademarks. Another potential ethical problem that may occur in online companies is unfair competition. Businesses risk engaging in unethical tactics such as price fixing, collusion, and monopolistic activity, which may be detrimental to customers and other companies. E-commerce companies are responsible for upholding ethical standards of competitiveness and abstaining from participating in acts that might be seen as unfair. E-business is facing many ethical challenges, one of which is environmental sustainability. The need for energy to power data centers, servers, and other infrastructure increases with the expansion of online commerce. E-commerce companies have to mitigate the adverse effects of their operations on the natural environment by adopting environmentally friendly business procedures. These procedures should include the use of renewable energy sources, the reduction of carbon emissions, and the promotion of environmentally friendly items. E-business also faces challenges when it comes to issues of social responsibility. E-commerce companies are morally obligated to contribute to the communities they operate constructively. This may be accomplished through giving money to charitable organizations, encouraging people to be open-minded and accepting of others' differences, and ethically doing business [3].

Last but not least, online firms are responsible for ensuring that they comply with all rules and regulations that apply to them. Failure to comply with rules and regulations may result in various adverse outcomes, including legal and financial ramifications, harm to reputation, and a reduction in customers' confidence. E-commerce companies are responsible for monitoring the development of new laws and regulations and guarantee that they comply with all applicable standards.

Ethical problems are becoming increasingly widespread due to the expansion and development of online businesses. To keep their consumers' faith and confidence, businesses must take the initiative to resolve ethical problems. They need to do so to maintain their trust and confidence. In this tutorial, we will go through the most effective methods for dealing with ethical concerns that arise in online business.

A company's beliefs and ideals need to be spelled out in a comprehensive ethical code, which should be the first thing the company does. This code of ethics should be made available to all stakeholders and workers. It should describe the company's position on various ethical problems, including data protection, intellectual property, and transparency. Companies may guarantee that everyone is on the same page when making ethical decisions by implementing a clear code of ethics into their business practices. Companies have to nominate an ethics officer or committee and have a set of ethical guidelines to abide by. The monitoring and resolution of any ethical problems that may crop up inside the organization should fall within the purview of this individual or group. They should be able to provide direction and assistance to workers who may need clarification about how to respond to a particular circumstance and need their help. When required, the ethics officer or committee should also be able to undertake investigations. Transparency should also be prioritized as an effective best practice for resolving ethical challenges in online business. Businesses need to communicate openly and honestly with their clients about the usage of their data and the individuals who have access to it. They need to be clear about any possible conflicts of interest or financial links that might influence the choices they make for their company. Businesses have the opportunity to earn their consumers' confidence and show their dedication to ethically conducting themselves when they place a priority on openness. Companies must also emphasize data privacy in e-business. They should establish stringent security measures to secure client data and only gather data essential to their organization's running. Before a company collects or uses its customers' data, it should obtain its express consent. Companies may show their commitment to ethical conduct and safeguard their consumers from possible data breaches by prioritizing the privacy of their customer's data. In addition to data privacy, corporations must also consider intellectual property rights. They need to take precautions to safeguard their intellectual property while avoiding infringing on the rights of others to their intellectual property and ensuring that they are not violating the rights of others. This can include registering trademarks and patents and keeping an eye on the Internet for any instances of intellectual property theft. Companies have a better chance of avoiding legal conflicts and demonstrating their commitment to ethical conduct if they prioritize protecting intellectual property rights [2].

In conclusion, businesses should prioritize ethical decision-making at every level of the company. This entails giving all personnel training in ethics and making it a point to ensure that ethical issues are considered whenever business choices are made. Establishing a culture inside a company that emphasizes ethical conduct and rewarding people who respect ethical standards should also be a priority for businesses. Companies can develop a sustainable business model that is based on trust and integrity if they place a high priority on the ethical decision-making process.

In conclusion, resolving ethical concerns arising while doing business online is very necessary to preserve the trust and confidence of consumers. The establishment of an ethical code, the appointment of an ethics officer or committee, the prioritization of data privacy and intellectual property rights, and the prioritization of ethical decision-making at all levels of an organization are all requirements for businesses. Companies can develop a sustainable business model based on trust and integrity if they follow these best practices and put them into effect. Figure 15.2 illustrates best practices for addressing ethical issues in e-business.

Summary

Intellectual Property Rights

- Intellectual property rights refer to legal protections granted to individuals or organizations for their creations or inventions, such as patents, trademarks, copyrights, and trade secrets. These rights safeguard original works and encourage innovation and creativity in e-business.

Privacy and Data Protection

- Privacy and data protection encompass the measures and regulations designed to safeguard individual's personal information in the digital realm. It involves collecting, storing, and sharing data, ensuring that it is handled securely and lawfully while respecting individuals' rights to privacy.

Cybersecurity and Fraud Prevention

- Cybersecurity protects computer systems, networks, and data from unauthorized access, damage, or disruption. It involves implementing preventive measures,

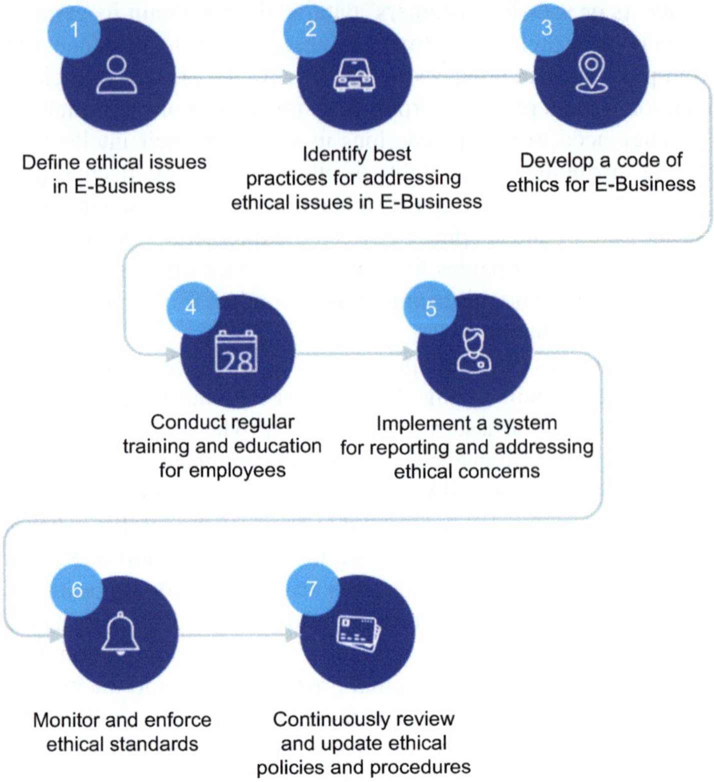

Fig. 15.2 Best practices for addressing ethical issues in e-business

such as firewalls, encryption, and intrusion detection systems, to counteract cyber threats and mitigate the risk of fraud and data breaches.

Consumer Protection

- Consumer protection focuses on safeguarding the rights and interests of consumers in e-business transactions. It involves regulations and practices that ensure fair and transparent dealings, accurate product information, secure payment methods, and resolving disputes to build trust and confidence among online consumers.

Ethical Issues in E-Business

- Ethical issues in e-business pertain to the moral considerations and dilemmas arising from online business activities. This includes issues such as the digital divide, intellectual property theft, online harassment, algorithmic bias, automation, job displacement, and the responsible use of emerging technologies. Addressing these ethical concerns promotes responsible and sustainable e-business practices.

Case Study

XYZ Corporation is an online retailer that distributes a variety of consumer goods. The company has recently expanded its operations and is now confronted with several legal, regulatory, and ethical issues regarding its e-business practices. The management team is concerned with ensuring compliance with applicable laws, protecting intellectual property rights, preserving customer privacy and data protection, preventing cybersecurity threats and fraud, adhering to consumer protection laws and advertising standards, and addressing potential ethical issues. They have asked students to analyze these issues and provide recommendations.

1. XYZ Corporation has created an original product design for one of its most popular commodities. How can the company prevent competitors from copying or imitating its design in the e-business environment? The XYZ Corporation desires to extend internationally. How should the company safeguard its trademarks and copyrights on international markets?
2. During the purchasing procedure, XYZ Corporation obtains personal information from its consumers. What legal responsibilities does the business have to safeguard this information and guarantee customer confidentiality? What steps should XYZ Corporation take in the event of a data breach to comply with applicable data protection laws and mitigate potential injury to its customers?
3. There has been an increase in phishing attacks against XYZ Corporation's customers. What steps can the business take to strengthen its cybersecurity and safeguard its consumers from such threats? XYZ Corporation accepts payments through its e-commerce platform. How can the company safeguard customer financial information and prevent fraudulent transactions?
4. The XYZ Corporation desires to conduct a promotional campaign offering discounts on specific products. What ethical and legal considerations should the business consider when advertising these discounts? XYZ Corporation receives a customer complaint regarding an online-purchased product. How should the company manage the complaint to assure compliance with consumer protection laws and customer satisfaction?
5. XYZ Corporation is contemplating using customer information for targeted advertising. What ethical considerations should the company take into account prior to implementing this practice, and how can they establish a balance between personalization and customer privacy? Several of XYZ Corporation's international suppliers have been implicated in labor rights violations. How can the business resolve these ethical concerns and ensure its supply chain reflects its values?

Discussion

1. Why is it important for e-businesses to consider legal, regulatory, and ethical considerations?
2. What are some common legal issues that e-businesses face?

3. How do regulations impact e-business operations?
4. What are some ethical considerations that e-businesses should keep in mind?
5. How can e-businesses ensure that they are complying with relevant laws and regulations?
6. What are some examples of intellectual property rights in e-business?
7. How can e-businesses protect their intellectual property rights?
8. What are some common privacy concerns in e-business?
9. How can e-businesses ensure that they are protecting customer privacy?
10. What are some common cybersecurity threats that e-businesses face?
11. How can e-businesses prevent fraud in their operations?
12. What are some common consumer protection laws that e-businesses must follow?
13. How can e-businesses ensure that their advertising is compliant with relevant regulations and guidelines?
14. What are some ethical issues that can arise in e-business operations?
15. How can e-businesses address ethical issues in their operations?
16. What are some best practices for ensuring legal, regulatory, and ethical compliance in e-business?
17. How can organizations stay up to date with emerging trends and changes in legal, regulatory, and ethical considerations for e-business?
18. What are some challenges associated with legal, regulatory, and ethical compliance in e-business?
19. How can organizations overcome these challenges to ensure compliance and ethical operations?
20. How can organizations balance the need for compliance with the need for innovation and growth in e-business?

Multiple-Choice Question

1. Which of the following is not an important consideration in e-business?
   (a) Legal considerations
   (b) Regulatory considerations
   (c) Ethical considerations
   (d) Financial considerations
2. Which of the following is not a key law or regulation relevant to e-business?
   (a) GDPR (General Data Protection Regulation)
   (b) CAN-SPAM Act
   (c) Sarbanes-Oxley Act
   (d) HIPAA (Health Insurance Portability and Accountability Act)

3. Intellectual property rights are important in e-business because they:
   (a) Protect customer data
   (b) Prevent cyberattacks
   (c) Ensure fair competition
   (d) Establish advertising standards
4. Trademarks, copyrights, and patents are types of:
   (a) Legal documents
   (b) Business contracts
   (c) Intellectual property rights
   (d) Consumer protection laws
5. Privacy and data protection laws impact e-business by:
   (a) Regulating online advertising
   (b) Preventing fraudulent activities
   (c) Safeguarding personal information
   (d) Ensuring fair pricing practices
6. Protecting personal data and customer privacy is important in e-business because it:
   (a) Builds customer trust
   (b) Reduces operational costs
   (c) Increases shareholder value
   (d) Enhances search engine optimization
7. Cybersecurity threats in e-business can include:
   (a) Social media advertising
   (b) Online surveys
   (c) Phishing attacks
   (d) Product pricing strategies
8. Fraud prevention measures in e-business involve:
   (a) Encryption techniques
   (b) Strategic partnerships
   (c) Competitive pricing
   (d) Product warranties
9. Consumer protection laws in e-business aim to:
   (a) Ensure product quality
   (b) Limit advertising opportunities
   (c) Encourage fair competition
   (d) Reduce corporate taxes

10. E-businesses must follow regulations and guidelines when advertising their products and services to:
    (a) Protect customer privacy
    (b) Prevent fraudulent transactions
    (c) Ensure fair competition
    (d) Minimize shipping costs
11. Ethical issues in e-business can arise in areas such as:
    (a) Logistics and supply chain management
    (b) Financial reporting
    (c) Product development
    (d) All of the above
12. Best practices for addressing ethical issues in e-business include:
    (a) Implementing transparent policies
    (b) Minimizing employee benefits
    (c) Ignoring customer feedback
    (d) Prioritizing profits over social responsibility
13. Which of the following is not an important consideration in e-business?
    (a) Legal considerations
    (b) Regulatory considerations
    (c) Ethical considerations
    (d) Financial considerations
14. Which of the following is not a key law or regulation relevant to e-business?
    (a) GDPR (General Data Protection Regulation)
    (b) CAN-SPAM Act
    (c) Sarbanes-Oxley Act
    (d) HIPAA (Health Insurance Portability and Accountability Act)
15. Intellectual property rights are important in e-business because they:
    (a) Protect customer data
    (b) Prevent cyberattacks
    (c) Ensure fair competition
    (d) Establish advertising standards

References

1. Yang, Y. and Z. Yin, *Accountancy for E-Business Enterprises Based on Cyber Security*. International Journal of Data Warehousing and Mining (IJDWM), 2023. **19**(6): p. 1–17.
2. Reynolds, G.W., & Stair, R. M, *Ethics in information technology*, ed. C. Learning. 2014.
3. Chesher, M., R. Kaura, and P. Linton, *Legal Aspects of e-Business*, in *Electronic Business & Commerce*, M. Chesher, R. Kaura, and P. Linton, Editors. 2003, Springer London: London. p. 319–331.
4. Klosek, J., *Data privacy in the information age*. 2000.

Appendix 1 (Discussion Answers)

Chapter 1

1. The term "e-business" refers to the electronic conduct of business activities using the Internet and other digital technologies.
2. E-business differs from conventional business models in that it depends predominantly on digital platforms and technology for operations, communication, and transactions, as opposed to physical locations and face-to-face interactions.
3. Online retail stores, digital marketplaces, online banking, e-commerce platforms, and online service providers are examples of e-business in action.
4. The advantages of e-business for businesses include increased market reach, increased efficiency, cost savings, enhanced consumer targeting and personalization, improved customer experiences, and streamlined business processes.
5. The digital landscape has changed substantially as a result of advances in technology, Internet connectivity, and consumer behavior over time. It has expanded to incorporate multiple industries and become a fundamental component of contemporary business operations.
6. Technology and infrastructure play a crucial role in e-business by providing the required tools, platforms, and networks to support digital operations, data administration, online transactions, and consumer communication.
7. Some essential benefits of e-business include access to global markets, availability 24 hours a day, 7 days a week, scalability, quicker and more efficient transactions, enhanced consumer insights, and the ability to leverage data for strategic decision-making.
8. E-businesses can increase an organization's efficacy by automating processes, reducing bureaucracy, facilitating quicker communication, enhancing supply chain management, facilitating real-time data analysis, and optimizing resource allocation.

H. Taherdoost, *E-Business Essentials*, EAI/Springer Innovations in Communication and Computing, https://doi.org/10.1007/978-3-031-39626-7

9. Costs of initial investment, cyber threats, data privacy concerns, technological complexity, resistance to change, and the need for skilled personnel are some of the obstacles organizations may face when implementing e-business strategies.
10. E-business can influence customer experiences by providing convenience, personalized interactions, simple access to information, seamless transactions, prompt customer support, and the capacity to provide customized products or services based on customer preferences.
11. Cybersecurity intrusions, data theft, identity fraud, online frauds, system malfunctions, legal and regulatory compliance issues, and reputational harm are risks associated with e-business.
12. Organizations can reduce these risks by instituting comprehensive security measures, utilizing encryption and secure payment mechanisms, routinely updating software and systems, educating employees and customers about online safety, and adhering to applicable legal and privacy regulations.
13. Trust is crucial in e-business because it establishes credibility, dependability, and confidence between businesses and their customers. It ensures that online transactions and sharing of personal information are secure for customers.
14. Secure and encrypted transactions, transparent privacy policies, clear communication, dependable customer support, positive customer reviews and ratings, an established brand reputation, and adherence to ethical business practices all contribute to building trust in an e-business environment.
15. A lack of trust in e-business can result in diminished customer confidence, lower conversion rates, the loss of business opportunities, a negative brand perception, an increase in customer attrition, and harm to the organization's overall reputation.
16. Organizations can build customer trust in e-commerce by prioritizing data security and privacy, providing transparent and accurate product information, providing secure payment options, ensuring prompt and reliable customer support, and cultivating a positive online reputation through social proof and customer testimonials.
17. Security is essential for establishing trust in e-business. Robust security measures, including encryption, firewalls, secure sockets layer (SSL) certificates, and authentication protocols, assist in protecting sensitive data, preventing unauthorized access, and demonstrating a dedication to customer safety.
18. Organizations can address customer concerns regarding data privacy and security in e-business transactions by implementing stringent data protection policies, obtaining customer consent for data usage, utilizing secure data storage and transmission methods, complying with applicable data protection regulations, and providing clear information on privacy practices.
19. Organizations need technology such as e-commerce platforms, secure payment gateways, customer relationship management (CRM) systems, and inventory management software to support e-business operations.
20. Cloud computing provides scalable and flexible hosting solutions, enables remote access to data and applications, facilitates data storage and backup, and supports collaborative work environments.

21. Critical e-business infrastructure components include secure payment gateways for online transactions, robust supply chain management systems for inventory and logistics management, customer relationship management (CRM) systems for customer data and interactions, and web hosting services for online presence and data storage.
22. Integration with existing systems, compatibility issues, scalability concerns, technical expertise requirements, infrastructure costs, and the need for ongoing updates and maintenance are some of the obstacles that organizations may encounter when implementing e-business infrastructure and technology.
23. By regularly updating and patching software, implementing strong access controls and authentication mechanisms, conducting regular security audits and testing, monitoring network traffic for anomalies, and investing in cybersecurity measures, organizations can ensure the security and dependability of their e-business infrastructure.
24. IT professionals play a vital role in sustaining the infrastructure and technology of e-businesses. They are accountable for system design, implementation, maintenance, and troubleshooting, as well as ensuring data security, managing networks and servers, providing technical support to users, and keeping abreast of emergent technologies and trends.

Chapter 2

1. E-business models are the frameworks and structures that define how businesses conduct online business operations. Due to technological advancements and alterations in consumer behavior, they have evolved over time, transitioning from basic online presence to complex platforms and ecosystems.
2. E-business models include Business-to-Business (B2B), Business-to-Consumer (B2C), Consumer-to-Consumer (C2C), Peer-to-Peer (P2P), and Platform-based. These models vary according to the target market, the nature of transactions, and the value proposition they provide.
3. Yes, a company can use multiple e-business models simultaneously. This enables them to target distinct consumer segments, diversify revenue streams, and broaden their market presence. Managing multiple models is however complex and requires effective coordination.
4. E-business models have transformed how businesses operate by facilitating global reach, efficient transactions, personalized marketing, enhanced consumer interactions, and streamlined supply chains. They have also disrupted conventional business models and created new innovation opportunities.
5. B2B e-business models emphasize transactions between businesses, whereas B2C e-business models emphasize transactions between businesses and individual consumers. B2B models typically involve greater transaction volumes, lengthier sales cycles, and more intricate purchasing procedures than B2C models.

6. In terms of customer behavior, B2B and B2C e-business models differ, as B2B customers are driven by rational decision-making and long-term relationships, whereas B2C customers are influenced by emotions and individual requirements. In B2B models, the purchasing process involves negotiation and procurement cycles, whereas in B2C models, decision-making processes are typically shorter. B2B marketing strategies emphasize relationship-building, whereas B2C marketing strategies emphasize widespread marketing and customer engagement.
7. The benefits of B2B e-business models include higher-order values, recurring business relationships, and the ability to provide customized products. B2B models face lengthier sales cycles, more complex negotiations, and a smaller customer base. B2C models can benefit from a larger customer base, quicker transactions, and impetuous purchases, but they may encounter intense competition and reduced profit margins.
8. Successful implementation of B2B and B2C e-business models requires a clear comprehension of the target audience, effective digital marketing strategies, a robust technological infrastructure, a seamless user experience, and effective customer relationship management.
9. E-business revenue models include subscription-based, transaction-based, advertising-based, freemium, and data-based models. Transaction fee-based models earn revenue from transactional commissions, whereas subscription-based models charge consumers a recurring fee for access to products or services. Advertising-based models generate revenue through the positioning of advertisements, whereas freemium models provide free fundamental services with optional premium enhancements. Data monetization models incorporate leveraging customer data for revenue generation.
10. Revenue models influence a company's pricing strategy by determining how consumers are charged and how offerings are structured. In addition to influencing customer acquisition, pricing models and value propositions attract distinct customer segments.
11. Subscription-based revenue models provide predictable recurring revenue, but customer retention may present challenges. Transaction fee-based models generate revenue per transaction, but profitability requires a high volume of transactions. Free services are provided by advertising-supported models, which rely on high website traffic and ad engagement. Depending on the business context, each model has its pros and cons.
12. Companies select the optimal revenue model for their e-business based on factors such as their target market, the character of their products or services, market dynamics, and competitive advantage. They must evaluate the revenue potential, consumer preferences, and scalability in order to make an informed choice.
13. Companies must develop an e-business strategy to align their online initiatives with their overall business objectives, obtain a competitive advantage, and remain competitive. It enables businesses to specify their target market, value proposition, marketing strategies, and operational processes.

14. Key components of an e-business strategy include defining business objectives, identifying target markets, developing a value proposition, designing customer experiences, establishing a digital presence, optimizing logistics and supply chains, implementing marketing and sales strategies, and continuously adapting to new technologies. These factors contribute to the success of a company by facilitating efficient operations, consumer satisfaction, and sustainable growth.
15. Companies segment their customers in order to identify distinct consumer categories with shared characteristics and preferences. They create a value proposition by comprehending consumer requirements, providing distinctive benefits, and setting themselves apart from competitors. Competitive analysis assists businesses in evaluating market trends, identifying rivals, and formulating strategies to obtain a competitive advantage.
16. Amazon, Alibaba, eBay, Airbnb, Uber, and Netflix are examples of companies that have devised and successfully implemented e-business strategies. These businesses have achieved substantial growth and market dominance in their respective industries by utilizing e-business models, implementing effective marketing strategies, delivering seamless customer experiences, and capitalizing on digital platforms.

Chapter 3

1. The e-environment is the digital space where electronic business activities occur, whereas the physical environment refers to the conventional offline world. The e-environment is distinguished from brick-and-mortar settings by its virtual interactions, online platforms, and digital transactions.
2. Multiple factors make the e-environment crucial for enterprises and consumers. It provides opportunities for global reach, cost-effective operations, enhanced consumer targeting, and enhanced efficiency for businesses. Convenience, access to a vast array of products and services, personalized experiences, and the ability to readily compare prices and reviews are advantageous to consumers.
3. The e-environment has revolutionized access to information and communication. It enables real-time, instantaneous communication via multiple channels, including email, messaging applications, social media, and videoconferencing. It has also facilitated the availability and accessibility of information through search engines, websites, online databases, and social networks.
4. Hardware (computers, mobile devices), software (applications, operating systems), networks (Internet, intranets), online platforms (websites, e-commerce platforms), digital content (text, images, and videos), and security systems (firewalls, encryption) are the primary components of the e-environment.
5. By facilitating data exchange, online transactions, automation, connectivity, and collaboration, digital technologies and online communication contribute to the e-environment. They provide the infrastructure and tools that enable

businesses and consumers to interact, exchange information, conduct business, and engage in a variety of online activities.

6. The e-environment's components have evolved substantially over time. Hardware, software, and networking technology advancements have resulted in quicker and more reliable Internet connections, enhanced user interfaces, and the proliferation of mobile devices. These developments have had a profound effect on society, transforming the ways in which people work, communicate, access information, and conduct business.
7. Since its inception, the e-environment has undergone continuous evolution. The Internet has evolved from simple text-based websites to interactive web applications, social media platforms, mobile applications, and the Internet of Things (IoT). In addition, advances in artificial intelligence, cloud computing, and data analytics have contributed to the evolution of the e-environment.
8. The creation of the World Wide Web, the advent of e-commerce, the rise of social media, the widespread adoption of smartphones, the emergence of cloud computing, and the advancements in data analytics and machine learning are significant turning points in the evolution of the e-environment.
9. The historical evolution of the e-environment has influenced its current state by influencing technological innovations, user behaviors, and regulatory frameworks. The accumulated knowledge and experiences from earlier phases have paved the way for future developments and enhancements in the e-environment.
10. The e-environment is distinguished by its connectivity, interactivity, global reach, personalization, scalability, and data-driven decision-making. These characteristics distinguish the e-environment from the limitations of the physical environment and define its digital nature.
11. These characteristics have various effects on enterprises and consumers. Utilizing connectivity and global reach, businesses can expand their customer base and enter new markets. Interactivity and personalization allow businesses to engage with consumers on a more personal level, delivering customized experiences and targeted marketing. Data-driven decision-making enables businesses to make well-informed decisions based on user data-derived insights.
12. The characteristics of the e-environment present both difficulties and possibilities. The need for comprehensive cybersecurity measures, addressing privacy concerns, managing information inundation, and adapting to swiftly evolving technologies are all obstacles. The ability to reach a global audience, collect enormous quantities of data for analysis, automate processes, and deliver personalized experiences creates opportunities.
13. The e-environment has substantially altered consumer behaviors and expectations. Consumers now anticipate convenience, round-the-clock accessibility, personalized experiences, prompt responses, and seamless interactions. Before making a purchase, they actively conduct online research and rely on online reviews, social media influencers, and recommendations.
14. Businesses operating in the e-environment are presented with opportunities such as increased market reach, enhanced consumer targeting, cost savings through automation, and the capacity to collect and analyze data for informed

decision making. However, they also face obstacles such as intense competition, managing their online reputation, ensuring data privacy and security, adapting to shifting consumer behaviors, and maintaining customer trust.

15. The e-environment has altered marketing and advertising techniques. Digital channels such as search engine optimization (SEO), social media marketing, influencer marketing, content marketing, email marketing, and targeted online advertising have supplemented and sometimes replaced traditional methods. These methods permit precise targeting, personalized communication, and quantifiable results.
16. Best practices for online reputation management include monitoring online mentions, interacting with customers via social media, promptly addressing customer concerns, encouraging positive reviews, managing online ratings and comments, and maintaining authenticity in online communications.
17. Businesses should establish a strong online presence via their websites, social media accounts, and other relevant online platforms in order to effectively engage with consumers online. They should actively attend to and respond to customer feedback, provide valuable content, offer personalized experiences, and facilitate customer participation and co-creation.
18. Use of web analytics tools, social media monitoring platforms, sentiment analysis, keyword tracking, and customer relationship management (CRM) systems are techniques for monitoring and analyzing online activity. These tools assist businesses in monitoring website traffic, measuring campaign effectiveness, monitoring social media mentions, gaining consumer insights, and identifying trends.
19. Data encryption software, secure payment gateways, firewall and intrusion detection systems, virtual private networks (VPNs), identity and access management solutions, backup and recovery systems, and data privacy compliance frameworks are available for managing and protecting online data. These tools aid businesses in protecting sensitive data, ensuring secure transactions, and adhering to data protection regulations.

Chapter 4

1. The earliest examples of e-business on the Internet were online marketplaces and directories, such as eBay and Craigslist, which facilitated individual-to-individual transactions.
2. Due to technological advancements, enhanced connectivity, and the development of secure online payment systems, the Internet has become an indispensable instrument for e-commerce. This enabled business to reach a global audience, conduct online transactions, and provide customized services.
3. The creation of secure protocols for online transactions (such as SSL), the emergence of search engines like Google, the growth of social media platforms,

the advent of mobile technology, and the rise of cloud computing are significant developments that have shaped the Internet and e-business.

4. The advent of mobile technology and the ubiquity of mobile devices have had a significant impact on e-business. It has resulted in the creation of mobile applications, responsive websites, and location-based services, allowing businesses to reach consumers at any time and in any location.
5. Existing e-business applications include online retail (e-commerce), online marketplaces, digital services, software as a service (SaaS), online finance, social media platforms, and online advertising.
6. E-commerce and online purchasing have revolutionized the way consumers browse and acquire products. It provides convenience, a larger selection of products, competitive prices, personalized suggestions, and the ability to compare and read reviews.
7. Pristine targeting, real-time analytics, the ability to reach a global audience, personalized content, and interactive ad formats are essential aspects of digital marketing and advertising that make it effective for e-business.
8. The influence of social media and user-generated content on e-business is significant. They offer businesses platforms for engaging with consumers, fostering brand loyalty, gathering feedback, and leveraging user-generated content for marketing purposes.
9. The advantages of cloud computing and software as a service (SaaS) for e-business include scalability, cost-effectiveness, adaptability, accessibility from any device, automatic updates, and centralized data storage and management.
10. E-business has a global reach, reduced administrative costs, 24/7 availability, the ability to collect and analyze consumer data, and the potential for rapid scalability.
11. Security threats (e.g., data intrusions), privacy concerns, online payment fraud, intellectual property infringement, competition, and the need to adapt to swiftly evolving technology are major risks and challenges associated with e-business.
12. By implementing robust cybersecurity measures, encrypting sensitive data, regularly updating software, providing secure payment options, educating employees and customers about online security, and adhering to privacy regulations, businesses can mitigate these risks and address these challenges.
13. Businesses must ensure compliance with relevant laws and regulations in the e-business space by keeping abreast of legal requirements, implementing appropriate data protection measures, obtaining required licenses and permits, and collaborating with legal professionals.
14. Amazon, Alibaba, Google, Facebook, Uber, Airbnb, and Netflix are examples of companies that have effectively utilized the Internet for e-commerce.
15. Innovative business models, customer-centric approaches, seamless user experiences, effective use of data and analytics, continuous technological advancement, and strong brand development are the key factors that contributed to the success of these companies.

16. Lessons learned from these e-business success stories include the significance of embracing technology, focusing on customer experience, leveraging data for insights and personalization, being adaptable to changing market conditions, and establishing a solid online presence and brand reputation.

Chapter 5

1. E-enterprise security and control refer to the measures and practices implemented to safeguard electronic business operations against illicit access, data intrusions, and other cyber threats. It differs from traditional security and control measures in that it concentrates on securing online systems, networks, and transactions, which may require different technologies and approaches than physical security.
2. Hacking and unauthorized access to systems, malware and ransomware attacks, phishing and social engineering schemes, data breaches, identity theft, and denial-of-service (DoS) attacks are examples of e-business security threats.
3. It is crucial for businesses to prioritize security and control in their e-business operations because the digital landscape poses substantial risks. Data breaches and assaults can result in financial losses, reputational harm, legal repercussions, and a loss of consumer confidence. By placing a premium on security, businesses can safeguard sensitive data, maintain the integrity of their operations, and mitigate potential risks.
4. Due to the growing complexity and interdependence of digital systems, the evolution of e-business has had an impact on security and control measures over time. As e-business has expanded, so have the sophistication and frequency of cyber threats, necessitating the development of more robust security protocols, technologies, and strategies.
5. Hacking, malware, phishing, internal threats, data intrusions, and web application vulnerabilities are typical threats to e-business security. They differ from traditional security threats in that they target digital assets primarily, exploit online vulnerabilities, and frequently employ techniques unique to the digital environment.
6. Various methods, such as security audits, vulnerability assessments, penetration testing, monitoring network traffic, analyzing system records, and staying informed about emergent threats and security best practices, allow businesses to identify potential vulnerabilities in their e-business systems.
7. In recent years, notable security vulnerabilities include the Equifax data breach, the Facebook-Cambridge Analytica scandal, and the Marriott data breach. These intrusions have had significant effects on businesses and consumers, resulting in monetary losses, compromised confidential information, a loss of trust, regulatory investigations, and legal repercussions.
8. Using a combination of technical measures such as firewalls, encryption, multifactor authentication, intrusion detection systems, and routine software

updates, businesses can secure themselves from cyberattacks and e-business security threats. In addition to employee training, security policies, incident response plans, and partnerships with cybersecurity specialists, protection can be enhanced through employee education, security policies, and incident response plans.

9. Best practices for securing e-business systems and networks include implementing strong access controls and user authentication, conducting regular security audits and risk assessments, encrypting sensitive data, maintaining backups, educating employees about security risks, and establishing incident response procedures.
10. By establishing a comprehensive security framework, defining roles and responsibilities, conducting regular security training, enforcing strict access controls, monitoring system activity, and conducting periodic security assessments, businesses can implement security protocols and procedures to protect their e-business operations.
11. Encryption and other security technologies play an essential role in assuring the security and control of e-businesses. Encryption safeguards sensitive information during transmission and storage by encrypting it in a manner that only authorized parties can decrypt and access. Other security technologies, including secure sockets layer (SSL), virtual private networks (VPNs), and intrusion detection systems (IDS), also contribute to the security of e-business operations.
12. Businesses can remain current with the most recent security and control measures for e-business by actively monitoring cybersecurity news and industry trends, participating in relevant conferences and workshops, collaborating with cybersecurity experts, maintaining partnerships with technology vendors, and engaging in ongoing professional development.
13. Data protection and privacy laws, industry-specific regulations, cybersecurity standards, and intellectual property laws are among the laws and regulations relevant to e-business security and control. These laws have an effect on businesses because they impose legal obligations regarding data security, breach notification, consumer privacy rights, and industry-specific compliance requirements.
14. When instituting security and control measures for e-business, companies should take ethical considerations such as preserving consumer privacy and assuring transparency into account, obtaining informed consent, minimizing data collection and retention, and addressing the potential impact of security measures on user experience and trust in data collection and utilization.
15. Uber, which faced penalties for concealing a data breach, and Yahoo, which suffered reputational damage and financial losses due to multiple large-scale data breaches, are examples of companies that have faced legal or ethical repercussions for inadequate e-business security measures.
16. Businesses can achieve a balance between the need for security and control and user privacy and transparency in their e-business operations by adopting privacy-by-design principles, providing clear and concise privacy policies, obtaining consent for data collection and use, implementing strong security

measures while minimizing data collection, and communicating regularly with users about security practices and updates.

Chapter 6

1. In the context of e-business, billing and payment systems refer to the mechanisms and procedures used to manage financial transactions online. They operate electronically, using the Internet and digital technologies to effectuate payments, as opposed to relying on physical currency or checks.
2. Effective billing and payment systems are crucial for multiple reasons for e-businesses. They provide consumers with a seamless and convenient payment experience, resulting in increased customer satisfaction and sales. In addition, they improve operational efficiency by automating payment processes and minimizing human error. Additionally, dependable and secure payment systems cultivate client confidence and credibility, which is essential for establishing lasting relationships.
3. Over time, billing and payment systems have evolved to satisfy the specific requirements of e-businesses. Initial online payment gateways allowed consumers to make purchases using credit or debit cards. New payment methods, such as digital wallets, mobile payments, and cryptocurrencies, have emerged as a result of technological advancements. These systems have become more complex with the addition of recurring invoicing, subscription management, and fraud prevention mechanisms.
4. Effective billing and payment systems for e-businesses include user-friendliness, secure transaction processing, multiple payment options, currency support, integration with other business systems (such as inventory management), robust reporting and analytics, and scalability to accommodate increasing transaction volumes.
5. Credit and debit card payments, electronic funds transfers (EFT), digital wallets (e.g., PayPal), mobile payments (e.g., Apple Pay, Google Pay), and cryptocurrencies (e.g., Bitcoin) are the most prevalent forms of payment systems used in e-business. These systems transmit payment data between the consumer, the merchant, and the financial institutions involved in the transaction in a secure manner.
6. Different payment systems have their benefits and drawbacks. Credit and debit cards are convenient and widely accepted, but processing fees may apply. Customers must have an account with the wallet provider in order to use digital wallets, which allow for a streamlined transaction experience. Convenient for on-the-go transactions, mobile payments are dependent on smartphone availability. Cryptocurrencies provide security and global accessibility, but they can be volatile, and their adoption is limited.
7. Fees charged for transactions and processing periods required to complete payments can vary between payment systems. Credit card transactions typically

incur transaction fees, whereas digital wallet and mobile payments may incur reduced or no fees. The processing period for digital wallets can be instantaneous, while certain bank transfers can take several days.

8. When selecting a payment system for e-business operations, companies should consider customer preferences, security features, transaction costs, integration capabilities with existing systems, scalability, international payment support, as well as the reputation and dependability of the payment service provider.
9. Implementing secure sockets layer (SSL) encryption, using tokenization to protect sensitive data, adopting two-factor authentication, monitoring transactions for suspicious activity, regularly updating and patching payment software, and complying with industry security standards (e.g., PCI DSS) are best practices for ensuring secure transactions and preventing fraud in billing and payment systems.
10. Using tools like real-time transaction monitoring, machine learning algorithms to identify fraudulent activity patterns, and manual review processes for high-risk transactions, businesses can implement fraud detection and prevention measures in their payment systems. In addition, they can employ address verification, CVV verification, and risk assessment mechanisms to determine the legitimacy of a transaction.
11. Encryption and other security technologies play a vital role in the security of invoicing and payment systems. Encryption ensures that payment information transmitted between parties cannot be read by unauthorized parties. Secure protocols such as SSL and transport layer security (TLS) create secure connections between the customer's device, the merchant, and the payment service provider, preventing data interception and tampering.
12. Businesses can strike a balance between the need for security and the desire to offer a seamless payment experience by instituting user-friendly payment interfaces and optimizing transaction processes, minimizing customer friction and clearly communicating security measures. This can be accomplished by implementing transparent privacy policies, visible trust seals, and proactive customer service.
13. Payment disputes, chargebacks, transaction failures, unauthorized transactions, and technical errors are typical problems and challenges in invoicing and payment systems. Various factors, including communication errors, processing delays, fraudulent activities, and consumer dissatisfaction, can cause these problems.
14. Businesses can effectively address these issues and challenges by having clear refund and dispute resolution policies, providing prompt customer support, maintaining accurate transaction records, leveraging fraud detection mechanisms, conducting thorough investigations when disputes arise, and collaborating with payment service providers to resolve issues promptly.
15. Successful invoicing and payment systems include characteristics such as dependability, security, seamless integration with other business systems, a variety of payment options, user-friendly interfaces, comprehensive reporting

and analytics, and solid customer support. PayPal, Stripe, Square, and Braintree are four examples of successful payment systems.

16. In order to acclimate to shifting trends and technologies in the invoicing and payment system landscape, businesses must remain abreast of industry developments, track customer preferences, and utilize emergent payment methods. They should be proactive in incorporating new technologies such as mobile payments, digital wallets, and cryptocurrencies, while assuring compatibility with existing systems and maintaining a strong emphasis on security and consumer experience.

Chapter 7

1. In e-business, supply chain management incorporates digital technologies and online platforms to manage the flow of products and services, in contrast to traditional businesses. It accentuates the incorporation of online marketplaces, virtual communication, and automated processes in order to improve efficiency and consumer satisfaction.
2. Effective supply chain management offers numerous advantages to e-businesses, including improved inventory control, reduced costs, speedier order fulfillment, enhanced customer service, increased transparency, and enhanced collaboration with suppliers and partners.
3. E-businesses can use supply chain management to gain a competitive advantage by streamlining operations, reducing lead times, providing flexible delivery options, optimizing inventory levels, establishing strong partnerships with dependable suppliers, and utilizing data analytics to make informed decisions.
4. In the overall supply chain management procedure for e-businesses, sourcing plays a vital function. It entails identifying and selecting suppliers who can satisfy quality standards, offer competitive pricing, provide dependable delivery, and align with the e-business's values and goals. Effective sourcing enhances the effectiveness and dependability of the supply chain.
5. In e-business supply chain management, inventory management is crucial. It entails optimizing inventory levels to meet customer demand, minimizing stockouts and overstocks, utilizing forecasting techniques, employing just-in-time practices, and instituting efficient warehousing and fulfillment processes to guarantee timely order fulfillment and customer satisfaction.
6. Implementing robust transportation management systems, utilizing route optimization algorithms, adopting real-time tracking technologies, establishing strategic partnerships with logistics providers, and continuously monitoring and improving delivery processes are best practices for optimizing logistics in e-business supply chain management.
7. The function of technology in enhancing supply chain management in e-businesses is crucial. It permits automation, data integration, real-time visibility, and stakeholder collaboration. It is possible to use cloud computing, the

Internet of Things (IoT), artificial intelligence (AI), and blockchain to optimize operations, improve forecasting accuracy, increase inventory visibility, and facilitate seamless communication.

8. Enterprise resource planning (ERP) systems, warehouse management systems (WMS), transportation management systems (TMS), demand planning software, electronic data interchange (EDI), and advanced analytics tools are examples of supply chain management technologies. By automating processes, optimizing inventory, facilitating data exchange, and enabling data-driven decision-making, these technologies increase efficiency and effectiveness.
9. E-businesses can ensure the security and dependability of the technology used in supply chain management by implementing robust cybersecurity measures, conducting regular audits and vulnerability assessments, encrypting sensitive data, establishing secure communication protocols, and collaborating with reputable technology vendors and service providers.
10. Common challenges encountered by e-businesses in supply chain management include managing logistics complexities, coordinating multiple stakeholders, mitigating risks associated with data security and privacy, handling demand fluctuations, adapting to changing customer expectations, and addressing product quality and returns issues. Through effective communication, collaboration, process optimization, and the utilization of technology solutions, these obstacles can be overcome.
11. E-businesses can ensure quality control throughout the entire supply chain management process by implementing quality assurance protocols, conducting regular inspections and audits, enforcing supplier compliance with quality standards, implementing stringent testing procedures, and establishing effective feedback and monitoring mechanisms.
12. E-businesses can select the best suppliers and vendors for their supply chain management requirements using a variety of methods. These strategies consist of conducting comprehensive supplier assessments, evaluating supplier capabilities and performance, taking into account factors such as pricing, reliability, responsiveness, and sustainability, establishing clear contractual agreements, fostering collaborative relationships, and periodically reviewing and optimizing the supplier base.

Chapter 8

1. E-procurement is the use of electronic technology, such as the Internet, to expedite and automate the business's procurement processes. It is closely related to e-business because online platforms and digital tools are utilized to manage purchasing activities.
2. E-procurement offers a number of advantages to businesses, including cost savings, enhanced efficiency, increased transparency, enhanced precision, and access to a larger pool of suppliers. It streamlines procurement processes by

automating duties, reducing documentation, providing real-time information, enabling quicker communication with suppliers, and facilitating improved negotiation and sourcing.

3. The implementation of e-procurement systems can present businesses with challenges, including resistance to change, lack of technological infrastructure, security concerns, data integration issues, supplier adoption, and ensuring user acceptability. Effective change management strategies, investment in a robust IT infrastructure, cybersecurity measures, supplier engagement and training, and addressing user concerns through appropriate training and support can help to overcome these obstacles.
4. Businesses can ensure the effectiveness and efficiency of their e-procurement systems by establishing clear goals and objectives, aligning the system with business needs, selecting the appropriate technology solution, regularly evaluating performance, optimizing workflows and processes, fostering collaboration between stakeholders, ensuring data accuracy and integrity, and continuously enhancing the system based on feedback and best practices.
5. E-catalogs (electronic catalogs for sourcing products), e-tendering (online bidding and proposal submission), e-invoicing (electronic invoice generation and processing), e-reverse auctions (online auctions where suppliers compete to offer the lowest price), and supplier relationship management systems (digital platforms for managing supplier interactions) are the most common types of e-procurement.
6. In e-procurement, online auctions involve consumers publishing their needs and suppliers bidding to win the contract by offering the lowest price. Increased competition, potential cost savings, and transparency are among the benefits. Quality concerns, a lack of relationship building, and time-consuming tendering processes can be disadvantages.
7. A request for proposal (RFP) is a document used in e-procurement to solicit proposals from potential suppliers. It specifies the buyer's needs, evaluation criteria, and contract conditions. Using RFPs provides advantages such as standardized evaluation, plain communication, and competitive tendering. Possible disadvantages include protracted response periods, limited innovation, and potential selection process biases.
8. A reverse auction is a form of e-procurement in which the consumer publishes their specifications and suppliers compete by submitting progressively reduced proposals. Reduced prices, increased competition, and time savings are all advantages. Challenges may include supplier dissatisfaction, a limited emphasis on quality, and the possibility of supplier collusion.
9. Best practices for implementing e-procurement systems include conducting a thorough needs assessment, involving key stakeholders, selecting the appropriate technology solution, providing comprehensive training, ensuring supplier participation and integration, monitoring and evaluating performance, and continuously optimizing processes based on industry benchmarks.
10. Businesses can adopt compatible technologies, establish data-sharing mechanisms, align workflows and information flows, implement robust APIs

(application programming interfaces) for system integration, and utilize enterprise resource planning (ERP) or procurement management systems to achieve seamless coordination between e-procurement, supply chain management, and inventory management.

11. Resistance to change, lack of organizational support, data security concerns, supplier engagement, system integration complexities, and ensuring user adoption are key challenges in e-procurement implementation. Effective change management strategies, strong leadership and support, investment in cybersecurity measures, supplier orientation and training, dedicated integration efforts, and the provision of user-friendly interfaces and support can be used to overcome these obstacles.
12. Organizations can evaluate the success of their e-procurement implementation by analyzing key performance indicators (KPIs) such as cost reductions realized, process cycle times, and supplier performance metrics. Order accuracy, contract compliance, user satisfaction, system availability, and return on investment (ROI) are key performance indicators. These metrics aid in assessing the efficacy, effectiveness, financial impact, and overall performance of an organization.
13. E-procurement offers advantages such as cost savings through efficient sourcing, reduced administrative expenses, and optimized supplier negotiations. It enhances productivity by automating processes, minimizing human error, providing real-time data, and streamlining workflows. In addition, it increases transparency, facilitates regulatory compliance, improves supplier management, and promotes data-driven strategic decision-making.
14. By implementing effective supplier relationship management (SRM) practices, fostering open communication and collaboration, providing timely feedback, establishing mutually beneficial partnerships, offering prompt payment terms, ensuring fair and transparent bidding processes, and recognizing and rewarding high-performing suppliers, businesses can use e-procurement to improve supplier relationships.
15. Challenges and limitations of e-procurement include supplier selection and qualification, legal and regulatory compliance, data security and privacy concerns, the possibility of supplier resistance, limited supplier diversity, and the requirement for ongoing monitoring and auditing. Businesses can address these challenges by implementing robust supplier evaluation and onboarding processes, remaining current with legal requirements, adopting secure technology solutions, fostering supplier trust and transparency, promoting supplier diversity, and conducting routine compliance audits.
16. Businesses can determine the efficacy of their e-procurement systems by monitoring metrics such as cost savings, process efficiency, cycle times, error rates, supplier performance, user satisfaction, and system reliability. Other evaluation techniques include user surveys, feedback from stakeholders, benchmarking against industry standards, and periodic audits to ensure compliance and identify enhancement opportunities.

Chapter 9

1. The rise of e-commerce and mobile technology has had a significant impact on the expansion of digital marketing by providing new platforms and channels for businesses to engage with their target audience. It has widened the scope of digital marketing campaigns and made it simpler for businesses to interact with consumers at any time, from any location.
2. Nike's "Just Do It" campaign, which focused on inspiring athletes and promoting their brand values, and Coca-Cola's "Share a Coke" campaign, which personalized their product packaging and encouraged customers to share their experiences on social media, are examples of successful digital marketing campaigns. These campaigns contributed to the success by generating brand awareness, stimulating consumer engagement, and boosting sales.
3. The COVID-19 pandemic has substantially increased the value of digital marketing for businesses. As a result of lockdowns and social distancing measures, businesses had to rely more on online channels to reach consumers. Digital marketing has become indispensable for maintaining consumer relationships, generating revenue, and adapting to the shifting market dynamics.
4. Influencer marketing, voice search optimization, chatbots and AI-powered customer support, interactive content, and personalized marketing are some emerging trends in digital marketing. Businesses should be aware of these trends and consider incorporating them into their strategies in order to remain competitive and engage their target audiences effectively.
5. To ensure that their digital marketing efforts align with their overall marketing strategy, businesses should establish clear goals and objectives, identify their target audience, conduct market research, develop a cohesive brand message, select appropriate digital marketing channels, create consistent and relevant content, and analyze performance metrics on a regular basis.
6. The most effective digital marketing channels for reaching a particular target audience depend on business type, demographics of the target audience, and marketing objectives. Popular channels, however, include social media platforms such as Facebook and Instagram for targeting a broad audience, email marketing for reaching a more segmented audience, and search engine marketing for targeting users actively seeking for specific products or services.
7. By monitoring key performance indicators (KPIs) such as website traffic, conversion rates, cost per acquisition (CPA), return on ad spend (ROAS), and customer lifetime value (CLV), businesses can effectively measure the ROI of their digital marketing efforts. Utilizing analytics tools, configuring conversion monitoring, and conducting A/B testing can assist businesses in measuring and optimizing the effectiveness of their digital marketing campaigns across multiple channels.
8. Common social media marketing errors include inconsistent branding, lack of audience targeting, excessive self-promotion, disregarding customer feedback, and failure to engage followers. To avoid these errors, businesses should develop

a social media strategy, understand their target audience, provide valuable content, interact actively with followers, and track social media metrics to determine the success of their efforts.

9. By creating personalized and relevant email campaigns, segmenting their email lists based on customer preferences, providing valuable content and offers, using compelling subject lines, and regularly analyzing email performance metrics to optimize campaigns, businesses can use email marketing to build relationships with customers and increase sales.
10. Best practices for optimizing a website's SEO to increase its visibility in search engine results include conducting keyword research and incorporating relevant keywords into website content, optimizing meta tags and descriptions, improving website loading speed and mobile-friendliness, creating high-quality and shareable content, obtaining backlinks from credible sources, and routinely monitoring and analyzing website analytics.
11. A lack of clear goals and identification of the target audience, insufficient market research, inconsistent messaging across channels, failure to track and measure campaign performance, and resistance to adapt to changing trends are common errors made by businesses when developing digital marketing strategies. To avoid these errors, businesses must define specific objectives, conduct exhaustive market research, develop a unified and consistent brand message, monitor and analyze performance metrics, and keep abreast of industry trends.
12. Businesses can achieve a balance between automation and personalization in their digital marketing strategies by utilizing automation tools for repetitive duties, such as email scheduling, while still providing personalized and customized content based on customer preferences and behaviors. It is essential to maintain a human touch in communication and ensure that automation enhances rather than detracts from the consumer experience.
13. Utilizing audience segmentation based on demographics, interests, and behaviors, leveraging social media advertising platforms' targeting capabilities, retargeting website visitors with relevant ads, creating personalized content and offers, and conducting market research to understand the needs and preferences of the target audience are all effective ways to target specific audiences with digital marketing campaigns.
14. Businesses can use content marketing to build brand awareness and attract new customers by creating valuable and informative content that resonates with their target audience, using formats such as blog posts, videos, infographics, and podcasts, promoting content via social media and email marketing, optimizing content for search engines, and measuring content performance using metrics such as engagement and conversion rates.
15. Storytelling is essential to the success of digital marketing campaigns because it captures the attention and emotions of the intended audience. It helps establish a relationship between the brand and the consumer, conveys the brand's identity and values, and makes the marketing message more memorable and relatable. Storytelling can be integrated into a number of digital marketing

channels, including website content, social media posts, and video advertisements.

16. Analyzing data from multiple sources, such as website analytics, social media insights, email marketing metrics, and consumer behavior monitoring, businesses can use data analytics to identify trends and optimize their digital marketing strategies. Data analytics can provide businesses with insights into consumer preferences, campaign performance, and market trends, allowing them to make data-driven decisions and improve their marketing strategies.
17. When collecting and utilizing customer data for digital marketing purposes, businesses should consider ethical considerations such as obtaining proper consent for data collection, protecting customer privacy and security, being transparent about data usage, offering opt-out options, and adhering to applicable data protection regulations. Maintaining a positive brand image and fostering long-term customer relationships requires a commitment to consumer trust and the responsible use of data.
18. Businesses can use A/B testing to improve the effectiveness of their digital marketing campaigns by comparing and measuring the performance of two versions of a campaign element (such as a landing page, email subject line, or call to action). A/B testing allows businesses to make data-driven decisions and optimize campaigns accordingly.
19. Open rates, click-through rates, conversion rates, unsubscribe rates, exit rates, and return on investment (ROI) are key metrics that businesses should monitor to determine the success of their email marketing campaigns. By tracking these metrics, businesses can assess the efficacy of their email marketing campaigns, identify areas for development, and optimize their email marketing strategies for maximum results.
20. By monitoring metrics such as website traffic, visitor demographics, popular content, conversion rates, and referral sources, Google Analytics enables businesses to gain insight into their website's performance and user behavior. Google Analytics provides valuable data for understanding how users interact with a website, identifying areas for optimization, measuring the efficacy of digital marketing initiatives, and making informed decisions to improve the overall user experience.

Chapter 10

1. In the context of e-business, CRM refers to customer relationship management, a strategy and technology used to manage and analyze interactions with consumers throughout their lifecycle.
2. CRM in e-business differs from conventional business models in that it uses digital channels and technology to collect and analyze client data, thereby facilitating personalized interactions and targeted marketing campaigns.

3. Effective CRM is essential for e-businesses because it facilitates the development and maintenance of strong customer relationships, the improvement of customer satisfaction, the increase of customer retention, and the stimulation of business growth through personalized marketing and sales efforts.
4. Customer data management, sales force automation, marketing automation, customer service and support, and analytics and reporting tools are common CRM components in e-business.
5. Customer data management is essential to effective CRM because it collects and organizes customer data from multiple touchpoints, allowing businesses to gain insight into customer behaviors, preferences, and requirements. This information facilitates the delivery of customized experiences and targeted marketing campaigns.
6. The advantages of marketing automation in CRM include streamlining marketing processes, automating repetitive duties, and facilitating personalized messaging. However, the risk of impersonal communication and the need for cautious planning and customization to prevent consumer alienation are disadvantages.
7. Customer service can be incorporated into a CRM system by collecting customer interactions and inquiries, monitoring support tickets, and ensuring information flows seamlessly between customer service representatives and other departments.
8. Best practices for implementing CRM systems and processes in e-businesses include aligning CRM strategy with business objectives, selecting the appropriate CRM software, ensuring proper training and user adoption, and measuring and analyzing CRM performance on a regular basis.
9. Through data integration, API (application programming interface) integrations with other software solutions, and cross-functional collaboration between departments such as marketing, sales, and customer service, CRM can be effectively incorporated with other systems and processes.
10. Common obstacles encountered during the implementation of CRM in e-business include data quality issues, employee resistance to change, integration complexities with existing systems, and assuring data privacy and regulatory conformance.
11. The advantages of implementing CRM in e-business include increased customer satisfaction and loyalty through personalized experiences, enhanced customer engagement, increased sales effectiveness, higher customer retention rates, and data-driven insights for strategic decision-making.
12. Data privacy and security can be addressed when implementing CRM in e-business by implementing robust data protection measures, complying with applicable privacy regulations, utilizing secure data storage and transmission methods, and providing customers with transparent information about data usage and consent.
13. Challenges or limitations of CRM in e-business include the requirement for ongoing data management and maintenance, the risk of data breaches or misuse, the potential for customer privacy concerns, and the need for continuous adaptation to changing customer preferences and technologies.

Chapter 11

1. Business intelligence (BI) differs from traditional data analysis by focusing on providing insights and actionable information for decision-making. Unlike traditional data analysis, which primarily involves examining historical data, BI involves the use of advanced analytics techniques to analyze data in real time, identify patterns and trends, and make informed predictions.
2. Examples of how business intelligence has been used to drive business success include:

   - Identifying customer preferences and behaviors to optimize marketing strategies and improve customer satisfaction
   - Analyzing sales data to identify product performance and market opportunities
   - Monitoring operational processes and supply chain to enhance efficiency and reduce costs
   - Forecasting demand and optimizing inventory management to minimize stockouts and overstocks.
   - Analyzing financial data to improve budgeting, forecasting, and financial decision-making

3. The growth of big data has significantly increased the importance of business intelligence for e-businesses. Big data provides a vast amount of structured and unstructured data, which can be challenging to manage and analyze without advanced BI tools. Business intelligence helps e-businesses extract valuable insights from big data, understand customer behavior, make data-driven decisions, and gain a competitive edge.
4. Some emerging trends in business intelligence that e-businesses should be aware of include:

   - Artificial intelligence and machine learning algorithms for advanced analytics and predictive modeling
   - Natural language processing and conversational analytics for easier data exploration and querying
   - Self-service BI tools that empower business users to perform their own data analysis
   - Integration of BI with cloud computing for scalability and cost-effectiveness
   - Real-time analytics and dashboards for instant data monitoring and decision-making

5. E-businesses can ensure that their business intelligence efforts align with their overall business strategy by:

   - Clearly defining their business goals and objectives and identifying key performance indicators (KPIs) to measure progress
   - Aligning BI initiatives with specific business needs and focusing on areas that have the most significant impact on achieving strategic objectives

- Involving stakeholders from different departments to ensure that BI solutions address their specific requirements
- Regularly evaluating and refining BI strategies to adapt to changing business conditions and technological advancements

6. Data warehousing is the process of collecting, organizing, and storing large volumes of structured and semi-structured data from various sources in a central repository. It contributes to business intelligence by providing a consolidated and integrated view of data, which enables efficient analysis, reporting, and decision-making. Data warehouses are designed for query and analysis purposes and often include data cleansing, transformation, and aggregation steps to improve data quality and usability.
7. Data mining helps e-businesses uncover valuable insights from their data by applying statistical and machine learning techniques to identify patterns, relationships, and trends. It involves extracting useful information from large datasets and transforming it into actionable knowledge. Data mining can be used to understand customer behavior, segment customers, detect fraud, optimize processes, and make predictions based on historical data.
8. Predictive analytics involves using historical and current data to make predictions about future events or outcomes. E-businesses can use predictive analytics to make more informed decisions by analyzing patterns and trends in data, identifying potential risks and opportunities, and forecasting future performance. It can be applied in various areas such as sales forecasting, demand planning, customer churn prediction, and inventory optimization.
9. Advantages of using different components of business intelligence:

   - Enhanced decision-making based on data-driven insights
   - Improved operational efficiency through process optimization
   - Increased competitive advantage by identifying market trends and opportunities
   - Better understanding of customer preferences and behaviors
   - Improved collaboration and communication across departments

   Disadvantages of using different components of business intelligence:

   - High initial investment and ongoing maintenance costs
   - Complexity in implementing and managing BI systems
   - Data quality and integration challenges
   - Privacy and security concerns related to data storage and usage
   - Potential reliance on technical expertise for data analysis and interpretation

10. Best practices for implementing a business intelligence system in an e-business include:

    - Clearly define business objectives and align BI initiatives accordingly.
    - Ensure data quality and consistency through data cleansing and validation processes.

- Involve stakeholders from different departments to understand their requirements and prioritize their needs.
- Select appropriate BI tools and technologies that align with business goals and user capabilities.
- Provide training and support to users to promote adoption and effective utilization of the BI system.
- Regularly monitor and evaluate the performance of the BI system to identify areas for improvement.

11. E-businesses can ensure that their business intelligence system is integrated with other systems and processes by:

    - Establishing data integration mechanisms to extract data from various sources and consolidate it in the BI system
    - Integrating the BI system with existing business applications and systems to enable seamless data flow and reporting
    - Implementing data governance practices to ensure data consistency, accuracy, and security across different systems
    - Promoting collaboration and communication between the BI team and other departments to understand their needs and provide relevant insights

12. Common challenges that e-businesses face when implementing a business intelligence system include:

    - Data quality issues, such as inconsistent or incomplete data
    - Data integration challenges when dealing with multiple data sources and formats
    - Lack of user adoption and resistance to change
    - Technical complexities and resource requirements
    - Ensuring scalability and flexibility as the business grows

    These challenges can be overcome by:

    - Implementing data quality processes and data governance practices
    - Utilizing data integration tools and techniques to streamline data consolidation
    - Providing training and support to users to encourage adoption
    - Collaborating with IT teams and involving business stakeholders in the implementation process
    - Planning for scalability and flexibility by selecting scalable BI architectures and platforms

13. Business intelligence helps e-businesses make better decisions and improve operational efficiency by providing timely and accurate insights based on data analysis. It enables e-businesses to:

    - Identify trends, patterns, and correlations in data to make informed strategic decisions

- Monitor key performance indicators (KPIs) in real time to identify performance gaps and take corrective actions
- Optimize operational processes by identifying bottlenecks, inefficiencies, and areas for improvement
- Understand customer behaviors and preferences to personalize marketing efforts and enhance customer experience
- Anticipate market changes and trends to stay competitive and seize new opportunities

14. Potential drawbacks of using business intelligence include concerns around data quality and privacy. These include:

    - Data quality issues, such as incomplete or inaccurate data, leading to erroneous insights and decisions
    - Privacy concerns related to the collection, storage, and usage of customer data
    - Compliance with data protection regulations, such as GDPR or CCPA
    - Vulnerabilities in data security, leading to potential data breaches or unauthorized access
    - Overreliance on data-driven decision-making, neglecting qualitative factors and human judgment

15. E-businesses can address these challenges and ensure ethical and responsible use of business intelligence by:

    - Implementing robust data governance practices to ensure data quality, privacy, and security
    - Complying with relevant data protection regulations and industry standards
    - Being transparent about data collection and usage practices and obtaining consent when required
    - Applying anonymization or pseudonymization techniques to protect personally identifiable information
    - Establishing ethical guidelines and frameworks for data usage, including ethical considerations in decision-making processes
    - Regularly reviewing and auditing BI processes to identify and address potential ethical or privacy concerns

Chapter 12

1. E-business adoption is the incorporation and utilization of electronic technologies for undertaking business operations. It is essential for modern businesses because it allows them to reach a larger audience, increase efficiency, cut costs, and remain competitive in the digital age.
2. The availability and affordability of Internet infrastructure, the development of secure online payment systems, the emergence of mobile technologies, and the

evolution of data analytics and cloud computing are technological factors that influence e-business adoption.

3. E-business adoption is significantly influenced by organizational factors. These factors include management support and leadership, employee availability and willingness to adopt digital technologies, the presence of a competent IT workforce, and effective change management strategies.
4. Environmental factors such as legal and regulatory frameworks, industry standards, market demand, and competitive pressures can have a substantial effect on e-business adoption. Organizations must navigate these factors in order to maintain compliance and capitalize on opportunities.
5. Typically, the stages of the e-business adoption process are awareness, interest, evaluation, trial, and adoption. Organizations must guide prospective users through these phases by supplying information, demonstrating value, and providing assistance.
6. Organizations can increase e-business awareness by engaging in marketing and promotional activities, leveraging digital channels, educating stakeholders on the benefits, and showcasing success stories and case studies.
7. Best practices for generating interest in e-business include offering incentives, providing training and support, demonstrating success through experimental projects, nurturing an innovative culture, and involving key decision-makers.
8. Tracking key performance indicators (KPIs) such as website traffic, conversion rates, customer satisfaction, and cost savings enables organizations to evaluate the efficacy of their e-business adoption efforts. Regular evaluation and data analysis help identify improvement opportunities.
9. Common obstacles to e-business adoption include resistance to change, a lack of digital skills, security and privacy concerns, cost concerns, and a lack of understanding of the benefits. A comprehensive strategy involving communication, training, addressing concerns, and providing support is required to overcome these obstacles.
10. Organizations can surmount resistance to e-business adoption by providing employees with training and education, involving them in decision-making, addressing concerns about job security, and demonstrating the positive influence of digital technologies on their work.
11. Organizations can address e-business adoption security concerns by instituting comprehensive cybersecurity measures, utilizing encryption and authentication mechanisms, ensuring compliance with data protection regulations, and routinely updating security protocols.
12. Organizations can ensure that their e-business initiatives align with their overall business strategy by undertaking comprehensive planning and analysis, establishing clear goals and objectives, involving key stakeholders, and routinely reviewing and adapting their strategies in response to market dynamics.
13. Increased cyber threats, reliance on technology infrastructure, disruption of traditional business models, digital divide, and information saturation are potential drawbacks of e-business adoption. Organizations must effectively manage these hazards and challenges.

14. By assessing financial data, comparing costs and benefits, conducting customer surveys, and monitoring key metrics such as sales growth, customer acquisition, and customer retention, organizations can calculate the return on investment (ROI) of their e-business initiatives.
15. Emerging trends in e-business adoption include the rise of mobile commerce, the incorporation of artificial intelligence and machine learning, the increasing significance of personalized customer experiences, the adoption of blockchain technology, and the expansion of the Internet of Things (IoT) in business operations.
16. Organizations can remain current with e-business technology advancements by actively monitoring industry trends and developments, participating in pertinent conferences and events, nurturing partnerships with technology providers, and investing in research and development activities.

Chapter 13

1. Change management in e-business is the process of planning, implementing, and directing organizational changes associated with electronic business operations. It is crucial because it enables organizations to adjust to the dynamic nature of e-business, maximize the benefits of technological advances, and minimize disruptions.
2. Change management can have a substantial effect on the success of an e-business implementation. It facilitates transitions, reduces resistance, and aligns the organization with the changes. Effective change management increases employee adoption and engagement, operational efficiency, and the likelihood of achieving desired results.
3. Common change management in e-business challenges include employee resistance, lack of organizational readiness, deficient communication, insufficient training and education, technological complexity, and the need to balance change with ongoing business operations.
4. Organizations can prepare for change management in e-business by conducting comprehensive assessments of their current state, establishing clear goals and objectives, developing a comprehensive change strategy, fostering a culture of openness and innovation, and allocating adequate resources and support.
5. The change management process includes stages such as planning, analyzing the impact of change, communicating with and engaging stakeholders, implementing the changes, monitoring their progress, and making any necessary adjustments. Change planning, stakeholder analysis, communication strategies, training and education, and evaluation are its main components.
6. Setting clear objectives, establishing a timeline, identifying potential risks and mitigation strategies, procuring the necessary resources, and aligning the change effort with the organization's overall strategy all contribute to effective change management in E-Business.

7. Best practices for communication during the change management process in e-business include maintaining open and transparent communication channels, promptly addressing concerns and questions, utilizing multiple communication methods, customizing messages for various stakeholders, and providing regular updates on progress and milestones.
8. Training and education are essential to supporting change management in e-business. Organizations should provide relevant training programs, seminars, and other resources to assist employees in acquiring the skills and knowledge necessary for effective adaptation.
9. Common sources of resistance to change in e-business include fear of job loss, a lack of understanding or awareness, perceived loss of control, cultural barriers, and skepticism regarding the proposed benefits of the changes.
10. Organizations can identify and address resistance to change in e-business by involving employees actively in the change process, addressing their concerns and anxieties, providing clear and consistent communication, providing training and support, and recognizing and rewarding positive behaviors.
11. Strategies for overcoming resistance to change in e-business include developing a compelling case for change, involving influential leaders and change champions, providing continuous support and feedback, creating a supportive and inclusive work environment, and addressing resistance individually.
12. Leadership support is essential for effective e-business change management. Leaders should articulate a clear vision, communicate the rationale for change, empower and engage employees, set an exemplary example, and allocate the necessary resources to support the change initiative.
13. Effective leadership strategies for managing change in e-business include active and visible leadership, nurturing a culture of innovation and continuous learning, establishing trust and collaboration, creating a shared purpose, and celebrating successes.
14. Organizations can determine the efficacy of their change management efforts in e-business by measuring key performance indicators, conducting surveys and feedback sessions, analyzing employee engagement and adoption rates, and comparing actual results to predetermined objectives.
15. The advantages of effective change management in e-business include improved operational efficiency, increased employee morale and engagement, enhanced customer satisfaction, faster adoption of new technologies, improved alignment with market demands, and the ability to capitalize on emerging opportunities.
16. Potential disadvantages of change management in e-business include temporary declines in productivity during the transition period, increased costs associated with training and implementation, potential resistance and conflicts within the organization, and the possibility of failure if the change process is poorly managed.
17. Organizations can strike a balance between the need for change and the potential risks associated with change management in e-business by conducting comprehensive risk assessments, developing contingency plans, involving

stakeholders in decision-making, adopting a phased approach to implementation, and continuously monitoring and adapting to changing circumstances.

18. The future of change management in e-business is likely to evolve as new digital trends emerge and technology advances. It may entail a greater emphasis on agility, flexibility, and adaptability, the utilization of artificial intelligence and automation for change implementation, and the incorporation of change management practices into agile development methodologies.
19. Organizations can stay current on best practices and emerging trends in change management for e-business by actively participating in industry conferences and events, joining professional networks and communities, engaging with consultants and experts, and reviewing relevant research and publications on a regular basis.
20. Aligning change efforts with business strategy, engaging stakeholders early and consistently, fostering a culture of change readiness, investing in employee development, leveraging technology to support change initiatives, and continuously learning and adapting based on feedback and lessons learned are important factors for organizations seeking to implement effective change management processes for e-business.

Chapter 14

1. Analysis and performance measurement are essential to e-business because they offer insight into the efficacy and efficiency of business operations. They aid in identifying development opportunities, optimizing strategies, and making data-driven decisions.
2. Conversion rates, website traffic, exit rates, average order value, customer acquisition cost, customer lifetime value, return on investment (ROI), and customer satisfaction are important e-business performance metrics.
3. Tracking and analyzing website data, user behaviors, and marketing campaigns, Web analytics contributes to the success of e-businesses. It facilitates the comprehension of consumer preferences, the improvement of website design, the optimization of marketing efforts, and the enhancement of the overall user experience.
4. Website traffic, page views, unique visitors, time spent on site, click-through rates, conversion rates, and exit rates are key metrics used in web analytics.
5. Social media metrics can be used to evaluate e-business performance by analyzing engagement, reach, followers, likes, shares, remarks, and conversions generated via social media platforms. They provide insights regarding brand visibility, consumer engagement, and campaign efficacy.
6. Follower growth, engagement rate, reach, impressions, click-through rate, conversion rate, and social media mentions are key social media metrics used in e-business.

7. Measurement of customer contentment is essential in e-business because satisfied consumers are more likely to become repeat buyers and brand advocates. It assists in identifying areas for improvement, enhancing the consumer experience, and fostering long-lasting relationships.
8. Customer surveys, feedback forms, online reviews and evaluations, net promoter score (NPS), and analysis of customer support interactions are common methods for measuring customer satisfaction in e-business.
9. By analyzing revenue, profitability, cash flow, return on investment, and other financial indicators, financial analysis can be used to evaluate e-business performance. It helps evaluate the company's financial health and success.
10. Revenue, gross profit margin, net profit margin, return on investment (ROI), customer acquisition cost (CAC), customer lifetime value (CLV), and average order value are essential e-business financial metrics.
11. Financial statement analysis software, budgeting and forecasting tools, cash flow management software, and financial performance dashboards are e-business financial analysis tools.
12. By monitoring key performance indicators (KPIs), tracking progress toward objectives, and identifying areas for improvement, performance measurement tools can be used to drive the success of e-business. They offer real-time data and insights for making intelligent decisions.
13. Examples of e-business performance measurement tools include analytics platforms (such as Google Analytics), customer relationship management (CRM) systems, marketing automation tools, and business intelligence software.
14. Dashboards and scorecards can be used to measure e-business performance by displaying key metrics and KPIs in a format that is visually appealing and simple to comprehend. They offer a comprehensive perspective of performance and enable data-driven decision-making.
15. Best practices for using performance measurement tools in e-business include defining clear objectives and KPIs, accumulating accurate and trustworthy data, routinely analyzing and interpreting the data, and utilizing the insights to drive strategic actions and improvements.
16. Organizations can better their e-business operations through analysis and performance measurement by identifying their strengths and weaknesses, optimizing processes, refining marketing strategies, enhancing customer experience, and remaining competitive in the digital marketplace.
17. E-business analysis and performance measurement challenges include data quality and accuracy, data integration from multiple sources, selecting relevant metrics, aligning metrics with business objectives, and keeping up with swiftly evolving technologies and trends.
18. Organizations can surmount these obstacles by ensuring data veracity and integrity, investing in comprehensive analytics tools, aligning metrics with business objectives, routinely evaluating and refining measurement processes, and remaining current on industry standards and best practices.
19. Organizations can stay current with emerging trends and best practices in analysis and performance measurement for e-business by actively participating in

industry forums and conferences, following credible blogs and publications, networking with industry professionals, and participating in pertinent training and education.

20. Key considerations for organizations seeking to implement effective analysis and performance measurement processes for their e-business include defining clear objectives and KPIs, selecting appropriate metrics and tools, establishing data governance and quality control processes, fostering a data-driven culture, and routinely reviewing and adjusting measurement strategies based on insights and evolving business needs.

Chapter 15

1. It is essential for e-businesses to take legal, regulatory, and ethical considerations into account because it enables them to operate within the confines of the law, maintain the trust of customers and stakeholders, and avoid potential legal and reputational risks.
2. Common legal issues faced by e-businesses include intellectual property infringement, contract disputes, violations of privacy, data protection, and compliance with consumer protection statutes.
3. Regulations impact e-business operations by providing e-businesses with guidelines and rules to follow in order to ensure fair competition, protect consumer rights, safeguard data and privacy, and maintain ethical business practices.
4. Ethical considerations for e-businesses include advertising and marketing transparency, treating customers and employees fairly, safeguarding user privacy, using data responsibly, and promoting sustainability and social responsibility.
5. E-businesses can ensure compliance with applicable laws and regulations by conducting regular audits, seeking legal counsel, implementing internal policies and procedures, providing employee training, and keeping abreast of changes to the legal and regulatory environment.
6. Trademarks, copyrights, patents, and trade secrets are examples of intellectual property rights in E-business that protect original creations, brand identities, inventions, and confidential information.
7. E-businesses can secure their intellectual property rights by registering trademarks and copyrights, securing patents, employing confidentiality agreements, monitoring and enforcing their rights, and pursuing legal action for infringements.
8. Common e-business privacy concerns include unauthorized data collection, security vulnerabilities, third-party access to personal information, and a lack of transparency in data management procedures.
9. E-businesses can protect consumer privacy by instituting robust data security measures, obtaining informed consent for data collection and use, providing

explicit privacy policies, facilitating opt-out options, and adhering to applicable data protection regulations.

10. Data breaches, hacking attempts, phishing attacks, malware infections, and ransomware are typical cybersecurity hazards faced by e-businesses.
11. E-businesses can prevent fraud by instituting secure payment systems, undertaking regular risk assessments, employing fraud detection technologies, monitoring transactions, and educating customers and employees about potential risks.
12. E-businesses must adhere to common consumer protection laws, including those pertaining to truthful advertising, accurate product information, consumer rights, refund policies, and dispute resolution mechanisms.
13. E-businesses can ensure advertising compliance by adhering to advertising regulations, avoiding deceptive or misleading claims, providing explicit disclosures, and substantiating all marketing claims.
14. Ethical issues that can arise in e-business operations include deceptive marketing practices, unjust competition, exploitation of consumer data, violations of labor rights, and environmental impact.
15. E-businesses can address ethical concerns by establishing ethical guidelines and policies, nurturing a culture of integrity and transparency, promoting responsible business practices, undertaking routine ethical audits, and pursuing external ethical certifications.
16. Best practices for ensuring legal, regulatory, and ethical compliance in e-business include establishing a legal and compliance team, conducting regular risk assessments, implementing robust internal controls, educating employees on legal and ethical obligations, and fostering a compliance-oriented organizational culture.
17. Organizations can stay abreast of emerging trends and changes in legal, regulatory, and ethical considerations for e-business by actively monitoring industry news, participating in relevant trade associations, engaging legal and compliance experts, attending conferences and seminars, and conducting ongoing research.
18. The challenges of legal, regulatory, and ethical compliance in e-business include navigating complex and changing regulations, adapting to different legal jurisdictions, managing data privacy and security risks, and ensuring consistent compliance across geographies and channels.
19. Organizations can overcome these obstacles by investing in legal and compliance expertise, conducting regular compliance audits, implementing robust risk management frameworks, leveraging technology for monitoring and reporting, and fostering an organization-wide culture of ethics and compliance.
20. Organizations can achieve a balance between the need for compliance and the need for innovation and growth in e-business by incorporating compliance considerations into the innovation process, conducting risk assessments for new initiatives, promoting ethical and responsible innovation, and seeking legal counsel to ensure compliance while pursuing growth opportunities.

Appendix 2 (Multiple-Choice Question Answers)

Chapter 1

1. (c) The use of electronic methods and technologies for business processes
2. (c) Information Age
3. (d) Morse code
4. (b) Trust is essential for building customer relationships and encouraging online transactions.
5. (c) Social media advertising
6. (c) Alibaba
7. (a) Search engine optimization
8. (b) eBay
9. (a) Cryptocurrency
10. (a) Increasing customer loyalty
11. (c) Airbnb
12. (a) To track customer behaviors and preferences
13. (c) Email marketing
14. (c) LinkedIn
15. (c) Ability to engage with a large audience

Chapter 2

1. (b) The process of selling products and services over the Internet
2. (d) B2G (Business-to-Government)
3. (d) Dell
4. (c) Direct interaction with end consumers

H. Taherdoost, *E-Business Essentials*, EAI/Springer Innovations in Communication and Computing, https://doi.org/10.1007/978-3-031-39626-7

5. (b) B2B models typically involve larger transaction volumes than B2C models.
6. (d) Shopify
7. (c) The different ways an e-business generates revenue
8. (c) Netflix
9. (c) Transaction fee-based model
10. (b) Facebook
11. (c) To differentiate from competitors
12. (c) Competitive analysis
13. (c) Customer segmentation
14. (c) Amazon
15. (b) The unique benefits or advantages an e-business provides to its customers

Chapter 3

1. (b) The digital landscape created by online platforms
2. (d) Digital technology and online communication
3. (c) The e-environment has evolved rapidly over time.
4. (d) Speed and accessibility
5. (c) The sharing and creation of content by online users
6. (d) Personalization and customization
7. (b) Consumers have higher expectations for personalized experiences.
8. (c) Enhanced targeting and customer insights
9. (c) Personalized and targeted advertising is more feasible.
10. (c) Responding promptly and appropriately to feedback
11. (c) Encouraging user-generated content and participation
12. (c) Leveraging web analytics tools and social listening platforms
13. (c) Implementing secure data storage and encryption
14. (c) Businesses collect more data to enhance personalization.
15. (c) To improve customer experiences and relationships

Chapter 4

1. (c) 1995
2. (c) Emergence of social media platforms
3. (d) Virtualization
4. (c) Supply chain management
5. (a) E-commerce
6. (a) Lower operating costs
7. (d) Security and privacy concerns
8. (a) Implementing strong encryption protocols
9. (c) Amazon

10. (a) Wide range of product offerings
11. (c) Increased adoption of virtual reality technology
12. (a) Machine learning
13. (c) Social media
14. (d) High shipping costs
15. (a) Alibaba

Chapter 5

1. (b) Protecting electronic transactions and data from unauthorized access
2. (d) To minimize the risk of financial loss and reputation damage
3. (b) It has shifted focus from physical security to virtual security.
4. (d) Cyberattacks and data breaches
5. (c) Weak passwords and authentication mechanisms
6. (a) Facebook's Cambridge Analytica scandal
7. (b) Regularly backing up data and storing it off-site
8. (d) To protect against unauthorized access and data breaches
9. (b) VPN (virtual private network)
10. (d) Data protection and privacy laws
11. (b) Ensuring transparency and user consent
12. (c) Legal and ethical consequences
13. (c) GDPR
14. (a) Equifax data breach
15. (b) Providing customers with clear and accessible privacy policies

Chapter 6

1. (b) To process payments for goods and services
2. (d) Cash on delivery
3. (a) High transaction fees
4. (c) Venmo
5. (c) Implementing encryption and other security technologies
6. (b) Implementing fraud detection measures
7. (a) SSL (secure sockets layer)
8. (a) Delayed order fulfillment
9. (b) A refund requested by a customer
10. (d) Implementing a clear and transparent dispute resolution process
11. (c) Quick and seamless payment processing
12. (a) Credit card payments
13. (c) Quick and secure payment processing
14. (c) It protects sensitive payment data from unauthorized access
15. (c) A system that provides 24/7 customer support and resolves issues promptly

Chapter 7

1. (b) Managing the flow of goods and services in online businesses
2. (a) Increases customer satisfaction
3. (c) Sourcing
4. (c) Managing stock levels and replenishment
5. (a) Managing the physical flow of goods and services
6. (c) Radio frequency identification (RFID)
7. (a) Reducing the need for human labor
8. (a) Ensure product quality control
9. (d) Supplier selection difficulties
10. (a) Implementing regular quality control checks
11. (c) Automating order fulfillment processes
12. (c) Selecting reliable suppliers
13. (c) Improved inventory accuracy
14. (d) Optimize the flow of goods and services
15. (d) Increased employee turnover

Chapter 8

1. (a) The process of purchasing goods or services through electronic means
2. (a) The use of technology to automate procurement processes
3. (b) It enables better inventory management
4. (a) Online auctions
5. (a) A formal document requesting suppliers to submit proposals for a specific project or requirement
6. (c) Increased competition leading to lower prices
7. (a) Limited supplier participation due to confidentiality concerns
8. (b) Ensuring clear communication and training for all stakeholders involved
9. (b) By integrating e-procurement systems with enterprise resource planning (ERP) systems
10. (c) Cost savings through improved efficiency
11. (a) Difficulties in supplier selection and evaluation
12. (c) By streamlining procurement processes and reducing manual tasks
13. (b) Difficulties in evaluating supplier performance and reliability
14. (c) By automating processes and reducing errors
15. (c) Implementing proper data security measures to protect sensitive information

Chapter 9

1. (c) Promoting e-businesses through digital channels
2. (a) It helps increase brand awareness and reach a larger audience.
3. (c) Print advertising
4. (a) Social media
5. (b) Enhancing a website's visibility in search engine results
6. (d) Pay-per-click (PPC) advertising
7. (a) Targeting and personalization
8. (d) Using different strategies for each channel independently
9. (a) Metrics that show customer engagement and conversion rates
10. (a) By identifying customer preferences and behavior patterns
11. (d) Using digital channels to promote, advertise, and sell products or services
12. (b) Social media
13. (b) Driving immediate website traffic and conversions
14. (b) Customizing marketing messages for specific customer segments
15. (a) By providing real-time data on customer preferences and behaviors

Chapter 10

1. (a) Customer relationship management
2. (c) To improve customer satisfaction and loyalty
3. (c) Customer data management
4. (d) Streamlined lead generation and nurturing
5. (b) Limited access to customer information
6. (a) Centralizing customer data
7. (c) By integrating CRM with ERP and e-commerce platforms
8. (c) Improved customer loyalty
9. (c) Compliance with data protection regulations
10. (b) By providing personalized experiences
11. (c) Customer service
12. (c) Inability to handle large volumes of data
13. (b) By providing excellent customer service
14. (b) Improved lead generation
15. (b) Resistance from employees to adopt new processes

Chapter 11

1. (b) The process of analyzing business data to gain insights and make informed decisions
2. (b) It enables e-businesses to analyze customer behaviors and preferences.

3. (d) All of the above
4. (a) Collecting and storing large amounts of structured and unstructured data
5. (a) It helps in identifying customer preferences and behavior patterns.
6. (a) Analyzes historical data to make future predictions and forecasts
7. (a) Improved decision-making and operational efficiency
8. (a) Defining clear business objectives and goals
9. (a) By sharing data and insights across departments and teams
10. (a) Improved customer service and satisfaction
11. (a) Data quality and accuracy concerns
12. (a) By providing real-time insights and analytics
13. (a) Data mining
14. (b) Increased complexity in data storage and management
15. (b) Inconsistent and inaccurate data across multiple sources

Chapter 12

1. (c) The process of integrating electronic technologies into business processes
2. (d) All of the above
3. (c) Financial factors
4. (d) All of the above
5. (d) Awareness
6. (a) Conducting a cost-benefit analysis of implementing e-business
7. (d) All of the above
8. (a) Providing comprehensive training and support to employees
9. (d) All of the above
10. (d) All of the above
11. (b) It will become the primary mode of business operations.
12. (d) All of the above
13. (b) Trial
14. (a) Partnering with technology providers or outsourcing IT services
15. (b) Creating informative and engaging content about e-business benefits

Chapter 13

1. (b) A process of managing organizational change in the implementation of e-business
2. (b) It ensures smooth transitions and minimizes disruptions.
3. (c) Training and development
4. (d) Implementing the change immediately without testing
5. (b) Employee motivation and engagement
6. (b) Providing proper training and support

7. (b) To communicate and inspire a shared vision
8. (b) Promoting open communication and collaboration
9. (a) Increased customer loyalty and retention
10. (b) Resistance and pushback from employees
11. (c) Continued evolution and adaptation to technological advancements
12. (a) Planning
13. (b) By providing ongoing training and support
14. (c) Ability to inspire and motivate employees during the change process
15. (a) Identifying areas for improvement and making necessary adjustments

Chapter 14

1. (d) To optimize e-business operations
2. (d) Website design
3. (b) Optimize website performance
4. (d) Website traffic
5. (c) Engagement rates
6. (d) Evaluate social media marketing efforts
7. (d) Improve customer retention
8. (c) Surveys
9. (c) Profit margins
10. (a) Dashboards
11. (c) Optimize marketing strategies
12. (c) Surveys
13. (d) Drive e-business success
14. (a) Setting realistic goals and targets
15. (d) Conversion rates

Chapter 15

1. (d) Financial considerations
2. (d) HIPAA (Health Insurance Portability and Accountability Act)
3. (c) Ensure fair competition
4. (c) Intellectual property rights
5. (c) Safeguarding personal information
6. (a) Builds customer trust
7. (c) Phishing attacks
8. (a) Encryption techniques
9. (c) Encourage fair competition
10. (a) Protect customer privacy
11. (d) All of the above

12. (a) Implementing transparent policies
13. (d) Financial considerations
14. (d) HIPAA (Health Insurance Portability and Accountability Act)
15. (c) Ensure fair competition

Index

H. Taherdoost, *E-Business Essentials*, EAI/Springer Innovations in Communication and Computing, https://doi.org/10.1007/978-3-031-39626-7

MIX
Papier aus verantwortungsvollen Quellen
Paper from responsible sources
FSC® C105338

If you have any concerns about our products,
you can contact us on
ProductSafety@springernature.com

In case Publisher is established outside the EU,
the EU authorized representative is:
Springer Nature Customer Service Center GmbH
Europaplatz 3, 69115 Heidelberg, Germany

Printed by Libri Plureos GmbH
in Hamburg, Germany